Mechanical Insulating

Level Three

Trainee Guide

Pearson

Boston Columbus Indianapolis New York San Francisco Amsterdam
Cape Town Dubai London Madrid Milan Munich Paris Montreal Toronto Delhi
Mexico City Sao Paulo Sydney Hong Kong Seoul Singapore Taipei Tokyo

NCCER

President: Don Whyte
Vice President: Steve Greene
Chief Operations Officer: Katrina Kersch
Mechanical Insulating Project Managers: Elizabeth Schlaupitz, Jamie Carroll
Senior Development Manager: Mark Thomas
Senior Production Manager: Tim Davis
Quality Assurance Coordinator: Karyn Payne
Desktop Publishing Coordinator: James McKay
Permissions Specialists: Kelly Sadler, Adrienne Payne
Production Specialist: Kelly Sadler
Editors: Graham Hack, Adrienne Payne

Writing and development services provided by Topaz Publications, Liverpool, NY

Lead Writer/Project Manager: Thomas Burke
Desktop Publisher: Joanne Hart
Art Director: Alison Richmond
Permissions Editor: Andrea LaBarge
Writers: Troy Staton, Thomas Burke, Terry Egolf, Veronica Westfall, Carol Hebert, Darrell Wilkerson

Pearson
Director of Alliance/Partnership Management: Andrew Taylor
Editorial Assistant: Collin Lamothe
Program Manager: Alexandrina B. Wolf
Assistant Content Producer: Alma Dabral
Digital Content Producer: Jose Carchi
Director of Marketing: Leigh Ann Simms
Senior Marketing Manager: Brian Hoehl
Composition: NCCER
Printer/Binder: LSC Communications
Cover Printer: LSC Communications
Text Fonts: Palatino and Univers

Credits and acknowledgments for content borrowed from other sources and reproduced, with permission, in this textbook appear at the end of each module.

Pearson

ISBN-13: 978-0-13-483379-8
ISBN-10: 0-13-483379-1

ScoutAutomatedPrintCode

Preface

To the Trainee

Insulation is a crucial element for the operation of mechanical systems and provides thermal protection for boilers, HVAC systems, duct work, and piping. Insulating these systems improves operating efficiency by managing system temperature, controlling and preventing the presence of condensation and mold, and protecting personnel from hot or cold surfaces.

Craft professionals in the mechanical insulating industry must learn both the fundamentals of the industry as well as specific skills and knowledge for their craft. Many are trained through craft apprenticeship programs, which provide both classroom instruction and on-the-job learning.

Workers in the industry are well paid and, according to the Bureau of Labor Statistics, constitute a workforce of over 25,000 workers. Regardless of where your career may take you, NCCER's Mechanical Insulating curriculum will provide you a solid foundation on your road to success.

New with *Mechanical Insulating Level Three*

NCCER is pleased to release *Mechanical Insulating* in full color, with new photographs and figures. This new version is now presented in NCCER's improved instructional systems design, in which the sections of each module are directly tied to learning objectives.

The curriculum has also been thoroughly redesigned, updated, and expanded. Level Three has been reorganized to provide advanced training of mechanical insulation installation. Modules included in this level cover topics such as board and block insulation, sheet-metal lagging, jacketing fabrication, and specialized insulation systems. Though the module on refrigeration and cryogenic systems has been deleted from this level, new modules developed include "Jacketing Systems" (Module ID 19313) and "Removable and Reusable Flexible Insulation Covers" (Module ID 19314). *Mechanical Insulating Level Three* also features the third edition of "Fundamentals of Crew Leadership" (Module ID 46101-17), which provides trainees with vital information about effective leadership, communication, and professionalism in the construction industry.

We wish you success as you progress through this training program. If you have any comments on how NCCER might improve upon this textbook, please complete the User Update form located at the back of each module and send it to us. We will always consider and respond to input from our customers.

We invite you to visit the NCCER website at **www.nccer.org** for information on the latest product releases and training, as well as online versions of the *Cornerstone* magazine and Pearson's NCCER product catalog.

Your feedback is welcome. You may email your comments to **curriculum@nccer.org** or send general comments and inquiries to **info@nccer.org**.

NCCER Standardized Curricula

NCCER is a not-for-profit 501(c)(3) education foundation established in 1996 by the world's largest and most progressive construction companies and national construction associations. It was founded to address the severe workforce shortage facing the industry and to develop a standardized training process and curricula. Today, NCCER is supported by hundreds of leading construction and maintenance companies, manufacturers, and national associations. The NCCER Standardized Curricula was developed by NCCER in partnership with Pearson, the world's largest educational publisher.

Some features of the NCCER Standardized Curricula are as follows:

- An industry-proven record of success
- Curricula developed by the industry, for the industry
- National standardization providing portability of learned job skills and educational credits
- Compliance with the Office of Apprenticeship requirements for related classroom training (*CFR 29:29*)
- Well-illustrated, up-to-date, and practical information

NCCER also maintains the NCCER Registry, which provides transcripts, certificates, and wallet cards to individuals who have successfully completed a level of training within a craft in NCCER's Curricula. *Training programs must be delivered by an NCCER Accredited Training Sponsor in order to receive these credentials.*

Special Features

In an effort to provide a comprehensive and user-friendly training resource, this curriculum showcases several informative features. Whether you are a visual or hands-on learner, these features are intended to enhance your knowledge of the construction industry as you progress in your training. Some of the features you may find in the curriculum are explained below.

Introduction

This introductory page, found at the beginning of each module, lists the module Objectives, Performance Tasks, and Trade Terms. The Objectives list the knowledge you will acquire after successfully completing the module. The Performance Tasks give you an opportunity to apply your knowledge to real-world tasks. The Trade Terms are industry-specific vocabulary that you will learn as you study this module.

00104-15
INTRODUCTION TO POWER TOOLS

Objectives

When you have completed this module, you will be able to do the following:

1. Identify and explain how to use various types of power drills and impact wrenches.
   a. Identify and explain how to use common power drills and bits.
   b. Identify and explain how to use a hammer drill.
   c. Identify and explain how to use pneumatic drills and impact wrenches.
2. Identify and explain how to use various types of power saws.
   a. Identify and explain how to use a circular saw.
   b. Identify and explain how to use saber and reciprocating saws.
   c. Identify and explain how to use a portable band saw.
   d. Identify and explain how to use miter and cutoff saws.
3. Identify and explain how to use various grinders and grinder attachments.
   a. Identify and explain how to use various types of grinders.
   b. Identify and explain how to use various grinder accessories and attachments.
4. Identify and explain how to use miscellaneous power tools.
   a. Identify and explain how to use pneumatic and powder-actuated fastening tools.
   b. Identify and explain how to use pavement breakers.
   c. Identify and explain the uses of hydraulic jacks.

Performance Tasks

Under supervision of your instructor, you should be able to do the following:

1. Safely and properly demonstrate the use of three of the following tools:
   - Electric drill
   - Hammer drill or rotary hammer
   - Circular saw
   - Reciprocating saw
   - Portable band saw
   - Miter or cutoff saw
   - Portable or bench grinder
   - Pneumatic nail gun
   - Pavement breaker

Trade Terms

Abrasive	Chuck key	Ground fault circuit interrupter (GFCI)	Revolutions per minute (rpm)
Alternating current (AC)	Countersink	Ground fault protection	Ring test
Arbor	Direct current (DC)	Kerf	Shank
Auger bit	Forstner bit	Masonry bit	Trigger lock
Carbide	Grit	Reciprocating	
Chuck			

Industry Recognized Credentials

If you are training through an NCCER-accredited sponsor, you may be eligible for credentials from NCCER's Registry. The ID number for this module is 00104-15. Note that this module may have been used in other NCCER curricula and may apply to other level completions. Contact NCCER's Registry at 888.622.3720 or go to **www.nccer.org** for more information.

Trade Features

Trade features present technical tips and professional practices based on real-life scenarios similar to those you might encounter on the job site.

Bowline Trivia

Some people use this saying to help them remember how to tie a bowline: "The rabbit comes out of his hole, around a tree, and back into the hole."

Figures and Tables

Photographs, drawings, diagrams, and tables are used throughout each module to illustrate important concepts and provide clarity for complex instructions. Text references to figures and tables are emphasized with *italic* type.

Notes, Cautions, and Warnings

Safety features are set off from the main text in highlighted boxes and categorized according to the potential danger involved. Notes simply provide additional information. Cautions flag a hazardous issue that could cause damage to materials or equipment. Warnings stress a potentially dangerous situation that could result in injury or death to workers.

NOTE
Nameplates must be posted on each material handling device. The nameplate must indicate

CAUTION
It is essential to note the revision designation on a construction drawing and to use only the latest

WARNING!
Saw teeth are very sharp. Use gloves and do not handle the saw teeth with bare hands. When cutting with a saw, ensure that your fingers remain clear of the teeth at all times.

Case History

Case History features emphasize the importance of safety by citing examples of the costly (and often devastating) consequences of ignoring best practices or OSHA regulations.

> **Case History**
>
> **Requesting an Outage**
>
> An electrical contractor requested an outage when asked to install two bolt-in, 240V breakers in panels in a data processing room. It was denied due to the 24/7 worldwide information processing hosted by the facility. The contractor agreed to proceed only if the client would sign a letter agreeing not to hold them responsible if an event occurred that damaged computers or resulted in loss of data. No member of upper management would accept liability for this possibility, and the outage was scheduled.
>
> **The Bottom Line:** If you can communicate the liability associated with an electrical event, you can influence management's decision to work energized.

Going Green

Going Green features present steps being taken within the construction industry to protect the environment and save energy, emphasizing choices that can be made on the job to preserve the health of the planet.

> GOING GREEN
>
> **Reducing Your Carbon Footprint**
>
> Many companies are taking part in the paperless movement. They reduce their environmental impact by reducing the amount of paper they use. Using email helps to reduce the amount of paper used,

Did You Know

Did You Know features introduce historical tidbits or interesting and sometimes surprising facts about the trade.

> **Did You Know?**
>
> **Safety First**
>
> Safety training is required for all activities. Never operate tools, machinery, or equipment without prior training. Always refer to the manufacturer's instructions.

Step-by-Step Instructions

Step-by-step instructions are used throughout to guide you through technical procedures and tasks from start to finish. These steps show you how to perform a task safely and efficiently.

> Perform the following steps to erect this system area scaffold:
>
> *Step 1* Gather and inspect all scaffold equipment for the scaffold arrangement.
>
> *Step 2* Place appropriate mudsills in their approximate locations.
>
> *Step 3* Attach the screw jacks to the mudsills.

Trade Terms

Each module presents a list of Trade Terms that are discussed within the text and defined in the Glossary at the end of the module. These terms are presented in the text with **bold, blue** type upon their first occurrence. To make searches for key information easier, a comprehensive Glossary of Trade Terms from all modules is located at the back of this book.

> During a rigging operation, the **load** being lifted or moved must be connected to the apparatus, such as a crane, that will provide the power for movement. The connector—the link between the load and the apparatus—is often a sling made of synthetic, chain, or **wire rope** materials. This section focuses on three types of slings:

Section Review

Each section of the module wraps up with a list of Additional Resources for further study and Section Review questions designed to test your knowledge of the Objectives for that section.

> **Additional Resources**
>
> *Materials Handling Handbook*, The American Society of Mechanical Engineers (ASME) and The International Material Management Society (IMMS), Raymond A. Kulwiec, Editor-in-Chief. 1985. New York, NY: Wiley-Interscience.
>
> *Manufacturing Facilities Design & Material Handling*, Matthew P. Stevens, Fred E. Meyers. 2013. West Lafayette, IN: Purdue University Press.
>
> **1.0.0 Section Review**
>
> 1. For material handling tasks, it is just as important to be mentally fit as it is to be _____.
>   a. physically fit
>   b. physically aggressive
>   c. closely supervised
>   d. over 200 pounds
>
> 2. Which of the following is a type of knot that is often used to join the ends of two ropes in non-critical, low-strain applications?
>   a. Bowline
>   b. Clove hitch
>   c. Half hitch
>   d. Square knot

Review Questions

The end-of-module Review Questions can be used to measure and reinforce your knowledge of the module's content.

> **Review Questions**
>
> 1. Identification tags for slings must include the _____.
>   a. type of protective pads to use
>   b. type of damage sustained during use
>   c. color of the tattle-tail
>   d. manufacturer's name or trademark
>
> 2. The type of wire rope core that is susceptible to heat damage at relatively low temperatures is the _____.
>   a. fiber core
>   b. strand core
>   c. independent wire rope core
>   d. metallic link supporting core
>
> 3. Synthetic slings must be inspected _____.
>   a. once every month
>   b. visually at the start of each work week
>   c. before every use
>   d. once wear or damage becomes apparent
>
> 4. An alloy steel chain sling must be removed from service if there is evidence that _____.
>   a. the sling has been used in different hitch configurations
>   b. replacement links have been used to repair the chain
>   c. the sling has been used for more than one year
>   d. strands in the supporting core have weakened
>
> 5. A piece of rigging hardware used to couple the end of a wire rope to eye fittings, hooks, or other connections is a(n) _____.
>   a. eyebolt
>   b. hitch
>   c. shackle
>   d. U-bolt
>
> 6. A lifting clamp is most likely to be used to move loads such as _____.
>   a. steel plates
>   b. piping bundles
>   c. concrete blocks
>   d. plastic tubing
>
> 7. Chain hoists are able to lift heavy loads by utilizing a _____.
>   a. rope and pulley system
>   b. rigger's strength
>   c. stationary counterweight
>   d. gear system
>
> 8. Before attempting to lift a load with a chain hoist, make sure that the _____.
>   a. hoist is secured to a come-along
>   b. load is properly balanced
>   c. tag lines are properly anchored
>   d. tackle is connected to its power source
>
> 9. A hitch configuration that allows slings to be connected to the same load without using a spreader beam is a _____.
>   a. double-wrap hitch
>   b. choker hitch
>   c. bridle hitch
>   d. basket hitch
>
> 10. To make the emergency stop signal that is used by riggers, extend both arms _____.
>   a. horizontally with palms down and quickly move both arms back and forth
>   b. directly in front and then move both arms up and down repeatedly
>   c. vertically above the head and wave both arms back and forth
>   d. horizontally with clenched fists and move both arms up and down
>
> 00106-15 Introduction to Basic Rigging Module Six 31

NCCER Standardized Curricula

NCCER's training programs comprise more than 80 construction, maintenance, pipeline, and utility areas and include skills assessments, safety training, and management education.

Boilermaking
Cabinetmaking
Carpentry
Concrete Finishing
Construction Craft Laborer
Construction Technology
Core Curriculum: Introductory Craft Skills
Drywall
Electrical
Electronic Systems Technician
Heating, Ventilating, and Air Conditioning
Heavy Equipment Operations
Heavy Highway Construction
Hydroblasting
Industrial Coating and Lining Application Specialist
Industrial Maintenance Electrical and Instrumentation Technician
Industrial Maintenance Mechanic
Instrumentation
Ironworking
Manufactured Construction Technology
Masonry
Mechanical Insulating
Millwright
Mobile Crane Operations
Painting
Painting, Industrial
Pipefitting
Pipelayer
Plumbing
Reinforcing Ironwork
Rigging
Scaffolding
Sheet Metal
Signal Person
Site Layout
Sprinkler Fitting
Tower Crane Operator
Welding

Maritime

Maritime Industry Fundamentals
Maritime Pipefitting
Maritime Structural Fitter

Green/Sustainable Construction

Building Auditor
Fundamentals of Weatherization
Introduction to Weatherization
Sustainable Construction Supervisor
Weatherization Crew Chief
Weatherization Technician
Your Role in the Green Environment

Energy

Alternative Energy
Introduction to the Power Industry
Introduction to Solar Photovoltaics
Power Generation Maintenance Electrician
Power Generation I&C Maintenance Technician
Power Generation Maintenance Mechanic
Power Line Worker
Power Line Worker: Distribution
Power Line Worker: Substation
Power Line Worker: Transmission
Solar Photovoltaic Systems Installer
Wind Energy
Wind Turbine Maintenance Technician

Pipeline

Abnormal Operating Conditions, Control Center
Abnormal Operating Conditions, Field and Gas
Corrosion Control
Electrical and Instrumentation
Field and Control Center Operations
Introduction to the Pipeline Industry
Maintenance
Mechanical

Safety

Field Safety
Safety Orientation
Safety Technology

Supplemental Titles

Applied Construction Math
Tools for Success

Management

Construction Workforce Development Professional
Fundamentals of Crew Leadership
Mentoring for Craft Professionals
Project Management
Project Supervision

Spanish Titles

Acabado de concreto: nivel uno (*Concrete Finishing Level One*)
Aislamiento: nivel uno (*Insulating Level One*)
Albañilería: nivel uno (*Masonry Level One*)
Andamios (*Scaffolding*)
Carpintería: Formas para carpintería, nivel tres (*Carpentry: Carpentry Forms, Level Three*)
Currículo básico: habilidades introductorias del oficio (*Core Curriculum: Introductory Craft Skills*)
Electricidad: nivel uno (*Electrical Level One*)
Herrería: nivel uno (*Ironworking Level One*)
Herrería de refuerzo: nivel uno (*Reinforcing Ironwork Level One*)
Instalación de rociadores: nivel uno (*Sprinkler Fitting Level One*)
Instalación de tuberías: nivel uno (*Pipefitting Level One*)
Instrumentación: nivel uno, nivel dos, nivel tres, nivel cuatro (*Instrumentation Levels One through Four*)
Orientación de seguridad (*Safety Orientation*)
Paneles de yeso: nivel uno (*Drywall Level One*)
Seguridad de campo (*Field Safety*)

Acknowledgments

This curriculum was revised as a result of the farsightedness and leadership of the following sponsors:

Brace Industrial Group
Brock Services LLC
Industrial Construction & Engineering Co.
Insulation Specialties Inc.
L&C Insulation
Marquis Construction Services, Inc.
National Insulation Association
Petrin Corporation
Southern Tier Insulations

This curriculum would not exist were it not for the dedication and unselfish energy of those volunteers who served on the Authoring Team. A sincere thanks is extended to the following:

Randy Beard
Tim Coughlin
Jeffrey DeGraaf
Peter Gauchel
David Gottlich
Wes Howard
Michele Jones
Ron King
Al Olvera
Gary Satterfield

NCCER Partners

American Council for Construction Education
American Fire Sprinkler Association
Associated Builders and Contractors, Inc.
Associated General Contractors of America
Association for Career and Technical Education
Association for Skilled and Technical Sciences
Construction Industry Institute
Construction Users Roundtable
Design Build Institute of America
GSSC – Gulf States Shipbuilders Consortium
ISN
Manufacturing Institute
Mason Contractors Association of America
Merit Contractors Association of Canada
NACE International
National Association of Women in Construction
National Insulation Association
National Technical Honor Society
National Utility Contractors Association
NAWIC Education Foundation
North American Crane Bureau
North American Technician Excellence
Pearson
Prov
SkillsUSA®
Steel Erectors Association of America
U.S. Army Corps of Engineers
University of Florida, M. E. Rinker Sr., School of Construction Management
Women Construction Owners & Executives, USA

NCCER Business Partners

Contents

Module Eight

Fundamentals of Crew Leadership

Covers basic leadership skills and explains different leadership styles, communication, delegating, and problem solving. Job-site safety and the crew leader's role in safety are also discussed. Explains details and procedures for project planning, scheduling, and estimating. (Module ID 46101; 22.5 Hours)

Glossary

Index

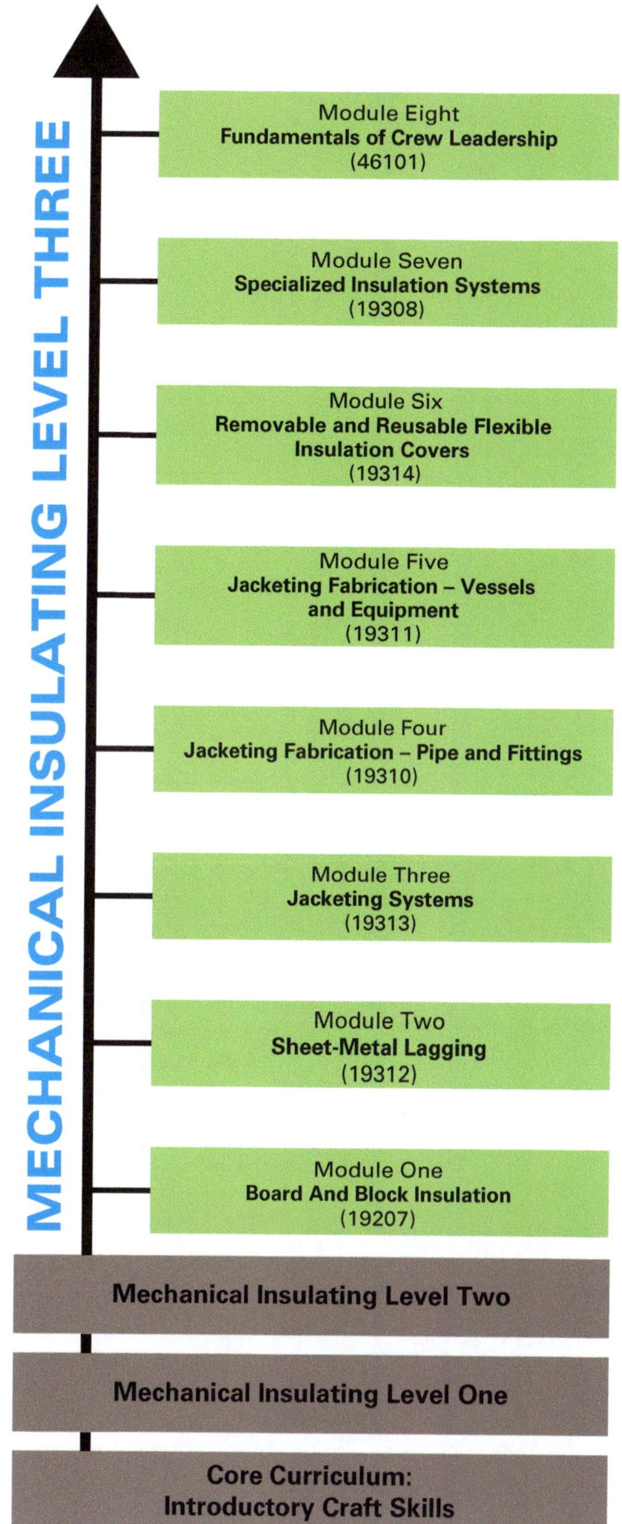

This course map shows all of the modules in *Mechanical Insulating Level Three*. The suggested training order begins at the bottom and proceeds up. Skill levels increase as you advance on the course map. The local Training Program Sponsor may adjust the training order.

Board and Block Insulation

Overview

Mechanical insulating differs from common building and residential insulating for a variety of reasons. One significant difference is the requirement to insulate pipes, ducting, and large vessels of various shapes. The demands on mechanical insulating materials include greater durability, rigidity, and thermal performance. This module presents methods for insulating large cylindrical vessels and other curved surfaces using rigid and semi-rigid materials.

Module 19207

Trainees with successful module completions may be eligible for credentialing through the NCCER Registry. To learn more, go to **www.nccer.org** or contact us at 1.888.622.3720. Our website has information on the latest product releases and training, as well as online versions of our *Cornerstone* magazine and Pearson's product catalog.

Your feedback is welcome. You may email your comments to **curriculum@nccer.org**, send general comments and inquiries to **info@nccer.org**, or fill in the User Update form at the back of this module.

This information is general in nature and intended for training purposes only. Actual performance of activities described in this manual requires compliance with all applicable operating, service, maintenance, and safety procedures under the direction of qualified personnel. References in this manual to patented or proprietary devices do not constitute a recommendation of their use.

19207 V2

19207
Board and Block Insulation

Objectives

When you have completed this module, you will be able to do the following:

1. Describe board and block insulation products and their applications.
   a. Describe common and special-use board insulation products.
   b. Describe common and special-use block insulation products.
   c. Describe the vapor retarder systems used for block and board insulation in below-ambient service.
2. Explain the layout and installation methods for board and block insulation.
   a. Explain how to estimate, lay out, and cut lags.
   b. Explain how to install board and block insulation on curved surfaces.
   c. Explain how to insulate the head of a vessel.

Performance Tasks

Under the supervision of your instructor, you should be able to do the following:

1. Assuming an insulation thickness of 2 inches (5 cm), determine the number of courses of block, the number of lags, and the bevel angle required on the lags for a given vessel.
2. Insulate the sidewall of a vessel using lags, curved segments, or scored block.
3. Insulate the head of a vessel.

Trade Terms

Billets
Block
Board
Buns
Convex
Course
Elliptical
Equipment
Floating ring
Friable
Gores
Juncture
Lags
Rolled angle irons
Scored
Spheroidal
Sub-girts
Tangent lines
Tanks
Tan-to-tan
Trapezoids
V-grooved
Vessels

Industry Recognized Credentials

If you are training through an NCCER-accredited sponsor, you may be eligible for credentials from NCCER's Registry. The ID number for this module is 19207. Note that this module may have been used in other NCCER curricula and may apply to other level completions. Contact NCCER's Registry at 888.622.3720 or go to **www.nccer.org** for more information.

Contents

Figures

Section One

1.0.0 Types of Board and Block Insulation

Objective

Describe board and block insulation products and their applications.

a. Describe common and special-use board insulation products.
b. Describe common and special-use block insulation products.
c. Describe the vapor retarder systems used for block and board insulation in below-ambient service.

Trade Terms

Billets: A block of material usually built up by cementing thinner layers together.

Block: A usually rigid form of insulating material where its thickness dimension relative to its width and length give it a blocky appearance.

Board: A usually semi-rigid form of insulation typically having an appearance more like a rectangular panel than a block.

Buns: A large uniform block of material from which thinner boards or smaller blocks are cut.

Equipment: In the context used here, the term generally refers to pumps, vessels, and irregularly-shaped mechanical devices.

Friable: The property of a brittle material that tends to crumble into small particles when crushed, rubbed, or scraped.

V-grooved: A term describing thicker and typically rigid board insulation that has manufactured or fabricated V-shaped slots cut into the material. These grooves allow the board to be bent around curved surfaces with minimal and allowable cracking.

Vessels: In the context of mechanical insulating, any fully- or partially-enclosed structures that can be used for storing or processing any type of substance, usually at pressures above or below atmospheric pressure. Vessels are used for many purposes, including containing chemical reactions, in addition to storing or transferring liquids and gases.

Insulation is manufactured from a wide variety of materials and in many forms that are appropriate for the type of service in which they are used. Examples include rolls, blankets, sheet, **board**, **block**, tubular, and sectional pipe, as well as pre-molded fitting covers, and even flexible removable covers. Each form of insulation has certain fabrication and installation properties suited to the form. Because the installation methods relate mainly to the form of insulation and not the material, the kind of material involved often doesn't matter. In mechanical insulating applications, especially in the industrial setting, board and block insulation are essential for **vessels** and other large, process-related structures.

Insulation often starts out as large pieces of material called **billets** or **buns**. A billet is a large block of material glued together from smaller pieces from which standard shapes, such as half-pipe insulation, are manufactured. Buns (*Figure 1*) are large, solid blocks of insulating material several feet wide and high, and up to 24' (7.3 m) long depending on the insulating material. Buns are sliced up into smaller pieces, or thin sheet- or board-like sections, depending on the application and material. Small, blocky pieces of rigid materials can be milled into special-purpose shapes.

1.1.0 Board Insulation

Board insulation is usually manufactured in rectangular pieces that are thin compared to their lengths and widths. Board insulation is typically rigid or semi-rigid compared to materials with similar dimensions, like closed-cell rubber foam insulation, which is quite flexible. Board insulation is usually supplied in widths of 24" or 48" (61 cm or 122 cm) and in thicknesses ranging between 1" and 4" (2.5 cm and 10 cm). Most manufacturers offer a range of densities. Higher density usually translates into a higher thermal resistance.

1.1.1 Fiberglass Board Insulation

Fiberglass board insulation is made from fine, inorganic glass fibers. The fibers are bonded together with a thermosetting resin to form a lightweight, semi-rigid or rigid board.

Specifications for fiberglass board insulation are provided in the *ASTM C 612* standards. Fiberglass board must comply with Types I, II, or III listed in the standard. The operating temperature of fiberglass board ranges from –20°F to 450°F (–29°C to 232°C) for Type I, up to 850°F (450°C) for Type II, and as high as 1,000°F (540°C) for Type 3. Fiberglass board is manufactured in a variety of densities from 1.5 pounds per cubic foot (lb/ft^3) to 7.0 lb/ft^3 (24–112 kg/m^3). The boards are produced plain (without a facing) or with various standard facings, and can be cut with a saw or knife. Faced fiberglass boards are typically limited to service temperatures less than 150°F (66°C) to prevent delamination of facing material. *Figure 2* is an example of fiberglass board insulating material.

1.1.2 Mineral Wool Board Insulation

Mineral wool is a general term for any fibrous material manufactured from molten rock, slag, glass, or ceramic spun into fibers. While fiberglass is also classified as a mineral fiber, in this module, the term *mineral wool* will refer to product obtained mainly from rock or slag. The fibers may be combined with a high temperature polymer binder. An oil may also be included to reduce dust production with handling. Mineral wool board may also be obtained with standard facings or unfaced (*Figure 3*). These products may also include a metal mesh reinforcement.

Mineral wool board insulation properties are also specified by the *ASTM C 612* standard. Mineral wool must meet the higher temperature specifications for Types IVA and IVB (1,200°F or 650°C). However, mineral wool is also suitable for below-ambient temperatures down to –20°F (–30°C). Mineral wool boards are generally available in nominal densities from 3 lb/ft^3 to 12 lb/ft^3 (48–192 kg/m^3). As with glass-fiber board, insulating properties improve as the density increases.

Mineral wool board is commonly available in thicknesses ranging from 1" to 4" (2.5–10 cm) in increments of $\frac{1}{2}$" (13 mm). Lengths and widths of standard board sizes are usually 24" × 48" (61 cm × 122 cm), with other sizes available by special order

CAUTION

Mineral wools are quite dusty, so appropriate breathing protection should be used.

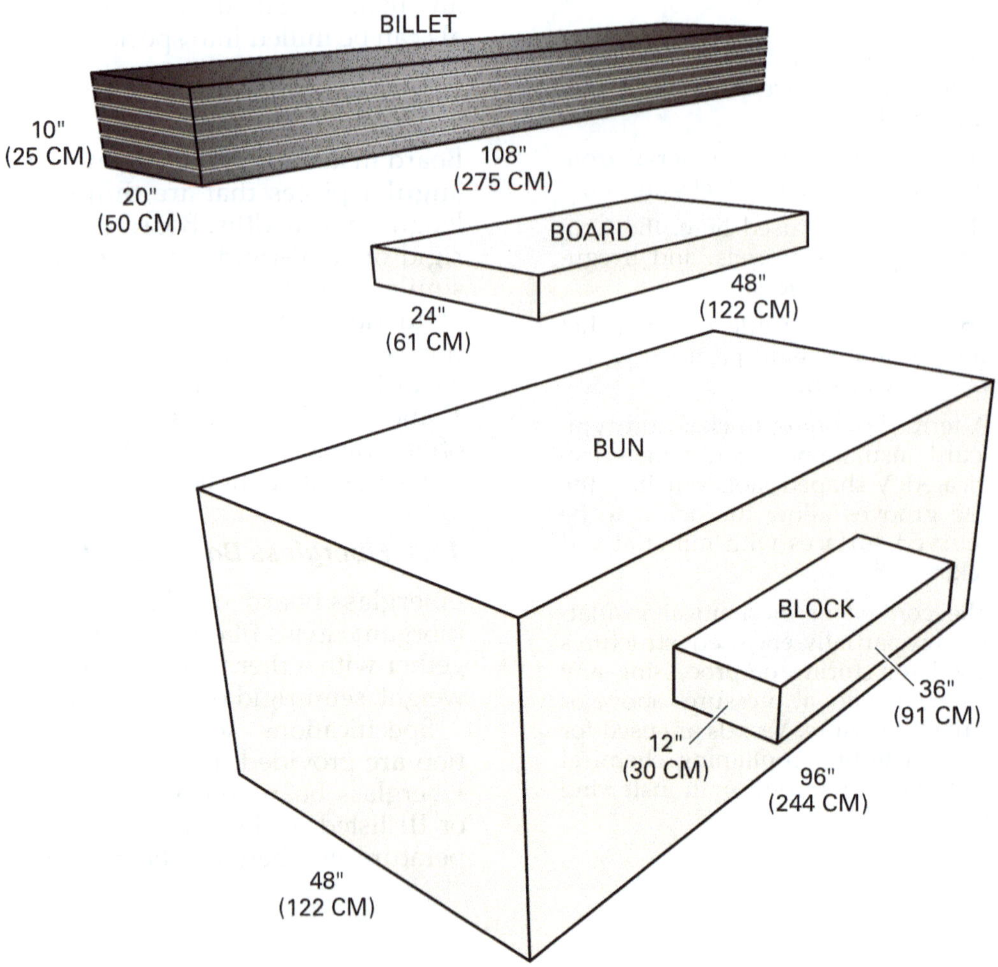

Figure 1 Comparison of the forms of rigid insulation products.

1.1.3 Rigid Polymeric Foam Board Insulation

Rigid polymeric foam boards are lightweight cellular insulation formed by expanding or extruding a plastic foam—phenolic, polystyrene (EPS or XPS), or polyisocyanurate (PIR). The manufacture and properties of these materials are thoroughly covered in the NCCER module 19206, "Rigid Foam and Cellular Glass Insulation". Therefore, their characteristics will only be briefly summarized here.

The polymeric foam materials are limited in their upper service temperatures but are ideal for cold-service applications:

- *Phenolic:* –290°F to 257°F (–179°C to 125°C)
- *Polyisocyanurate:* –297°F to 300°F (–183°C to 150°C)
- *Polystyrene board:* –65°F to 165°F (–54°C to 74°C)

> **NOTE**
> Polystyrene pipe insulation has a service temperature range of –297°F to 165°F (–183°C to 74°C) due to a different cellular structure compared to board materials.

Phenolic materials are typically manufactured in buns. The bun is cut into boards according to the dimensions for the fabricator's products. PIR is also initially manufactured as buns. The buns can be precut into boards by fabricators or sawn to size in the field.

Polystyrene (XPS) rigid foam is typically manufactured as rectangular billets. The manufacturer may offer the same product as boards in smaller widths.

In board form, the materials are generally available in 2' or 4' (0.6 m or 1.2 m) widths and various lengths, and in thicknesses from 1" to 4" (2.5–10 cm). Actual dimensions may vary among manufacturers and custom dimensions can be requested. Also, rigid polymeric foam boards can be supplied unfaced or with a variety of facing materials (*Figure 4*).

1.1.4 Special-Purpose Board Insulation Products

Many new board insulating products have been developed for the mechanical insulation market in recent years. The following is a brief summary of these products and their uses:

Figure 2 Fiberglass board insulation.

Figure 3 Unfaced mineral wool board insulation.

Figure 4 Faced and unfaced rigid foam board insulation.

- *Alumina-silicate board* – Alumina-silicate insulation is a glass-like fibrous ceramic, similar in form to mineral wool, but with much higher maximum operating temperatures. These materials are designed for refractory applications with maximum service temperatures between 2,600°F and 3,000°F (1,430°C and 1,650°C). The fibers are bonded by a water-soluble material that burns out at high temperatures.
- *Calcium silicate board* – Calcium silicate insulation is often called *Cal-Sil* or *CalSil* in the trade. CalSil is a common high-temperature insulating material. Some specialized manufacturers cut buns of CalSil into boards as well as smaller pieces. These are milled into unusual shapes required for special insulating needs inside machines and **equipment**. Type II CalSil refractory and fire-rated panels have a maximum service temperature of 1,700°F (930°C). CalSil boards are commonly used for insulating hot equipment supports, in foundries, and for creating passive fire-resistant surfaces.
- *Microporous composite board* – Microporous composite board insulation is made of inorganic fibers or powders compressed into dense boards, leaving very tiny pores between them. Materials can range from 18 lb/ft^3 to 25 lb/ft^3 (290–400 kg/m^3) for unfaced board. Microporous panels can also be given a glass facing to reduce dust during handling. These have lower densities in the range of 14 lb/ft^3 to 18 lb/ft^3 (225–290 kg/m^3). Maximum service temperatures depend on ASTM grades, ranging from about 1,650°F to 2,100°F (900°C to 1,150°C). Boards can be produced in thicknesses ranging from $\frac{1}{8}$" to 4" (3–100 mm). These materials can be used in hot appliances, flues, and as primary and backup insulation in furnaces and hot processes where high thermal resistance in a small space is required.

1.2.0 Block Insulation

Block insulation pieces are generally thicker and look more like blocks than rectangular boards. Board insulation is generally supplied in much larger lengths and widths than block, though not as a rule. Block insulation is often made of somewhat more rigid materials than board insulation, and it is generally required where thick layers of high-density insulation are specified. The smaller blocks make installing heavy insulation manageable for the average worker. Block insulation is also required for covering curved surfaces with a thick layer (*Figure 5*). The smaller dimensions shape to the curved surface better than large, flat boards.

1.2.1 Calcium Silicate/Expanded Perlite Block Insulation

Calcium silicate block is produced from heating calcium oxide (lime) and silica with cellulose reinforcing fibers. CalSil properties are specified in *ASTM C 533*. The standard identifies three types based on density and maximum service temperature. Types I and IA CalSil pertain to pipe and block insulations, which are rated up to 1,200°F (650°C). Type I block has a density of 15 lb/ft^3 (240 kg/m^3) and Type IA is 22 lb/ft^3 (352 kg/m^3).

CalSil block is typically manufactured in lengths of 36" (or 1 m), widths of 6", 12", and 18" (15 cm, 31 cm, and 46 cm), and thicknesses from 1" to 4" (25–100 mm) in increments of $\frac{1}{2}$" (13 mm). Besides the rectangular block form, it is available as curved radius block for pipes and **V-grooved** flat block (*Figure 6*). Calcium silicate can be cut by hand or with a table or band saw. When using a hand saw, a coarse-tooth cross-cut type is best. Trim cuts are made with a keyhole saw or boning knife.

Figure 5 A boiler steam drum being insulated with calcium silicate blocks.

Figure 6 Calcium silicate insulation products.

Expanded perlite is a natural form of volcanic glass that has been heated in the manufacturing process to produce a low density material through expansion. Its performance standards are specified under *ASTM C 610*. Perlite has many properties similar to calcium silicate and is often considered an equivalent insulating material. It has a service temperature range from 80°F to 1,200°F (27°C to 650°C).

Perlite block is manufactured in lengths of 36" (1 m), widths of 6" and 12" (15 cm and 31 cm), and in thicknesses from 1½" to 4" (3.8–10 cm) in increments of ½" (13 mm). Perlite block has a nominal density of 13 lb/ft^3 (210 kg/m^3). These properties may vary among manufacturers but are typical. In addition to the standard block forms, perlite has the ability to be cast in molds to produce insulation covers for standard pipe fittings, and can be machined into specialized shapes. The same tools used for CalSil can be used for perlite block.

1.2.2 Cellular Glass Block Insulation

Cellular glass block is a closed-cell glass composite material that is both fireproof and impervious to water absorption and diffusion. It also has excellent resistance to compression. It is easy to fabricate in the field with basic hand tools, and it provides excellent insulating performance.

Glass foam is manufactured by heating ground glass particles with a powdered agent, such as limestone, in molds. At the temperature when the glass particles soften and stick together, the limestone chemically breaks down, releasing a gas that produces tiny bubbles in the glass material. After cooling, the glass foam consists of enclosed cells of glass having no paths for liquids or vapors available through the material (e.g., no wicking).

Cellular glass performance specifications are provided in *ASTM C 552*. The ASTM Type I standard applies to cellular glass block and billets. It is designed for use on surfaces operating between about –450°F to 900°F (–268°C to 480°C). It is used more often on cold applications than on hot applications. Since it is a closed-cell material, it will not absorb moisture and has an extremely low water vapor permeance. It also is used in applications where absorption of the product contained in equipment could create a safety hazard. Cellular glass block has a nominal density of approximately 7.2 lb/ft^3 (115 kg/m^3). Loadbearing cellular glass for vessel bases have a higher density (8.7 lb/ft^3 or 140 kg/m^3).

Block is typically furnished by the manufacturer in 18" × 24" or 18" × 36" sections (*Figure 7*). The standard range of thicknesses is generally 2" to 8" in ½" increments. Metric blocks are 45 cm × 60 cm in size, and range in thickness from 4 cm to 18 cm in increments of 1 cm. Available sizes will vary by manufacturer and product line, and special sizes and shapes can be obtained by special order. Thicker cellular glass block is typically supplied as billets.

Cellular glass products do have some limitations. They conduct heat more readily (have slightly higher K-factors) compared to polymeric foams or mineral fiber insulation. While cellular glass does have excellent compressive strength, it will form stress cracks when exposed to high temperatures. It also must be set in mastic or covered with a protective coating if subjected to mechanical vibrations.

Cellular glass block is easily cut with a crosscut hand saw or keyhole saw. However, cellular glass is abrasive, and it will soon wear down the teeth of any saw used to cut it. Carbide-toothed saws are ideal for this purpose.

CAUTION

Always wear goggles, work gloves, and tight-weave, long-sleeved shirts when working with cellular glass products. Cellular glass is **friable** and can produce tiny, sharp particles that can work their way into skin and eyes.

1.2.3 Other Block Insulation Materials

More effective and easier-to-use mechanical insulation materials are continually being developed as new processes and applications are explored. The following is a summary of some other block insulation materials you may encounter in your work.

- *Alumina-silica fiber blocks* – Alumina-silica high-temperature fiber (HTF) ceramic blocks provide insulation for refractory and very high temperature applications. This block insulation has a maximum service temperature of up to 2,600°F (1,430°C). The relatively low density (nominally 15 lb/ft^3 or 240 kg/m^3) allows for reduced effort during installation. Sizes of blocks will vary among manufacturers; one offers blocks in 6" × 12" (15 cm × 31 cm) or 12" × 12" (31 cm × 31 cm) in various thicknesses between 3" and 12" (7.6–31 cm).
- *Microporous block insulation* – This material is similar in performance to microporous board insulation, but is available in block forms. *ASTM C 1676* sets specifications for this type of insulation. Type I applies to microporous board and block insulation. (See Section 1.1.4 for additional information.)

Figure 7 An example of a rigid glass foam block.

1.3.0 Vapor Retarder Systems

Board and block insulation materials are often specified for cold-service and below-ambient systems because of their excellent insulating characteristics. However, the seams between pieces of such insulation present a real challenge for maintaining an effective vapor retarder boundary that is necessary for cold-service systems. These must be sealed to assure proper performance. Vapor retarder systems consist of low- to zero-perm mastics, membranes, jackets, and appropriate accessories, and often require multiple layers to achieve the required performance.

> **NOTE**
> Recall that the lower the value of water-vapor permeance for a material, the better vapor retarder it is.

1.3.1 Mastics for Below-Ambient Service

Below-ambient systems can be classified by the following temperature ranges:

- *Cool or cold service*: between 32°F (0°C) and ambient temperature
- *Low-temperature service*: between –50° (–46°C) and 32°F (0°C)
- *Cryogenic service*: –50°F (–46°C) and below

Mastics are ideal for creating a vapor retardant system over board and block insulation. They can easily be applied by trowel, brush, or spraying, though troweling wis probably best for filling the seams between the sections of primary insulation. Mastics may also be used to protect the entire insulating system. Besides providing a vapor retardant boundary, mastics may also protect against absorption of liquid water, mechanical damage, and weather and other environmental conditions. Mastics can also provide a bedding compound for other accessories such as reinforcement meshes, finish fabrics, and jacket materials.

Not all mastics can be used as vapor retarders. Insulation specifications for cold service will require mastics with permeances of less than 0.5 perm and much lower when necessary. Manufacturers of mastics usually include permeance ratings in the product data sheets.

Water-based polymer mastics are typically used for low-temperature to ambient services. They are optimum for sealing seams, punctures in membranes, and terminating vapor retarder jackets and membranes. These mastics also have the benefit of easy cleanup during application, low personnel hazard, and good flame resistance and smoke generation. They can also be used in exterior and below-grade applications.

Solvent-based and resin mastics are suitable for cryogenic service, having permeance values less than 0.05 perm. Cryogenic service involves extreme conditions that require multiple layers of insulation and close attention to the construction of vapor dams throughout the insulation. While these mastics show good flame resistance and smoke generation properties, their use may be restricted due to flammability during application. Spills and drips are much harder to clean up when dry, so make it a point of cleaning up solvent-based mastics while still wet. Proper PPE is also required during application and cleanup.

1.3.2 Vapor-Retarder Jacket Systems

Low- and cryogenic-service applications must reduce the diffusion of water vapor toward the cold surface to the lowest possible extent. No primary insulation, even when installed in multiple layers and coated with low-permeance mastics, is able to meet this requirement. Engineers will therefore specify one or more layers of vapor retarder membranes, and sometimes an outer jacket, as key components of the complete insulation system.

Membranes may consist of thin polymer sheets of very low-perm material. Some products (such as SARAN™ products) have a lower service temperature limit of –40°F (–40°C). They can be applied over a mastic bedding and its edges sealed with associated tape. Other membranes (such as Polyguard's Zero-Perm® line) may be constructed of a thin, multi-ply, composite polyester-aluminum material, which also serves as an excellent environmental protective jacket (*Figure 8*).

Laminates can consist of a heavy composite-membrane product that combines a polymer film with a thick rubberized asphalt layer. Such materials are advertised as self-healing and usually have a self-adhesive feature. These membrane

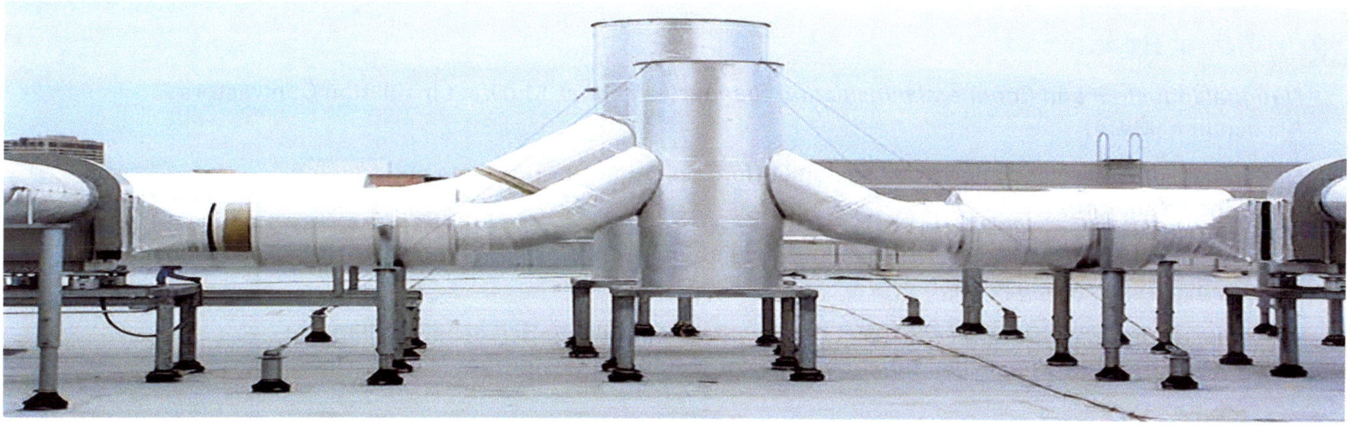

Figure 8 An aluminum-composite membrane installed on an outside ventilation duct system.

products are just a few examples of those available from various manufacturers, and their properties may change over time as they are improved or replaced. Newly-developed vapor retarder products are continually being introduced to the industry.

Most vapor retarder membrane materials perform their function only if they are properly sealed and intact. One small puncture can compromise the entire installation. For this reason, various kinds of rigid jacketing are required to protect the membrane system. These are commonly PVC, aluminum, or stainless steel. The metal jackets have low susceptibility to corrosion that can be caused by moisture condensation on the inner surfaces. Some metal jacket products include a low-permeability coating on the inside surface of the sheathing to minimize this possibility.

NOTE

A general overview of insulation jacketing systems is provided in NCCER Module 19313, "Jacketing Systems".

Additional Resources

National Industrial and Commercial Insulation Standards Manual, Midwest Insulation Contractors Association (MICA).

National Insulation Association (NIA) website offers resources for products and training. **www.insulation.org**.

"Exploring Insulation Materials: Microporous Insulation" (online article), National Insulation Association (NIA). July 2013. **www.insulation.org/io**

Mechanical Insulation Design Guide, National Institute of Building Sciences. Available at **www.wbdg.org/guides-specifications/mechanical-insulation-design-guide**.

1.0.0 Section Review

1. A billet of insulating material is a _____.
   a. large, solid, uniform block
   b. block glued up from layers
   c. flexible roll
   d. bag of particles

2. What characteristic of block insulation makes it different from board insulation?
   a. It's denser and more fibrous.
   b. It's more expensive.
   c. It's generally thicker and more rigid.
   d. It's generally thinner and more flexible.

3. Why are complex vapor retarder systems essential to the performance of board and block insulation?
   a. Board and block insulations tend to release water vapor at high temperatures.
   b. Board and block insulations aren't naturally water-repellent.
   c. Board and block insulations are difficult to make smooth and attractive.
   d. Board and block insulations have seams that must be sealed to keep out water vapor.

SECTION TWO

2.0.0 INSTALLING BOARD AND BLOCK INSULATION

Objective

Explain the layout and installation methods for board and block insulation.

a. Explain how to estimate, lay out, and cut lags.
b. Explain how to install board and block insulation on curved surfaces.
c. Explain how to insulate the head of a vessel.

Performance Tasks

1. Assuming an insulation thickness of 2 inches (5 cm), determine the number of courses of block, the number of lags, and the bevel angle required on the lags for a given vessel.
2. Insulate the sidewall of a vessel using lags, curved segments, or scored block.
3. Insulate the head of a vessel.

Trade Terms

Convex: A surface that appears to be outwardly curved or domed.

Course: As with masonry and similar trades, refers to one row of many rows of sections of insulation.

Elliptical: Shaped like an ellipse, which is a closed shape likened to a flattened circle. Curved lines and three-dimensional shapes formed from a portion of an ellipse are also referred to as being elliptical.

Floating ring: In the insulating trade, a ring formed of twisted wires, cable, or thin rod that is not attached to the underlying surface (i.e., it's floating) and is used to anchor insulation securement bands. Also called *chokers*, they are installed around manway covers, pipe flanges, and other obstructions on the surface to be insulated.

Gores: Sections of a curved surface, such as the surface of a sphere, that when flattened out, have little or no change of shape. The surface maps of model globes of the earth are created from gores glued to a ball.

Juncture: A line formed by the intersection of two surfaces. An edge of a cube is the juncture of two of its faces.

Lags: Narrow sections of insulation cut from rigid board or block material whose long sides have been beveled. The beveled sides minimize the air spaces between similar lags next to it when installed on a curved surface.

Rolled angle irons: Curved sections of standard angle iron that has been shaped in an industrial rolling mill. The rolled angle iron is welded to the curved sidewall of a vessel to form a horizontal shelf or ledge.

Scored: A type of grooving in rigid board insulation consisting of vertical slots or notches. The scoring gives some flexibility to the otherwise inflexible insulation to minimize fracturing of the material when bent. Compare with the term *V-groove*.

Spheroidal: Shaped like a sphere (or a part of a sphere). Spheroidal forms include spheres, hemispheres, and ellipsoids. In the insulating trade, it describes many types of cylindrical vessel heads.

Sub-girts: In industrial construction, the term girt refers to a secondary support iron (T-, H-, or L-bar) attached to primary structural components and surfaces for stiffening, reinforcement, or to support other equipment. A sub-girt is a similar bar attached to the underlying structure that supports an overlying cover or sheathing.

Tangent lines: When referring to cylindrical vessels, the lines where the vessel head begins to curve away from the straight vessel sidewall. On most vessels, these lines are visible as either weld beads or flanges.

Tanks: Fully- or partially-enclosed vessels mainly used for storing liquids or gases. Compare with the term vessel.

Tan-to-tan: A mechanical trade term referring to the distance between the tangent lines at opposite ends of a double-headed cylindrical vessel. Refer to the term tangent line.

Trapezoids: Four-sided figures in which two opposite sides are parallel and one of the parallel sides is shorter than the other.

Board insulation can be applied to any flat surface or equipment. The term *equipment* generally refers to pumps, **tanks**, vessels, and irregularly-shaped mechanical devices and machinery. Since a tank is a type of vessel, only the term *vessel* will be used throughout this module. The application of semi-rigid fiberglass and mineral wool board to ductwork and other flat surfaces is described in other NCCER *Mechanical Insulating* modules. This module will focus on using board and block materials to insulate curved walls and **convex** or **elliptical** vessel heads, which can be somewhat challenging until the methods are mastered.

2.1.0 Measuring and Cutting Lags

Molded pipe covering, curved segments, or scored block are the preferred shapes for insulating small-diameter vessel sidewalls. If these shapes are not available, flat block may be cut into beveled segments to fit. These segments, called lags (*Figure 9*), are beveled on the sides to fit together on the curved surface.

Lags are fabricated from standard-size block insulation by cutting with a hand or power saw. The following instructions will describe how to calculate lag size and how to cut lags by hand. While these instructions are time-consuming, they can be used in situations where prefabricated materials are not available or supplied by the vessel manufacturer.

2.1.1 Determine Lag Dimensions and Number

To estimate the number of lags required to cover a vessel sidewall and their dimensions, follow these general steps (refer to the example vessel in *Figure 10* while reading the steps):

Step 1 Determine the diameter of the vessel to be insulated by measuring its circumference and dividing by pi (π). If the vessel has a flat head, its diameter can be measured directly. The diameter may also be obtained from engineering plans.

Step 2 Add twice the thickness of the insulation to the diameter.

Step 3 Determine the outside circumference of the insulation (C_I). Multiply the total in Step 2 by π (3.14). The following equation summarizes Steps 1 through 3:

$$C_I = (D + T + T) \times \pi$$
$$C_I = (72" + 2" + 2") \times 3.14$$
$$C_I = 76" \times 3.14$$
$$C_I = 239.6" \textit{ (approx. 239")}$$

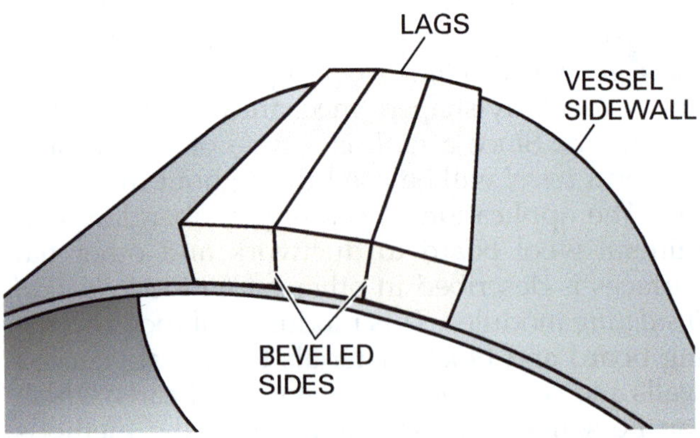

Figure 9 Beveled lags on a curved surface.

Step 4 Determine the number of lags (segments) required per course. Divide the insulation circumference by the approximate outside width of the lag as it lies snugly against the vessel. The smaller the vessel diameter the narrower the lag width. For estimation purposes, the approximate width should be in whole-number inches (in this example, 3"). The following equation shows how to calculate the number of lags:

Number of lags =
$$C_I \div \text{outside lag width}$$
$$239" \div 3"$$
$$79.66 \textit{ (approx. 80 lags per course)}$$

NOTE

This method will usually produce a usable estimate for lags per course. It is always a good practice to verify the outer lag width by dividing the insulation circumference by the whole number lags per course. If the resulting lag width varies more than $\frac{1}{16}$ inch from the assumed width, then use the revised outer width to minimize the amount of trimming for the last lag in the course.

Step 5 Determine the inside lag width. Use the vessel's uninsulated circumference (C_V) by multiplying the diameter of the vessel (D_V) by π.

$$C_V = D_V \times \pi$$
$$C_V = 72" \times 3.14$$
$$C_V = 226.1"$$

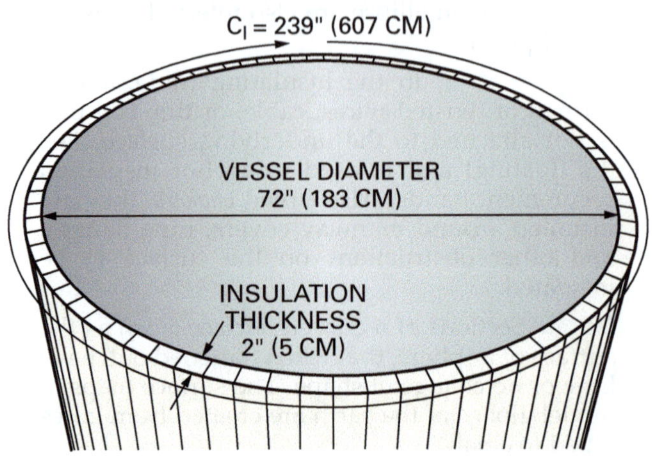

Figure 10 Dimensions required for estimating one course of lag materials.

Step 6 Divide the result by the number of lags determined in Step 4. This is done as follows:

Inside lag width = C_V ÷ number of lags
Inside lag width = 226.1" ÷ 80 lags
Inside lag width = 2.83" *or* $2^{13}/_{16}$"

Step 7 Determine the number of courses required. This requires knowing the distance between the vessel's upper and lower **tangent lines**. Measure the **tan-to-tan** height of the vessel (in this example, it is 12' or 144") and divide by the length of the material being cut (typically 36" for CalSil block). For example:

144" ÷ 36"/course = 4 courses

CAUTION

If the insulation specifications require overlapping the upper or lower vessel tangents by the thickness of the insulation, then be sure to include these extra lengths in the tan-to-tan measurement before determining the number of courses.

Step 8 Determine the total number of lags required. Multiply the number of lags per course by the number of courses required, as follows:

80 lags/course × 4 courses = 320 lags

NOTE

Figure 11 shows the courses of lags unstaggered, which is *not* the correct method of installation. The figure is designed to clearly illustrate how the quantity of insulation material is estimated. The quantity will not be affected the final arrangement of the lags.

To summarize, a vessel 12 feet high between tangent lines and 6 feet in diameter, with a 2" thickness of insulation, requires 80 lags per course and four courses. Each lag is 3 inches wide on the outside, $2^{13}/_{16}$" wide on the inside, and 36" long. A total of 320 lags is required. *Figure 11* visually summarizes this example.

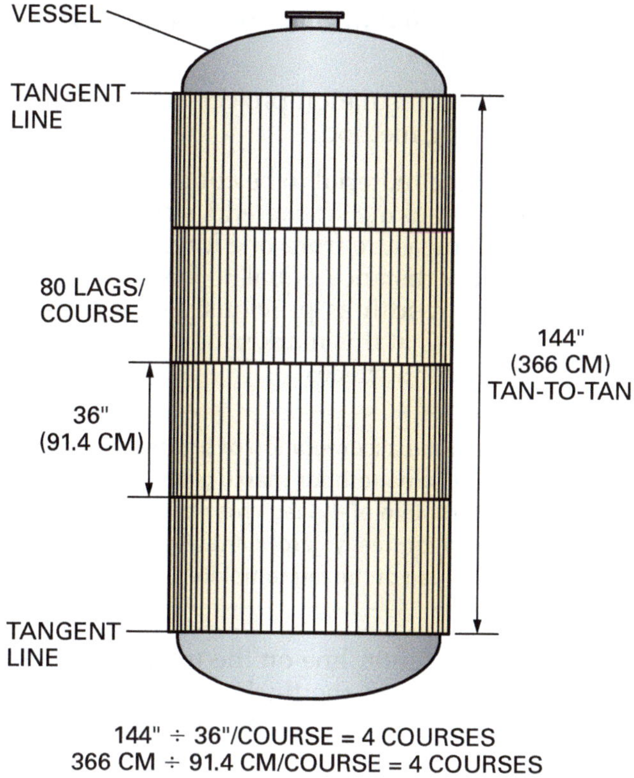

Figure 11 Estimation process for covering the sidewall surface of a cylindrical vessel with calcium silicate lags.

Think About It

How Many Lags?

Determine the number of lags required to insulate the sidewall of a vertical vessel. The vessel is supported on legs and has both upper and lower elliptical heads. The key dimensions required are as follows:

- Insulation is 3" (7.6 cm) board CalSil, 12" × 36" (30.5 cm × 91.4 cm)
- Vessel outer diameter is 120" (305 cm)
- Assume outer lag width is 4" (10 cm)
- Tan-to-tan height is 102" (259 cm)
- Insulation must overlap top and bottom covers to thickness of insulation

2.1.2 Cutting Calcium Silicate / Expanded Perlite Lags

Lags for a vessel insulation project can be cut using the following method:

Step 1 Place the material to be cut on the cutting table. Looking at the end surface of the piece, measure from one long edge of the block a distance equal to half the outer surface width as previously determined. Draw a C at this mark square to the inner and outer surfaces.

> **NOTE**
> To prevent the accumulation of errors by using a cut lag as a pattern for marking other lags, make a template of the lag cross-section from sheet metal or heavy cardboard. *Figure 12* shows the use of the lag template.

Step 2 Draw a center line on the template stock, then mark off the thickness of the lag on the center line. Center the outside and inside widths at the endpoints of the line and draw lines square to the center line with these widths. Using the examples dimensions:

- Outside Width = 3 inches
- Inside Width = $2^{13}/_{16}$ inches

Step 3 Connect the corresponding ends of the inner and outer widths to create the cutting bevels. Cut out the trapezoidal pattern. Place the pattern on the center line made on the end of the block made in Step 1 and trace the pattern. Mark the opposite end of the material to match.

Step 4 Using a straightedge, mark lines connecting the end marks on both surfaces to make the edges of the inner and outer faces of block.

Step 5 Hold the hand saw at a corresponding angle to the end mark and cut along the face mark. Monitor the alignment of the saw and continue cutting the bevel until the cut is complete.

Step 6 Reverse the pattern and mark the ends and faces on the leftover block of insulation for the next cut as described in Steps 3 through 5. Complete the cut and reverse the pattern and cuts repeatedly to obtain the largest number of whole pieces from each block. This practice minimizes waste.

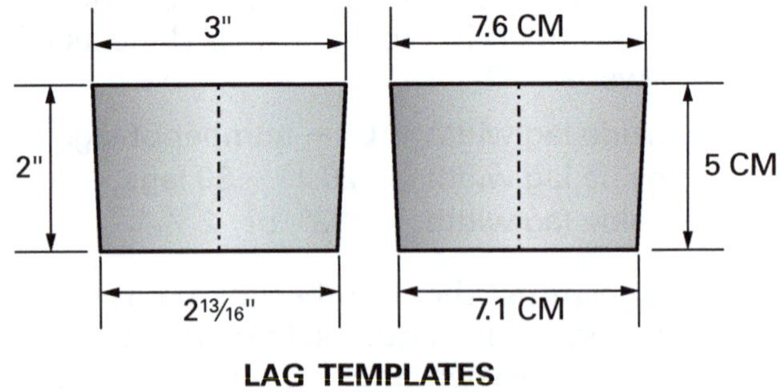

Figure 12 Patterns for transferring a lag cross section.

> **NOTE**
> If using a tilt-arbor table saw to cut lags, the proper side bevel can usually be cut more accurately by using some basic math.

Step 7 To determine the bevel angle for setting the saw, divide 360° by the number of lags per course. For example:

$$360° \div 80 \text{ lags} = 4.5°$$

Each side bevel on the lag is half this angle:

$$4.5° \div 2 = 2.25° \textit{ (approx. 3°)}$$

Step 8 Cut the lags from the flat block using the dimensions in Step 2. Check the cut bevel angle with a protractor and adjust the saw if needed. If more than one lag can be cut from a block, remember to reverse the block after each cut. Make the least number of cuts possible to reduce waste.

> **CAUTION**
> Power saws can create a lot of dust during repetitive sawing operations. Be sure to use the appropriate PPE when working in the vicinity of insulation sawing operations. Also use a dust collection system if available.

Step 9 Install the lags tightly spaced as described previously. Measure and cut the final pieces for a good fit.

The estimating and cutting methods just discussed can be used for covering the sidewall surface of any cylindrical vessel. Carefully follow the steps and a neat job will result.

2.2.0 Installing Block and Board Insulation on Curved Surfaces

Lags are one method for covering cylindrical vessels and other curved surfaces. Blocks or boards that have been scored or V-grooved may also be used. The cut or grooved side is placed against the curved surface, closing up the cuts to permit a custom fit (*Figure 13*). With rigid block or board, it is normal that the material cracks along the grooves. These can be filled with mastic if required by specifications.

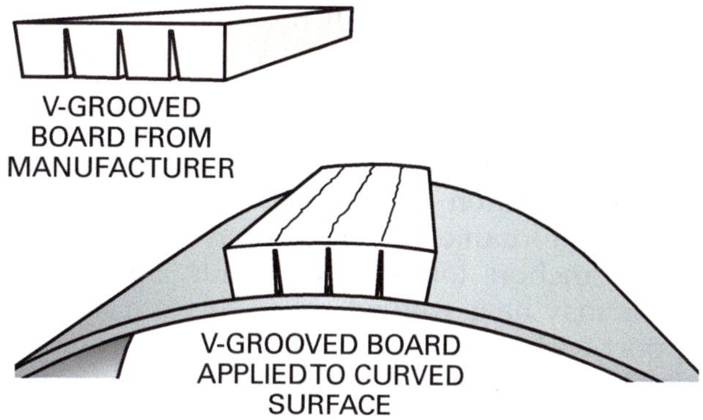

Figure 13 V-grooved board insulation.

2.2.1 Vertical Vessel Installations

Vessels may be either horizontal or vertical. This section covers the practices for insulating vertical (upright) vessels. Upright storage vessels may sit directly on a pad and not require the bottom to be insulated. Cold and cryogenic service vessels may be built setting on a layer of insulation. They may also be on legs or a skirt and require bottom insulation.

Heads of vessels may be flat, domed, spheroidal, or have a cone or pitched roof (*Figure 14*). How the sidewall insulation meets the vessel top insulation depends on the shape of the top. For flat-topped vessels, and those without head flanges or insulation support rings near the tangent line, the sidewall insulation should extend above the rim or tangent line to overlap the edges of the board insulation on top. For domed, conical, or elliptical heads, especially in larger vessels, the sidewall insulation may end at a support ring at or below the tangent line. In general, on vessels supported by legs, the sidewall insulation extends below the tangent line or rim of the lower head.

Joints in board or block insulation must fit tightly. There should be no grooves or joints larger than $\frac{1}{8}$" (3 mm) wide. Block and board insulation should lie tightly against the surface to which they are applied. Significant voids should not be allowed between insulation sections, and between the surface being insulated and the insulation. Also, all seams between insulation sections should be filled to the depth of the insulation with a compatible insulating cement or mastic.

Small vessels usually are not equipped with insulation anchors or support rings. However, many large upright storage vessels are manufactured with insulation support rings to bear the weight of block and board sections. They may also have a welded lip at the top of the sidewall that extends out over the sidewall insulation to which sheathing is attached. Some vessels may

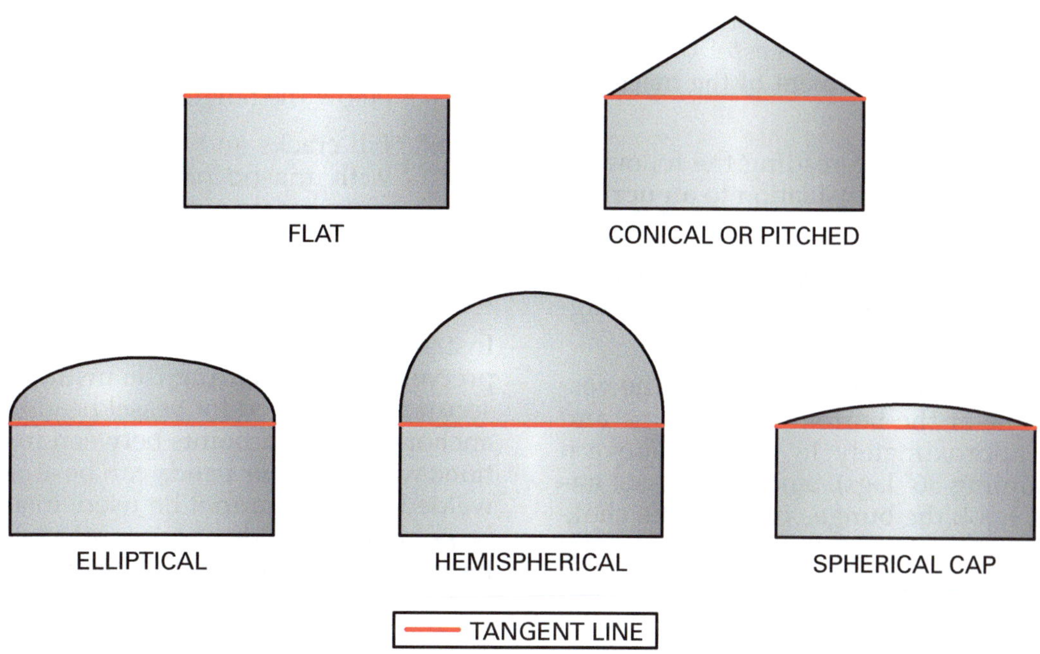

Figure 14 Examples of vessel head forms.

have welded insulation anchors installed on the heads of the vessels, especially the lower head for raised vessels.

Vessel sidewall insulation is generally secured by banding it in place. Vessel head, end, top, and bottom insulation is attached by anchors and wire. Large-diameter vessels may also require welded anchors for securing bands. Some large vessels may have welded **rolled angle irons** that support each course of insulation and provide a foundation for sheet metal jacket or sheathing.

The following is a description of installing block insulation to an upright process water vessel that is about 10' (3.0 m) high and 6' (1.8 m) in diameter. The vessel has straight sidewalls and elliptical bottom and top heads. The vessel is supported by legs, with the bottom of the vessel about 24" (61 cm) from the floor. This installation may use lags, molded curved sections, V-grooved blocks, or scored blocks.

Before beginning work, the following checks should be made:

- The mechanical foreman certifies that all piping connections are complete or at least accessible. Vessels are normally tested at the factory for leaks and may be insulated before all pipe connections are complete if pipe connections are left accessible.
- Determine whether any vessel leg connections protrude from the sides of the vessel. If so, place a double twist of wire (a choker) around the top part of the legs to form an anchor for the lower securement bands after the sidewall insulation is installed (*Figure 15*). An alternate for chokers is to weld lugs to legs or cut small holes through the leg to allow the passage of bands. Alterations to the vessel or its supports require the involvement of the mechanical foreman.

Refer to *Figure 16* while reading the following steps for installing block insulation to an upright process water vessel:

Step 1 Cut a sufficient number of lags in half for the first course in preparation for staggering the circumferential joints.

Step 2 Place two bungee cords around the vessel, one at the bottom tangent line and one approximately 18" (45 cm) above it (assuming 36" lags). Since this vessel has legs, hook the bungee cords to the chokers.

Step 3 Place one lag under the bungee cords, with the bottom end extending below the lower tangent line a distance equal to the thickness of the insulation (or according to specifications). Make sure the lag is plumb with a level.

Step 4 Place a half-section of lag to make a staggered end joint. Continue placing alternating full sections and half sections to complete the course. Make sure all joints are tight. Make cutouts for legs and other obstructions as necessary. (It will probably be necessary to trim the width of the final lag for proper fit. The lags should lie snugly against the vessel surface, with all joints tight.)

Step 5 Place a steel band near the lower ends of all the lags at the lower tangent line and another just below the top ends of the half-sections. Anchor the bands at the leg chokers, if required. Tighten the bands with a banding tool. Using a mallet, tap the bands to ensure a snug fit. Engage the ears of the band clip. Cut off excess banding.

Step 6 Move the bungee cords up the vessel and continue placing lags to complete the vessel sidewall insulation. Carry the sidewall insulation above the upper tangent line to a distance equal to the thickness of the insulation being used. Place bands for each course just below the middle stagger joint.

Step 7 For the top course, cut the full lags to the specified height and fill in the staggered sections to match. Place a top band just below the tangent line as in *Figure 17*.

Step 8 Fill cracks and voids at legs and cutouts with mastic or insulating cement and scrap pieces of insulation.

2.2.2 Insulating Vessel Heads

Installation of insulation on a vessel head requires preparation for securing the insulation. The best securement method for vessel heads uses welded anchors located at points between insulation sections where wire or bands can be attached. When welded anchors cannot be used, insulation must be secured by bands or wires attached to circumferential bands around the sidewall.

Figure 15 Example vessel with attachment wires around legs.

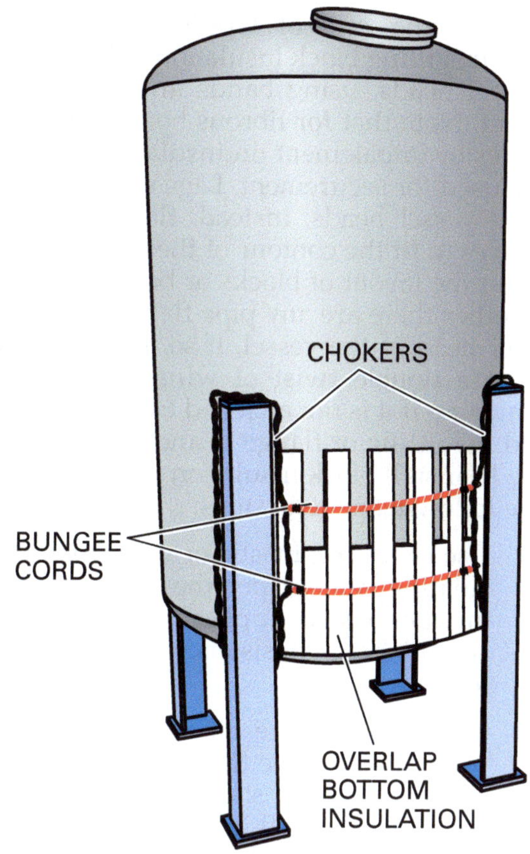

Figure 16 Installing the first course of insulation with bungee cords.

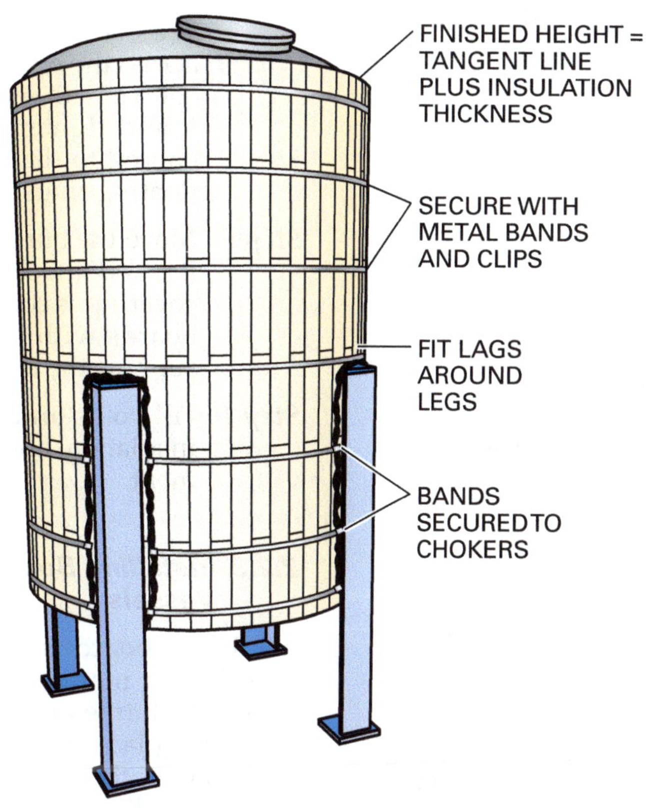

Figure 17 Completing the sidewall insulation.

The following series of steps is a general procedure for installing block insulation to bottom and top vessel heads, using bands and wire for attachment. Note that for fibrous board insulation, adhesives or impalement on insulation pins may also be used for securement. Lags usually are not used on vessel heads; instead, flat block is cut into pieces to fit the contour of the surface. When planning the layout of blocks or boards, check to see whether there are any pipe flanges or fittings in either head of the vessel. If so, place a choker made of a double twist of wire (also called a floating ring if it is not attached to any structure) around the fitting or flange to anchor the wire or bands. To install block insulation to vessel heads, use the following general steps:

Step 1 Use premixed mastic or mix a pail of insulating cement per the manufacturer's instructions, if applicable. The product should be a consistency that is suitable for troweling.

Step 2 Cut and trim the block or boards into pieces that will lie flat against the surface of the vessel end starting at the juncture with the sidewall insulation. The seams formed by the ends of the blocks/boards should be staggered as with the sides. Boards can be laid in one direction for flat-headed vessels. Boards for domed- or elliptical-headed vessels should be cut into gores (see Section 2.3.0). For conical or shallow-domed heads, a radial pattern may be appropriate as in *Figure 18.*

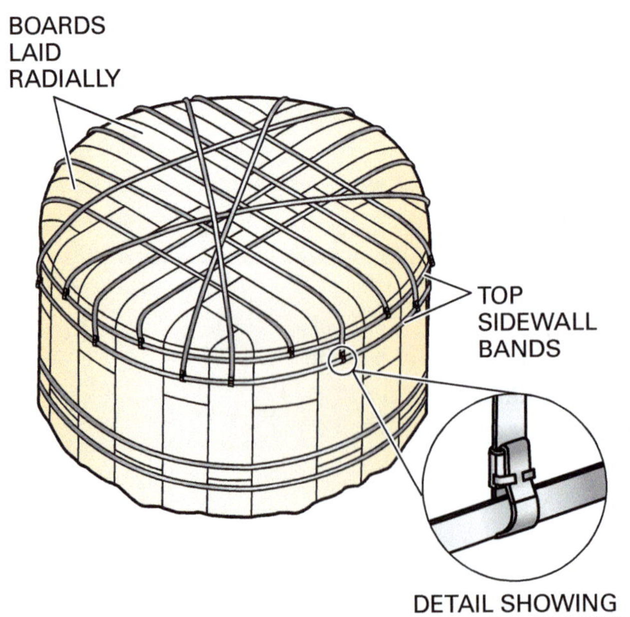

Figure 18 Insulating a shallow-domed vessel top.

Step 3 Lay the head segments closely, fitting one to the other. Fill voids and joints with mastic or insulating cement.

Step 4 After the head insulation is complete, secure the boards with tie wires from welded anchors or with bands attached to the top sidewall band. Cross the bands in a symmetrical pattern so that every board has at least one band securing it. Refer to *Figure 18.*

Step 5 To insulate the lower vessel head as in *Figure 19,* attach a bungee cord to the lower sidewall band, stretch it across the bottom, and attach it to the sidewall band on the other side of the vessel.

Step 6 Cut pieces of insulation in a similar pattern to those for the top, taking into account any fittings or penetrations. Fabricate floating ring attachments around larger flanges and fittings as needed.

Step 7 Coat the contact surface of a piece of insulation with adhesive mastic or insulating cement. Set each piece on the vessel bottom surface under the bungee cord and twist it slightly to ensure good contact. Do not move a piece once it has been placed.

Step 8 Continue placing pieces under the bungee cord until one course is completed across the bottom of the vessel. Secure the course with wires or stretch a band along it and tighten. The band should be anchored to the lower sidewall bands (and floating rings, if applicable).

Step 9 Move the bungee cord and continue placing and attaching pieces until the bottom covering is complete. Secure the blocks or boards with sufficient wires or bands to support the insulation per specifications.

Step 10 Fill voids and joints with scrap pieces of insulation and mastic or insulating cement.

2.2.3 Installing Board on Large-Diameter Vessels

Insulation board is installed on large-diameter vessels in a manner similar to installing block on small-diameter vessels. Generally, board is banded in place. If a vessel is more than 12' (3.7 m) in diameter, the specifications may call for band anchors to allow bands of shorter lengths to be installed. Expansion springs are included in bands

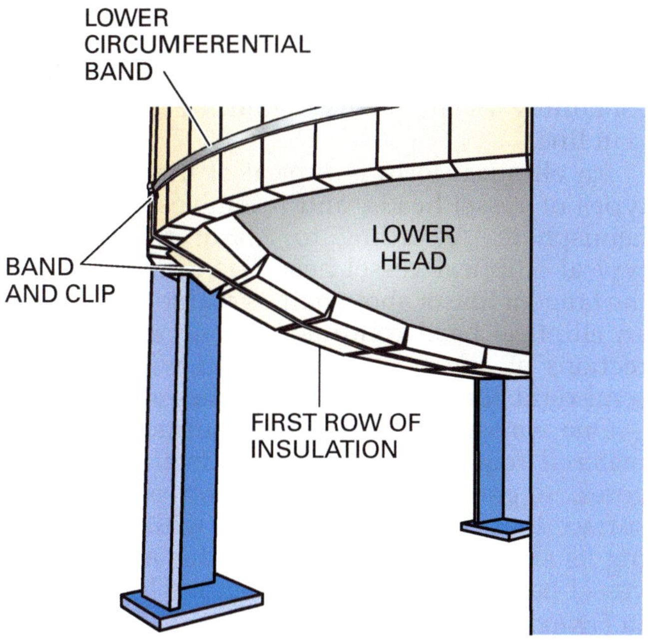

Figure 19 A method for insulating a raised-vessel bottom head.

to allow for vessel expansion and contraction as the vessel heats and cools, if required.

Band anchors may be ½" (13 mm) rod welded on vessels at a height appropriate for the specified insulation thickness. The band length should not exceed 40' (12 m) between anchors. Floating rings are installed around manholes or other large flanged fixtures as usual. Insulation support rings (for example, rolled angle irons) are usually fastened to vertical vessel sidewalls. The support ring may be a flange around the vessel and usually is not be wide enough to protrude beyond the insulation. *Figure 20* shows how band anchors are used for large vessels, as well as a choker securement around a vessel manway. Some job specifications might call for the bottom course of sidewall insulation to be cellular glass for support and the remainder of the insulation to be mineral or glass fiber board. The tops of vessels that could be walking surfaces may be insulated with rigid insulation such as calcium silicate or cellular glass. Fibrous board insulation on tops of large vessels is usually attached with weld or stick pins.

The following steps provide a description of how to install fiberglass board on a large-diameter upright vessel:

Step 1 Rig a rope and pulley so that material may be hoisted to the scaffold in a safe manner.

Step 2 Place a bungee cord around the vessel approximately 1' (30 cm) from the bottom. Place a second bungee cord around the vessel approximately 2' (60 cm) above the first cord.

NOTE

Bungee cords should be stretched so that some force is required to pull the cord away from the vessel but otherwise won't crush the insulation.

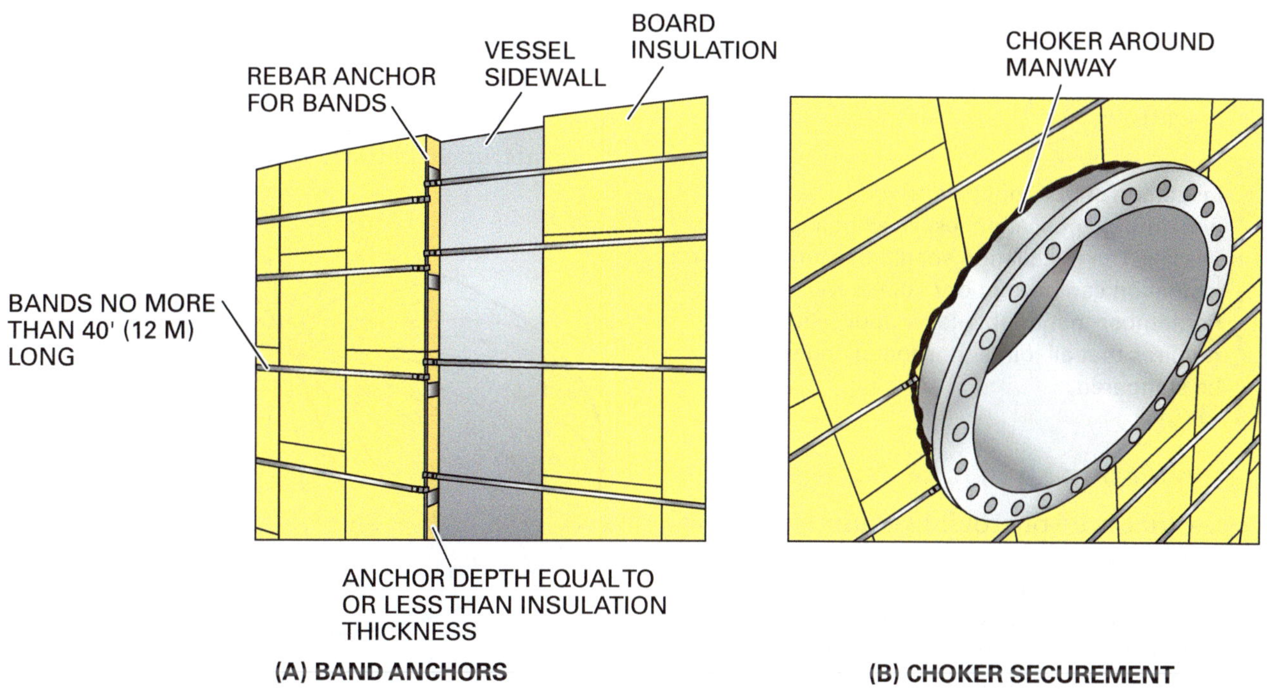

Figure 20 Band anchors and a choker securement.

Step 3 Cut a sufficient number of pieces of board in half to provide a staggered end joint for the lowest course.

Step 4 Place a full piece of board under the bungee cords. Check plumb with a level.

Step 5 Continue alternating half and full pieces until the first course is completed. Trim the last piece for a proper fit. Make cutouts for anchor rod supports and other interference. If band anchor rods interrupt the course, continue with the same course between rods until the course is complete around the vessel circumference.

Step 6 Check that all joints are tight and all pieces fit snugly against the vessel sidewall. Place a securement band on the first course of board just below the stagger joints and tighten. Secure with band seals.

Step 7 Move the bungee cords up to support the next course and continue placing full pieces of board and banding in place until the sidewall insulation is complete.

Step 8 Check that all joints are tight. Pack any cracks with scrap fiberglass.

Step 9 Apply insulation on the top vessel head as described in Section 2.3.0. Install weld pins on the vessel head on centers as called out per specifications. Working around **sub-girts** may be necessary if the head is to be finished with sheet metal.

Step 10 Install the specified vapor retarder and jacket as required before removing the scaffold.

CAUTION

Insulation should not be allowed to become wet during application, especially on tops of vessels. The installation should be started when insulation can be kept dry during application. Remember that double-layer application requires that all joints (horizontal and vertical) be staggered.

2.3.0 Insulating an Elliptical Vessel Top

Surfaces that are spheroidal require special attention when planning the installation of block or board insulation. The term *spheroidal* refers to any three-dimensional shape that is related to a sphere. A spheroidal shape includes true spheres, hemispheres, spherical caps, and ellipsoids. Ellipsoidal vessel heads are commonly called elliptical heads in the trade. Other curved vessel heads include flanged and dished heads. Certain forms of vessel heads are combinations of curved surfaces containing a sharply curved knuckle near the tangent line.

An elliptical form is typical of many of these types of vessel heads, and resembles a flattened hemisphere. According to ASME references, a typical elliptical vessel head has a height above the tangent line of about half its diameter. Because an elliptical head continuously curves in all directions at any point, covering it with rigid or semi-rigid board insulation becomes a challenge.

One way to cover a curving surface with flat material requires cutting pieces in the shape of gores. A gore is a narrow section of a curved surface that can be flattened out without changing its shape very much. A gore for an elliptical vessel head would look like the one highlighted in *Figure 21*.

2.3.1 Creating the Gore Pattern

The preferred method for generating a pattern for a vessel head gore is by the use of special engineering software. These programs are typically used by custom insulation fabricators. The insulation is delivered as stacks of numbered sections. Each number corresponds to a course starting at the edge of the vessel head. The insulation sections are shaped like rounded **trapezoids** and grow narrower as each course approaches the center of the head (*Figure 22*). Because the shapes are computer-generated for a specific vessel head shape, the pieces fit closely and make efficient use of the materials.

If a gore-pattern program is not available, the mechanical insulator will need to know how to draft up a pattern from field measurements.

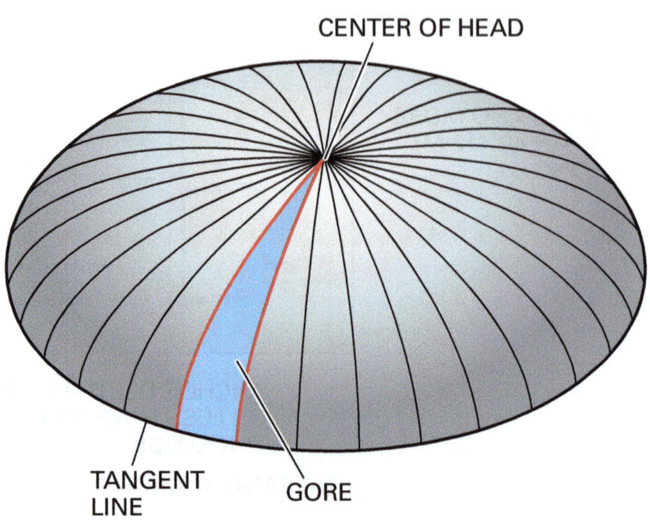

Figure 21 Gore of an elliptical vessel head.

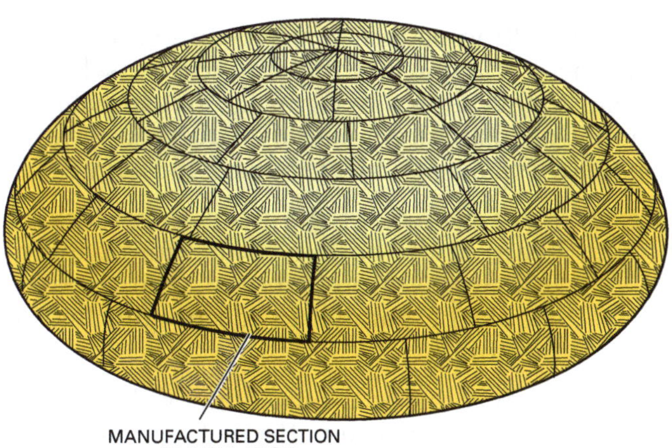

Figure 22 Computer-generated and manufactured insulating system.

There is no single correct way to insulate an elliptical head. Each method has its advantages relating to the type of material being used. Semi-rigid and V-grooved insulation can bend slightly and therefore has more options for covering a vessel head. The method discussed in this section can be used for a variety of such materials.

Before laying out the insulation patterns, the following dimensions and numbers should be obtained by direct measurement and calculation:

- *Circumference of the vessel* – Measure at the tangent line of the head.
- *Radial distance from the center of the head to the tangent line* – Measure along the vessel's surface. If there is a pipe or other obstruction at the head's center, then measure the radial distance from the obstruction's surface. Measure the obstruction's diameter or circumference to find its radius and add this value to the vessel's radial measurement.
- *Maximum gore width* – Place a straightedge against the tangent line and parallel to it. Note the size of the gaps between the straightedge and the vessel surface. To avoid too large a void under the insulation, these gaps on either side where the straightedge touches should be less than about $\frac{1}{2}$" (13 mm). Mark the vessel at the points where the gaps on both sides are $\frac{1}{4}$" to $\frac{1}{2}$" (7–13 mm), and measure this span.

NOTE

If the insulation is semi-rigid, such as mineral wool board, then a wider span can be used limited only by the material's tendency to buckle.

- *Number of gores* – Divide the circumference of the head by the width of the gore just measured. The result will likely not come out even. Round up to the next higher number, which will be the number of gores required to cover the head. Divide the head's outer circumference by this number to obtain the maximum width of the gore at the outer perimeter.

NOTE

The maximum gore width should not be wider than the maximum dimension of the insulation product to be used

- *Widths that determine shape of gore* – Mark the working span just determined on the vessel's head perimeter. Stretch a chalk line from these marks to the center of the vessel head and snap clean radial lines on the head. Measure the distances between the gore side lines at approximately equal intervals radially from the center of the vessel head to the tangent line. Depending on the size and shape of the vessel head, fewer measurements are needed where the vessel curvature is gentle; more measurements are needed where the curvature is sharp. Record the radial distance and associated measurement between the side lines.

NOTE

A valid gore pattern can be generated with at least four measurements, not including the knuckle, if applicable. At the vessel head knuckle, take a measurement just above the knuckle, at the middle at the sharpest curve, and below the knuckle. You can assume that the width of the gore below this point, if applicable, is constant. Again, record the radial surface distance from the head's center for each measurement.

The method outlined in the following steps assumes you are creating a full-sized pattern, which will allow direct measurements and estimates (refer to *Figure 23*); if you do not have the space to lay out a full-sized pattern, you will need to make an accurate scale drawing to provide cutting dimensions:

NOTE

Review NCCER Module 19212, "Trade Math and Layout", for a review of scale drawing methods.

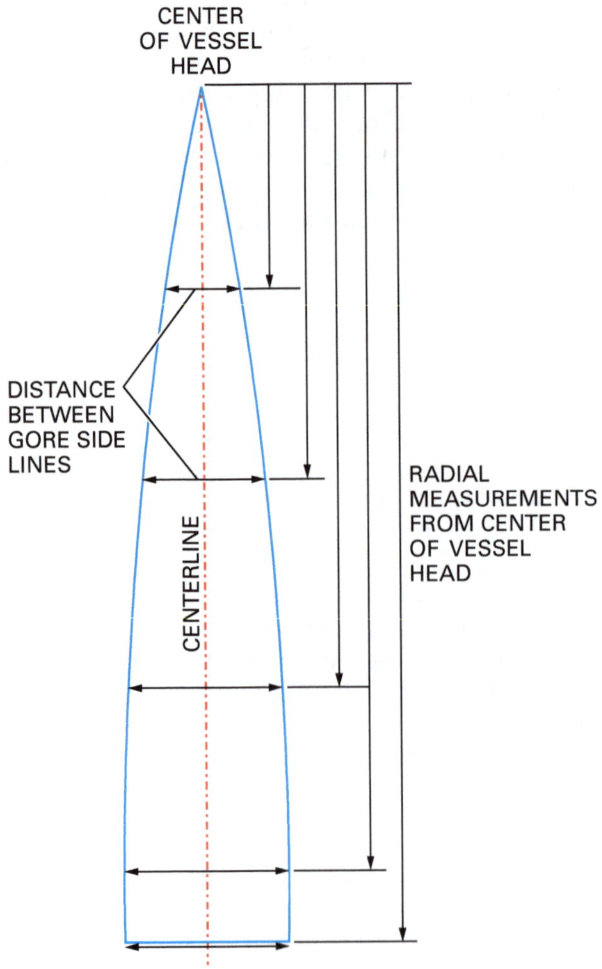

Figure 23 Layout of a gore pattern for a vessel's elliptical head.

Step 1 Draw a straight line representing the center line of the gore. It should be longer than any radial distance recorded in your field notes.

Step 2 Select one end of the center line to represent the center of the vessel head. Stretch a tape measure from the top of the center line to the first radial distance and mark the center line. Repeat for all the other radial distances you recorded.

Step 3 For each radial distance, draw a line square to the center line. Measure half the corresponding gore width on both sides of the center line. Mark these points on the squared line. Repeat for all the radial distance marks.

Step 4 Draw a smooth line from the top of the center line to the bottom mark on one side through the marks made in Step 3. Then repeat to create mirror-image curve on the other side of the center line.

The resulting pattern is a two-dimensional development of a gore for the elliptical vessel head similar to that shown in *Figure 23*. To estimate the amount of insulation material required, perform the following steps while referring to *Figure 24*:

Step 1 Place a piece of the insulation block or board at the bottom of the gore pattern. If using V-groove board, the grooves should be parallel to the bottom line of the gore.

Step 2 Draw a line at the top edge of the insulation on the gore pattern. Then move the insulation so that its bottom edge aligns with the line just drawn. Draw another line along its top edge. Repeat until you reach the top of the gore pattern.

Step 3 Count the number of trapezoidal insulation sections marked out on the pattern. This is the number of courses of insulation.

Step 4 It is likely you will be able to cut two or more pieces of insulation for a gore from a single section. Determine the number of pieces that can be obtained from a section starting at the bottom of the gore pattern, marking the width at the first course, then rotating the section and marking the dimensions of the second course. Note that a remnant too narrow for a lower course might be used near the top where the gore narrows.

Step 5 After determining the number of full insulation sections required to cover one gore, multiply this number by the total number of gores previously calculated. It is always a good practice to add 10–20 percent for waste and rework.

2.3.2 Cutting and Applying Insulation to an Elliptical Vessel Top

Once the gore pattern is laid out on the cutting table, and the section cut lines are drawn in, the sections can be rapidly measured and cut directly on the pattern. One person can be cutting while another is installing. It may be practical to cut all the pieces for one course first before beginning the next course. Straight cuts are made along the edges of the gore pattern between section lines. Waste pieces that are large enough to be used nearer the points of the gores should be set aside and trimmed to shape when needed.

As noted earlier, fibrous board or block insulation (e.g., mineral wool or fiberglass) can

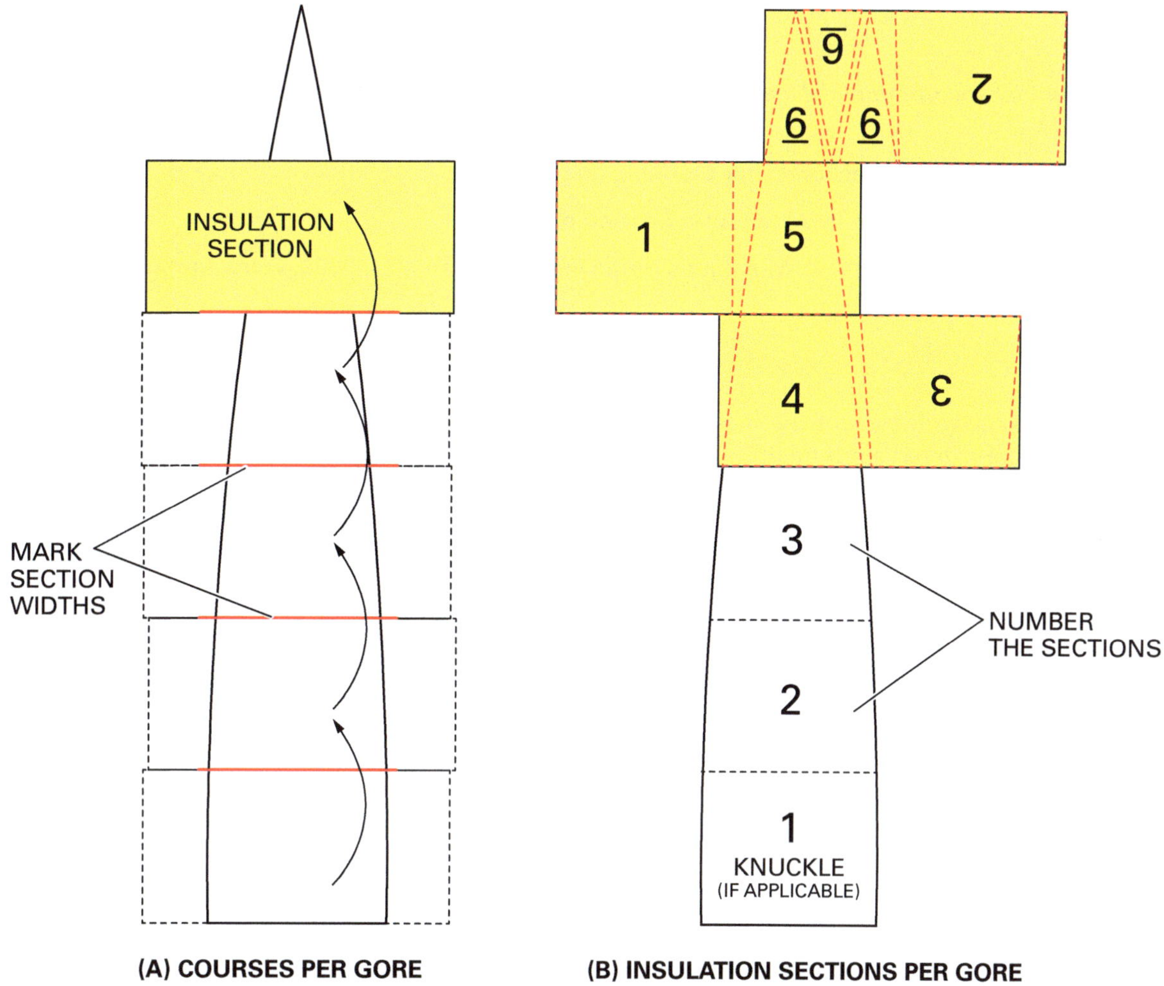

Figure 24 Determining the number of courses per gore and insulation sections per gore.

be impaled on weld or stick pins, or can be glued directly to the vessel with adhesive mastic. Insulation under walkways should be of a rigid cellular glass or calcium silicate/expanded perlite to prevent compression of the insulation. Rigid block or board insulation should be bedded in mastic or insulating cement and secured with wires or banding as described in Section 2.2.2. A single band down the center of each gore can suffice, secured at the top by a floating ring and at the bottom by the top band on the side wall.

Begin installation at the sidewall and work in one direction. Installing board on smaller-diameter vessels may require curving the circumferential edges to fit to the curve of the sidewall. After trimming the final piece in a course to fit, start the next course. Note that, since each course is the same height, joints between sections for one courses can be staggered with adjacent courses by simply rotating the pattern slightly. Fill in gaps with scrap insulation, mastic, or cement as described earlier in this module.

Additional Resources

National Industrial and Commercial Insulation Standards Manual, Midwest Insulation Contractors Association (MICA).

The following websites offer resources for products and training:

National Insulation Association (NIA). **www.insulation.org.**

"Dome Calculator," David South, Jr. Monolithic Dome Institute. March 24, 2009. **www.monolithic.org**

2.0.0 Section Review

1. In mechanical usage, which of the listed items would *not* be considered equipment?
   a. A tank or vessel
   b. A pipe elbow
   c. A mechanical structure
   d. A machine with wheels and levers

2. When staggering lags or boards of insulation on a sidewall of a vessel, the typical dimension of the cut sections in the staggered first course is _____.
   a. $\frac{1}{4}$ of the full section length
   b. $\frac{1}{2}$ of the section length
   c. $\frac{3}{4}$ of the section length
   d. 1 full section

3. Which of the listed vessel head shapes would *not* be considered a spheroidal head?
   a. Hemisphere
   b. Elliptical
   c. Conical
   d. Cap

Summary

In this module you have reviewed the key properties and characteristics of the common block and board insulating materials that are described with greater detail in other modules in the NCCER *Mechanical Insulating* curriculum. New or less common block and board insulating materials have been introduced for completeness.

This module introduces some scenarios that permit the trainee to develop estimating and takeoff work. Skills required for calcium silicate insulation are reinforced and expanded by learning the fabrication and installation of lags for vessel and vessel insulation. Several methods for installing insulation on broad curved surfaces are introduced.

The trainee's professional layout skills are augmented by learning a manual technique for developing gore patterns for spheroidal vessel heads. These methods may be used for insulating any type of surface related to a sphere.

Review Questions

1. A bun of insulating material is a(n) _____.
   a. large, solid, uniform block
   b. block glued up from layers
   c. flexible roll
   d. bag of granules

2. Which of the listed board insulation materials has the highest maximum service temperatures?
   a. Calcium silicate
   b. Alumina silicate
   c. Mineral wool
   d. Microporous composites

3. Calcium silicate is a common form of block insulation for which of the listed service temperature ranges?
   a. Cryogenic
   b. Cold
   c. Hot
   d. Refractory

4. What property makes cellular glass ideal for cold and cryogenic applications?
   a. Very low permeance
   b. High density
   c. Low density
   d. Absorbs liquids easily

5. Which of the listed mastics would be considered an ideal sealant for a cryogenic vapor retardant system?
   a. A mastic with a perm rating less than 0.05.
   b. A water-based mastic.
   c. A mastic with a perm rating of 1.0.
   d. Any available mastic will do.

6. A lag would be most commonly associated with the insulation of which item?
   a. Rectangular duct
   b. The wall of a power plant precipitator
   c. A cylindrical vessel
   d. A pipe elbow

7. The outside width of a lag determines which of the following values, everything else remaining the same?
   a. The insulation circumference (C_I)
   b. The vessel circumference (C_V)
   c. The number of lags per course
   d. The number of courses

8. When insulating vessel heads, which of the following statements is true?
   a. Insulation sections are never staggered on a vessel top.
   b. Elliptical vessel heads are*not*normally insulated in staggered rows.
   c. Insulation is secured only by mastic or cement, and gravity.
   d. Metal bands, wires, or pins are used to secure insulation on top and bottom heads.

9. When insulating the sidewalls of large-diameter vessels, the maximum length of securement banding should be _____.
   a. 40' (12 m)
   b. 30' (9 m)
   c. 25' (7.5 m)
   d. 20' (6 m)

10. Which of the following statements is *false* about cutting and applying insulation in gores on vessel heads?
    a. Numbered templates may be used to cut sections.
    b. Waste material may be used for sections near the point of the gore.
    c. Gaps between sections are filled with scrap insulation, and insulating cement or mastic.
    d. The sections are cut into rectangles of decreasing size based on pattern dimensions.

Trade Terms Introduced in This Module

Billets: A block of material usually built up by cementing thinner layers together.

Block: A usually rigid form of insulating material where its thickness dimension relative to its width and length give it a blocky appearance.

Board: A usually semi-rigid form of insulation typically having an appearance more like a rectangular panel than a block.

Buns: A large uniform block of material from which thinner boards or smaller blocks are cut.

Convex: A surface that appears to be outwardly curved or domed.

Course: As with masonry and similar trades, refers to one row of many rows of sections of insulation.

Elliptical: Shaped like an ellipse, which is a closed shape likened to a flattened circle. Curved lines and three-dimensional shapes formed from a portion of an ellipse are also referred to as being elliptical.

Equipment: In the context used here, the term generally refers to pumps, vessels, and irregularly-shaped mechanical devices.

Floating ring: In the insulating trade, a ring formed of twisted wires, cable, or thin rod that is not attached to the underlying surface (i.e., it's floating) and is used to anchor insulation securement bands. Also called *chokers*, they are installed around manway covers, pipe flanges, and other obstructions on the surface to be insulated.

Friable: The property of a brittle material that tends to crumble into small particles when crushed, rubbed, or scraped.

Gores: Sections of a curved surface, such as the surface of a sphere, that when flattened out, have little or no change of shape. The surface maps of model globes of the earth are created from gores glued to a ball.

Juncture: A line formed by the intersection of two surfaces. An edge of a cube is the juncture of two of its faces.

Lags: Narrow sections of insulation cut from rigid board or block material whose long sides have been beveled. The beveled sides minimize the air spaces between similar lags next to it when installed on a curved surface.

Rolled angle irons: Curved sections of standard angle iron that has been shaped in an industrial rolling mill. The rolled angle iron is welded to the curved sidewall of a vessel to form a horizontal shelf or ledge.

Scored: A type of grooving in rigid board insulation consisting of vertical slots or notches. The scoring gives some flexibility to the otherwise inflexible insulation to minimize fracturing of the material when bent. Compare with the term *V-groove.*

Spheroidal: Shaped like a sphere (or a part of a sphere). Spheroidal forms include spheres, hemispheres, and ellipsoids. In the insulating trade, it describes many types of cylindrical vessel heads.

Sub-girts: In industrial construction, the term girt refers to a secondary support iron (T-, H-, or L-bar) attached to primary structural components and surfaces for stiffening, reinforcement, or to support other equipment. A sub-girt is a similar bar attached to the underlying structure that supports an overlying cover or sheathing.

Tangent lines: When referring to cylindrical vessels, the lines where the vessel head begins to curve away from the straight vessel sidewall. On most vessels, these lines are visible as either weld beads or flanges.

Tanks: Fully- or partially-enclosed vessels mainly used for storing liquids or gases. Compare with the term vessel.

Tan-to-tan: A mechanical trade term referring to the distance between the tangent lines at opposite ends of a double-headed cylindrical vessel. Refer to the term tangent line.

Trapezoids: Four-sided figures in which two opposite sides are parallel and one of the parallel sides is shorter than the other.

Vessels: In the context of mechanical insulating, any fully- or partially-enclosed structures that can be used for storing or processing any type of substance, usually at pressures above or below atmospheric pressure. Vessels are used for many purposes, including containing chemical reactions, in addition to storing or transferring liquids and gases.

V-grooved: A term describing thicker and typically rigid board insulation that has manufactured or fabricated V-shaped slots cut into the material. These grooves allow the board to be bent around curved surfaces with minimal and allowable cracking.

Additional Resources

This module is presents thorough resources for task training. The following reference material is recommended for further study.

Mechanical Insulation Design Guide, National Institute of Building Sciences. Available at **www.wbdg.org/guides-specifications/mechanical-insulation-design-guide.**

National Industrial and Commercial Insulation Standards Manual, Midwest Insulation Contractors Association (MICA).

The following websites offer resources for products and training:

National Insulation Association (NIA). **www.insulation.org**

"Exploring Insulation Materials: Microporous Insulation" (online article), National Insulation Association (NIA). July 2013. **www.insulation.org/io**

"Dome Calculator," David South, Jr. Monolithic Dome Institute. March 24, 2009. **www.monolithic.org**

Figure Credits

Gary J. Bases – BRIL, Inc., Module opener, Figure 5

© iStock.com/Jarin13, Figure 3

Johns Manville Industrial Insulation, Figure 6

Pittsburgh Corning Corporation, Figure 7

Polyguard Products, Inc. – www.polyguard.com, Figure 8

Section Review Answer Key

Answer	Section Reference	Objective
Section One		
1. b	1.0.0	1a
2. c	1.2.0	1b
3. d	1.3.0	1c
Section Two		
1. b	2.0.0	2a
2. b	2.2.1	2b
3. c	2.3.0	2c

NCCER CURRICULA — USER UPDATE

NCCER makes every effort to keep its textbooks up-to-date and free of technical errors. We appreciate your help in this process. If you find an error, a typographical mistake, or an inaccuracy in NCCER's curricula, please fill out this form (or a photocopy), or complete the online form at **www.nccer.org/olf**. Be sure to include the exact module ID number, page number, a detailed description, and your recommended correction. Your input will be brought to the attention of the Authoring Team. Thank you for your assistance.

Instructors – If you have an idea for improving this textbook, or have found that additional materials were necessary to teach this module effectively, please let us know so that we may present your suggestions to the Authoring Team.

NCCER Product Development and Revision

13614 Progress Blvd., Alachua, FL 32615

Email: curriculum@nccer.org

Online: www.nccer.org/olf

❑ Trainee Guide ❑ Lesson Plans ❑ Exam ❑ PowerPoints Other ______________________

Craft / Level: Copyright Date:

Module ID Number / Title:

Section Number(s):

Description:

Recommended Correction:

Your Name:

Address:

Email: Phone:

Sheet-Metal Lagging

Overview

Insulation is critical to the efficiency and safety of many heavy industries, especially electrical power generation, dedicated industrial air pollution control installations, cement manufacturing, and any other manufacturing involving high temperature processes. These structures are usually large enough that they are directly exposed to the weather, which requires a protective sheathing strong enough to withstand the effects of wind and rain to protect the underlying insulation. In this module, you will learn about industrial insulation lagging and flashing.

Module 19312

Trainees with successful module completions may be eligible for credentialing through the NCCER Registry. To learn more, go to **www.nccer.org** or contact us at 1.888.622.3720. Our website has information on the latest product releases and training, as well as online versions of our *Cornerstone* magazine and Pearson's product catalog.

Your feedback is welcome. You may email your comments to **curriculum@nccer.org**, send general comments and inquiries to **info@nccer.org**, or fill in the User Update form at the back of this module.

This information is general in nature and intended for training purposes only. Actual performance of activities described in this manual requires compliance with all applicable operating, service, maintenance, and safety procedures under the direction of qualified personnel. References in this manual to patented or proprietary devices do not constitute a recommendation of their use.

19312 V2

19312
SHEET-METAL LAGGING

Objective

When you have completed this module, you will be able to do the following:

1. Describe various types of sheet-metal lagging and their basic installation considerations.
   a. Describe sheet-metal lagging and identify common power tools and specialty tools needed for fabrication and installation.
   b. Identify and describe various forms of sheet-metal lagging.
   c. Describe common lagging support methods.
   d. Describe how to install lagging and flashing.

Performance Tasks

Under the supervision of your instructor, you should be able to do the following:

1. Flash a top-to-side lagging juncture.
2. Flash a lagging inside vertical corner.
3. Flash a sidewall lagging offset detail.
4. Install sheet-metal lagging to supports.

Trade Terms

Flashing
Galvanic corrosion
Knurled
Lagging
Self-drilling
Self-tapping
Sheet-metal brake
Stiffeners
Sub-girts

Industry Recognized Credentials

If you are training through an NCCER-accredited sponsor, you may be eligible for credentials from NCCER's Registry. The ID number for this module is 19312. Note that this module may have been used in other NCCER curricula and may apply to other level completions. Contact NCCER's Registry at 888.622.3720 or go to **www.nccer.org** for more information.

Contents

Figures and Tables

SECTION ONE

1.0.0 INSULATION LAGGING AND FLASHING

Objective

Describe various types of sheet-metal lagging and their basic installation considerations.

a. Describe sheet-metal lagging and identify common power tools and specialty tools needed for fabrication and installation.
b. Identify and describe various forms of sheet-metal lagging.
c. Describe common lagging support methods.
d. Describe how to install lagging and flashing.

Performance Tasks

1. Flash a top-to-side lagging juncture.
2. Flash a lagging inside vertical corner.
3. Flash a sidewall lagging offset detail.
4. Install sheet-metal lagging to supports.

Trade Terms

Flashing: Narrow strips of sheet metal fabricated to protect joints in a covering as part of a weatherproofing system that may include caulking and neoprene washers for screw fasteners.

Galvanic corrosion: A form of corrosion (rusting) resulting from the creation of an electric current flowing between dissimilar metals in contact with each other.

Knurled: Having narrow milled ridges on a shaft or knob that are used to aid against slippage in a shaft hole or when gripped.

Lagging: With reference to insulation, heavy sheet metal sheathing over insulation, usually ribbed or corrugated, as weatherproofing and to protect insulation against industrial washdown operations.

Self-drilling: Describes a fastener (screw) that is equipped with a special drill-bit-like point that drills through the metalwork to be fastened and taps its own threaded hole. Compare with the term *self-tapping*.

Self-tapping: Describes a fastener equipped with a notched, tapered point that allows the screw to create its own threads in metals softer than the screw. These screws require a pre-drilled hole. Compare with the term *self-drilling*.

Sheet-metal brake: A heavily-constructed machine designed to create straight folds in sheet metal. The process of bending metal into a corner or fold is called *braking*.

Stiffeners: A piece of heavy formed steel, often with a T, H, I, or other cross section, welded to a flat surface to limit flexing. Also an angled ridge formed in sheet metal for the same purpose.

Sub-girts: In industrial construction, the term girt refers to a secondary support iron (T-, H-, or L-bar) attached to primary structural components and surfaces for stiffening, reinforcement, or to support other equipment. A sub-girt is a similar bar attached to the underlying structure that supports an overlying cover or sheathing.

More than half a century ago, the term **lagging** referred to just about any type of insulation material, especially on hot systems such as steam pipes and boilers. As large boiler system designs evolved around World War II, the term *cladding* became popular to identify sheet metal and other materials installed as protective sheathing over insulation, especially for outside surfaces. When sheet aluminum became a popular sheathing material, however, the cladding term was dropped by the industry because it referred to a metal bonded to another metal. Even later, sheet-metal unions, who installed most of the heavy metal sheathing over exterior industrial insulation, came to adopt the lagging identifier in the way it is used today.

1.1.0 Lagging Overview

Most sheet-metal lagging is formed into ribbed sheets provided with flat sheet metal as **flashing**. Sheets of lagging are installed by attachment to **sub-girts** or anchor plates. The sub-girts and anchor plates are usually fastened to the underlying structure before the insulation is applied to provide good support and stiffening for lagging. Flashing is generally attached only to the ribbed sheets or directly to the equipment surfaces at points where the insulation and lagging terminate.

Large flat surfaces are covered with ribbed lagging sheets (*Figure 1*). Flat sheet metal may occasionally cover large surfaces, but usually it is fabricated to enclose complex shapes, corners, and edges as flashing. The flashing is usually the same material type, thickness, and finish as used in the ribbed sheets.

Projects are often located in electric generating stations, power houses, paper mills, and other large industrial plants such as chemical and petrochemical plants, and refineries. Sheet-metal lagging can be installed on power and large industrial boilers, boiler flue structures, precipitators, bag houses, stacks, large ducts, and other apparatus. Much of the lagging installation work can be done at heights exceeding 100' to 200' (30 to 60 m) above ground.

Sheet-metal lagging work may involve some welding. The welding might be special heavy-duty stud type or it might be conventional arc welding used to attach sub-girts and angle iron supports. This type of welding requires welder certification and is usually performed by professional welders, or by the sheet-metal or boilermaker trades.

Scaffolding and rigging are major parts of most lagging jobs. Scaffolds may be built from the ground up, may be swinging scaffolds, or require boom lifts. Insulation and lagging are usually installed from the same scaffold. The lagging is installed as soon as practicable after the insulation to minimize exposure of the insulation to weather. There may be one crew installing insulation and another crew installing lagging, especially if separate trades are involved, or there may be only one crew installing both insulation and lagging.

Sheet-metal lagging work requires additional skills and safety practices compared with regular insulation work. In fact, in union-controlled work, sheet-metal tradesmen will likely perform the bulk of lagging installation. Because most lagging work is off the ground, the use of safety harnesses and safe scaffolds is mandatory. Safety glasses, goggles or face shields, and heavy gloves are necessary when cutting or drilling sheet metal with power tools. All power supplies must be GFCI protected and all power cords and power tools must be approved for its intended use. Work site supervision may require other safety practices and personal protective equipment (PPE).

Because installing sheet-metal lagging is different from other insulation work, different tools and equipment are used. For instance, sheet-metal tools and fabrication machinery are required for forming and installing lagging. Generally, the mechanics furnish their own hand tools. The contractor may furnish some required power tools and equipment such as the following:

Figure 1 Insulation lagging on parts of a power plant boiler island.

- Arc welding machine for welding sub-girts and angle iron
- Beaders, crimpers, and other sheet-metal forming tools for special shapes
- Metal-rated circular saws with metal-cutting blades
- Cutoff saws for cutting sub-girts
- **Sheet-metal brake** for bending and crimping flashing
- Lock former machine for use with a power-flanging machine
- Power-flanging (e.g., Pittsburgh®) machine for making watertight seams and for forming flat sheets used to flash **stiffeners** or complex shapes
- Reciprocating saws
- Sheet-metal roller for shaping metal for curved surfaces
- Sheet-metal shears for squaring sheet-metal work
- Slitting shears for cutting complex shapes in sheet metal
- Standard and impact drills with nut-driving sockets
- Stud welding machines

Some of these tools require special certifications to use. Do not use a tool or piece of equipment if you are not familiar with its operation and unless you are certain it is in proper working order.

1.2.0 Lagging Materials

One widely used type of sheet-metal lagging is stucco embossed 0.040" (1.0 mm) box rib aluminum. Thickness of aluminum sheet metal is usually stated in thousandths of an inch. Formed and flat aluminum sheet metal used in lagging work ranges from 0.032" to 0.063" (0.81 to 1.60 mm). Sheet-metal gauge is the measurement of galvanized or painted steel sheet-metal thickness. For heavy-duty installations, steel sheets with a 55% aluminum-45% zinc anticorrosion coating are common. The thickness of standard steel sheet metal used in lagging ranges from 20 gauge to 16 gauge (0.036" to 0.060" or 0.91 to 1.52 mm). Note that smaller gauge numbers means thicker steel. Aluminum thickness may also be stated in gauge, but the thicknesses of aluminum and the same gauge steel are not the same. For example, 18-gauge galvanized steel is 0.052" (1.32 mm) thick and 18-gauge aluminum is only 0.040" (1.02 mm) thick. *Table 1* lists the thicknesses and weights of different types of sheet metal that may be used for lagging and flashing.

Lagging sheet metal is identified by both the material and the way it is formed. The materials used are smooth aluminum, stucco-embossed aluminum, painted aluminum, galvanized steel, stainless steel, and painted steel. Common forms of lagging sheets are shown in *Figure 2*. The type, thickness, and finish of materials for flashing usually match the type of ribbed sheets used.

Ribbed sheet-metal lagging is used mainly for covering insulation over systems subject to expansion and contraction during operating cycles. However, ribbed materials are inherently more rigid in the direction of the ribs than flat materials, and can better withstand wind loads. Flat materials are used mainly as flashing for lagging, as well as covering circular, elliptical, and irregular structures, and in locations where the ribbed materials won't fit. Corrugated materials may also serve as lagging as long as they meet the metal thickness guidelines above. Generally, corrugated sheets are lighter in weight than box-rib sheets of the same size and thickness. In heavy industrial lagging work, corrugated sheets are mainly used as backing material instead of metal meshes to stabilize insulation under industrial pre-insulated lagging panels.

Sheets of lagging may be ordered to the required length and shipped on skids. Ribbed sheet metal needs to be protected from damage during storage and handling. Bent or damaged pieces will not properly mesh with adjacent pieces when installed and they detract from the

Table 1 Thicknesses and Weights of Typical Lagging Materials

Gauge	Galvanized Steel		Stainless Steel		Aluminum	
	Thickness inch (mm)	Weight lb/ft^2 (kg/m^2)	Thickness inch (mm)	Weight lb/ft^2 (kg/m^2)	Thickness* inch (mm)	Weight lb/ft^2 (kg/m^2)
20	0.040 (1.00)	1.62 (7.89)	0.038 (0.95)	1.56 (7.62)	0.032 (0.81)	0.45 (2.21)
19	0.046 (1.16)	1.86 (9.08)	0.045 (1.11)	1.82 (8.88)	0.036 (0.09)	0.51 (2.48)
18	0.052 (1.31)	2.11 (10.28)	0.050 (1.27)	2.08 (10.12)	0.040 (0.10)	0.57 (2.78)
17	0.058 (1.46)	2.35 (11.45)	0.056 (1.43)	2.34 (11.42)	0.045 (0.12)	0.64 (3.13)
16	0.064 (1.61)	2.59 (12.65)	0.063 (1.59)	2.60 (12.70)	0.051 (0.13)	0.72 (3.51)

*Aluminum sheet is dimensioned by inch or millimeter. The smaller values reflect the lower density of aluminum.

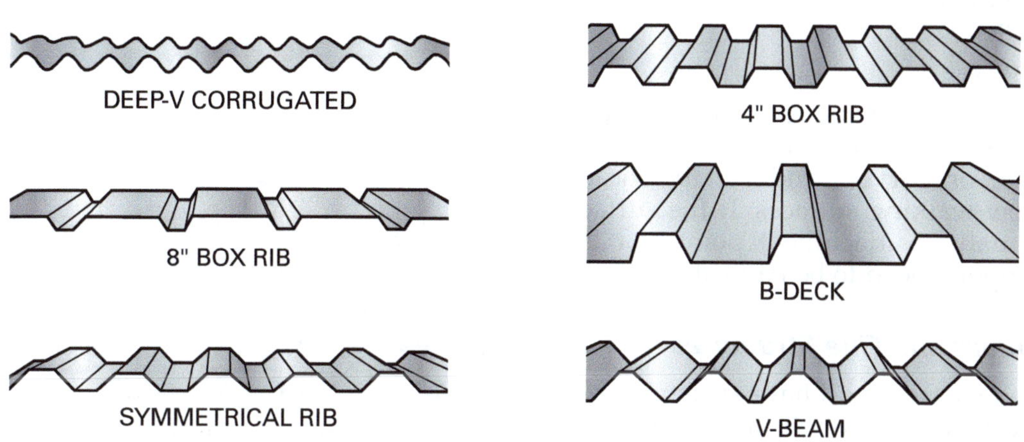

Figure 2 Common forms for industrial insulation lagging.

finished appearance of the facility. Both steel and aluminum sheets must be kept absolutely dry when stacked together. Any moisture allowed to contact the ends of the sheets might seep between them and cause corrosion or objectionable discoloration (*Figure 3*).

> **NOTE**
> Covering stacks of lagging sheet metal with clear plastic is not recommended, as it may cause condensation. Separating the cover from the sheet metal with 2" lumber to provide air circulation is recommended.

Pre-insulated lagging panel systems are available from manufacturers. They consist of a sheet-metal lagging outer skin, insulation, and the inner finish, which may be sheet metal, expanded-metal mesh, or hex-mesh wire. There are panel systems available with interlocking seams, smooth or ribbed skin, and fabricated for vertical or horizontal mounting on flat surfaces. While pre-insulated panels can be installed faster, they are quite heavy and require crane or other lifting services to handle them. They are also more difficult to modify to fit around openings, penetrations, and fittings than field-fabricated lagging. These panels are best installed on large flat surfaces that are relatively cool (less than 350°F or 180°C) because they do not flex as readily with thermal expansion and contraction as field-installed lagging.

1.3.0 Lagging Supports

Structures with surfaces of many shapes and configurations are encountered in insulation and sheet-metal lagging work. Industrial flues (sometimes called breechings) may be in any of the following forms:

- Stiffened rectangular duct
- Stiffened round duct
- Radiused elbows
- Square elbows

There may be transitions, reducers, or expansion joints. Lagged surfaces may be vertical, horizontal, sloped, or curved, and can be located above or below walking surfaces. If the surfaces do not have natural attachment points for lagging, then supports need to be fabricated.

1.3.1 Equipment and Flue Structures

Most lagged surfaces will have structural ribs, flanges, or stiffeners (*Figure 4*). If the surface to be insulated is similar to a boiler flue structure with flanged connecting joints, the insulation may be applied directly to the surface of the flue and built up to the height of the ribs and flanges, or even thicker. In these cases, the ribs, flanges, and stiffeners are buried out of sight under the lagging and insulation, forming a continuous, eye-pleasing surface, which can be important to facility owners.

For high-temperature installations, especially power plant boiler island components subject to continual vibrations from high-speed flue gas, the insulation will often be backed by expanded metal mesh or corrugated sheet metal to hold the insulation in place after the binder burns out.

Flat insulated surfaces may also be covered by standard lagging sheets and the exposed ribs may be covered individually with insulation and sheet metal. The actual approach to lagging will be determined by the engineering drawings and specifications.

Figure 3 Discolored aluminum lagging due to improper storage.

Figure 4 Typical equipment framing and stiffeners underlying insulation and lagging.

1.3.2 Lagging Support Attachments

On most heavy industrial applications, attachments for lagging or provisions for attachments must be installed before the insulation. Commonly-used attachments are H-bars, Z-bars, L-angle iron, and various types of sub-girts. Sub-girts are usually fabricated from 16-gauge to 22-gauge sheet metal. *Figure 5* shows different shapes of sub-girts that are used in lagging work.

Sub-girts are generally attached to stiffeners or flanges of ducts and casings with screws. However, sub-girts are not attached by screws directly to breechings or equipment—sub-girts must be welded to surfaces to be insulated or they are fastened to the insulation securements. The insulation is either installed around the support members flush to the outer flange of the sub-girts or the sub-girts are installed on top of the insulation. Lagging is attached to the sub-girt (*Figure 6*), never directly to the flue or equipment structure. In addition, any steel surface to which aluminum is installed must be protected from galvanic corrosion. This can be done by coating the steel at the points of contact with paint or by gluing a felt or woven glass fabric barrier to it. More commonly, most aluminum sheet-metal products can be obtained with an anticorrosion coating factory-applied to the underside.

Boiler wall lagging may be attached over insulation by means of anchor plates. Special $^5/_{16}$" (8

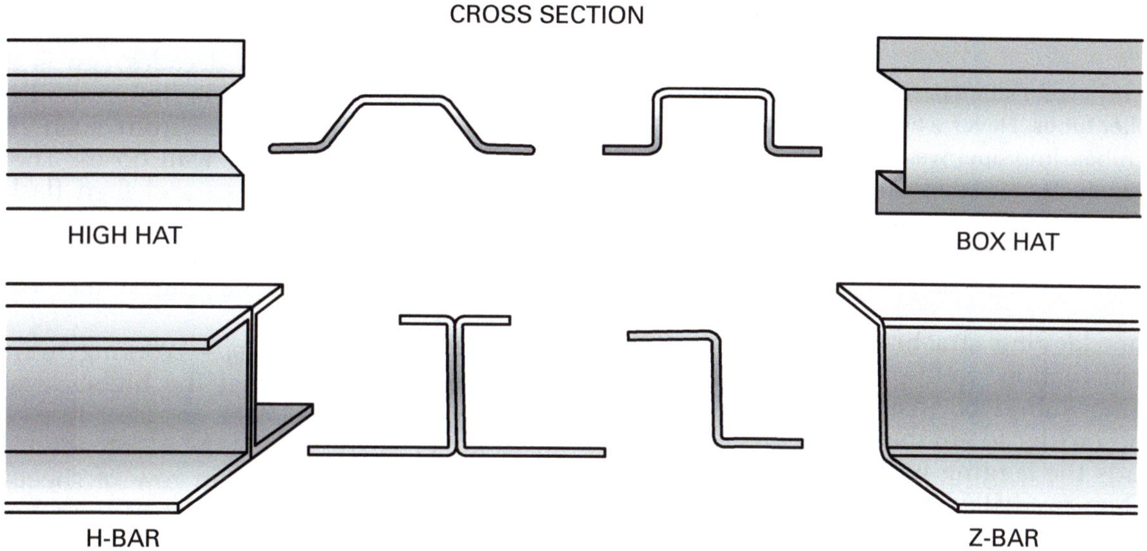

Figure 5 Examples of sub-girts for lagging.

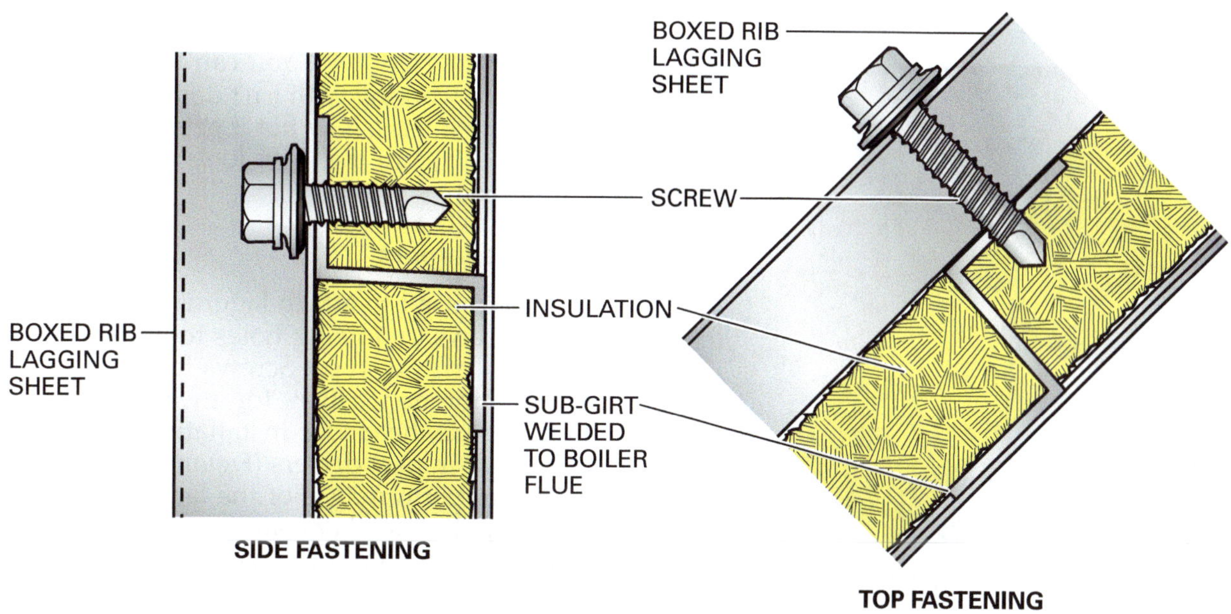

Figure 6 Typical method for attaching a box rib lagging section to a sub-girt.

mm) studs are welded to the webbing between the tubes in the boiler wall before positioning the insulation. The outer end of the stud is knurled. The insulation is installed over the stud, whose end should be nearly flush with the face of the insulation. A plate with a short section of $^{5}/_{16}$" (8-mm) pipe stock welded to its center is hammered over the knurled end of the stud. The lagging sub-girts are positioned on the anchor plates and holes are drilled and sized for self-tapping screws. Then sub-girts are secured to the plates and the lagging is attached by screws to the sub-girts. This method permits some flexing when the underlying equipment surface expands and contracts. It has the further advantage of keeping the support for the lagging separate from the insulation.

> **CAUTION**
>
> Always weld studs to the webbing between boiler tubes. Never weld studs to boiler pipes or tubes. Improper welding by unqualified individuals could permanently damage tubes and lead to a steam rupture, malfunction resulting in personal injury, or other casualty.

Another lagging attachment method uses the fibrous insulation weld-pin system to secure the lagging sub-girts. Insulation weld pins are spaced according to engineering plans to handle wind loads the lagging will be expected to withstand. Then predrilled sub-girt channels are installed onto the insulation surface and secured with speed clips (*Figure 7*). The lagging is fastened to the channels with screws as usual. This method is usually more economical than other approaches.

Figure 7 Attaching lagging sub-girts to insulation anchor pins with pin clips.

1.4.0 Installing Lagging

The installation of lagging may be done by mechanical insulators, but if the general contractor is using union workers, the lagging operation will normally be completed by members of a sheet-metal union. The approach to lagging should begin with some generally-accepted rules for a straight-forward installation.

- *Work direction*—In the industry, it is an accepted practice to work from left to right. The main advantage to this approach is that ribbed lagging starts at the same location on all walls and the same distances from corners. This allows work to start on multiple walls at the same time and have a similar appearance at corners.
- *Lag first and then flash*—A more efficient and time-saving method for flashing than the practice from even a few decades ago is to install the ribbed lagging sheets first, providing specified stand off distances from such things as penetrations, doors, or wall boxes. The lagging can be rapidly installed and then flashing can be fabricated afterward to the exact dimensions required. This approach doesn't rule out installing flashing before lagging at roof edges and wherever necessary.
- *Work from bottom to top*—Working from bottom to top guarantees that the higher-level of lagging overlaps the lower, ensuring that water and weather is shed properly as intended. Also, by making sure the bottom row of sheets is level and square makes the rest of the job easier to level.
- *Use all your senses*—When working with sheet metal, if you are having trouble with a tool binding, making excessive noise or vibration, or getting stuck, then it probably means you aren't using the right tool. You can tell if the material you are working with isn't doing what you want by how it feels, how hot it gets, or if it begins to smoke. Obtain the right tools and learn how to use them.

1.4.1 Fasteners

Fasteners used to attach lagging may be self-tapping screws, requiring holes to be predrilled, or the self-drilling type. Screws for lagging should have neoprene washers for a watertight application and hex heads for installation by power nut-drivers or impact drivers (*Figure 8*).

Proper performance of the lagging depends on ensuring the washers are not too loose or too tight. *Figure 9* shows the method for checking the

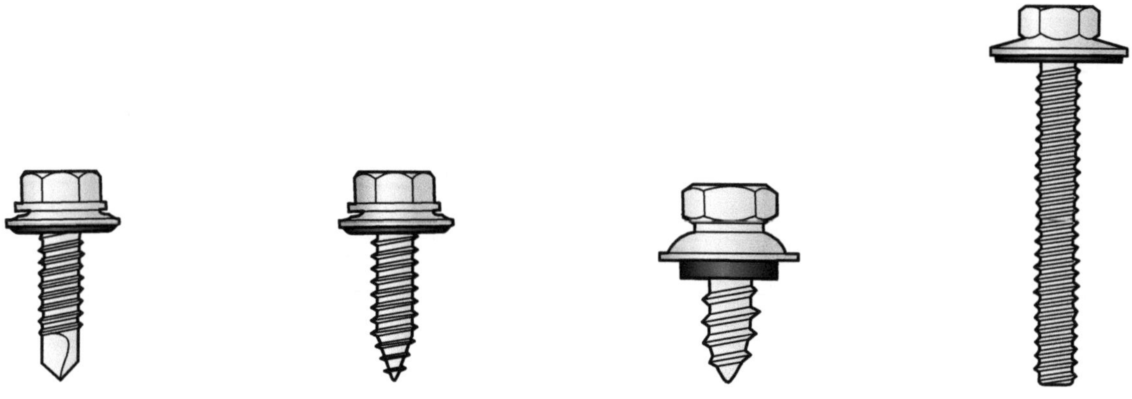

Figure 8 Examples of fasteners used for lagging installation.

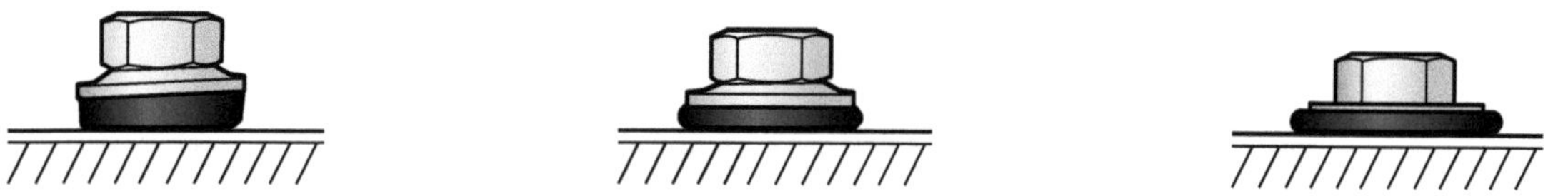

Figure 9 Evaluating the tightness of a lagging screw.

compression of the washer. The washer is too loose if you can spin it with your fingers.

A variety of fasteners may be used for installing lagging, but each type has a specific purpose. Using the wrong fastener can cause early failure of the lagging system. Engineering plans or specifications should be consulted for choosing the correct fasteners.

Sheet-metal screws can be used for installing flashing to lagging and to itself. Self-drilling screws are used to attach lagging sheets to sub-girts and will drill through most thicknesses of sheet metal and fabricated sub-girts. Self-tapping screws can be used to attach flashing sections to lagging sheets and to secure overlaps of lagging sheets. These screws will not drill through structural steel angles, channels, or beams satisfactorily. If required to attach lagging to such members, they must be predrilled with regular bits sized to the fasteners to be used and then secured with self-tapping screws. Review *Figure 10* to understand the range of options for fastener selection when working with sheet-metal materials.

Screws are available in stainless steel, galvanized steel, cadmium-plated steel, and aluminum. Self-tapping and self-drilling aluminum screws will drill only into aluminum and are only used to attach one piece of aluminum to another.

1.4.2 Installation and Flashing Considerations

Top surfaces should always be sloped to prevent ponding of water (*Figure 11*). This is accomplished by installing sub-girts of differing heights so that a slope is created, forcing the water to drain. The practice of humping lagging over structural frames and stiffeners to save money (*Figure 12*) often creates low spots where water can pond. This leads to corrosion and failure of the underlying insulation.

In locations where the open ends of box-ribbed lagging are exposed to the weather, specifications usually require the ribs be plugged with closure strips (*Figure 13*). Polymeric end closure strips are available for wide variety of lagging and corrugated sheets. The EPDM form of closure strips are resistant to a wide range of temperatures. They are usually supplied in 36" (1 m) strips. End closures are not intended to weatherproof ends oriented upward. Remember, flashing must always be used for proper weatherproofing.

Sheets installed on roofs or upward-facing insulated surfaces should overlap side finish. The ribs of sheets should run vertically and downslope so as to allow water runoff, regardless of the direction of the long dimension of the surface being covered. Sheets should be of sufficient

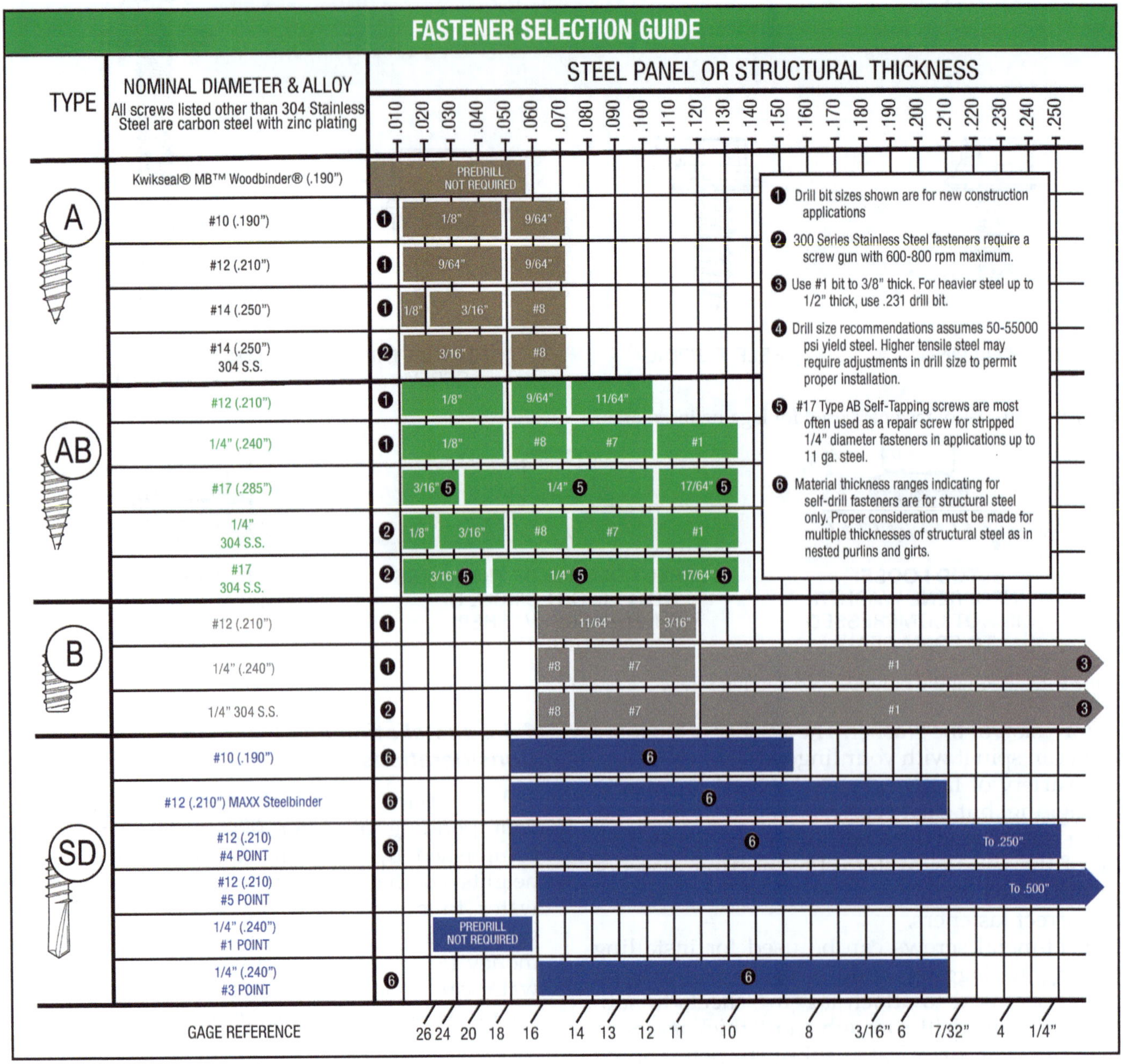

Figure 10 An example of a fastener selection guide for sheet-metal securement.

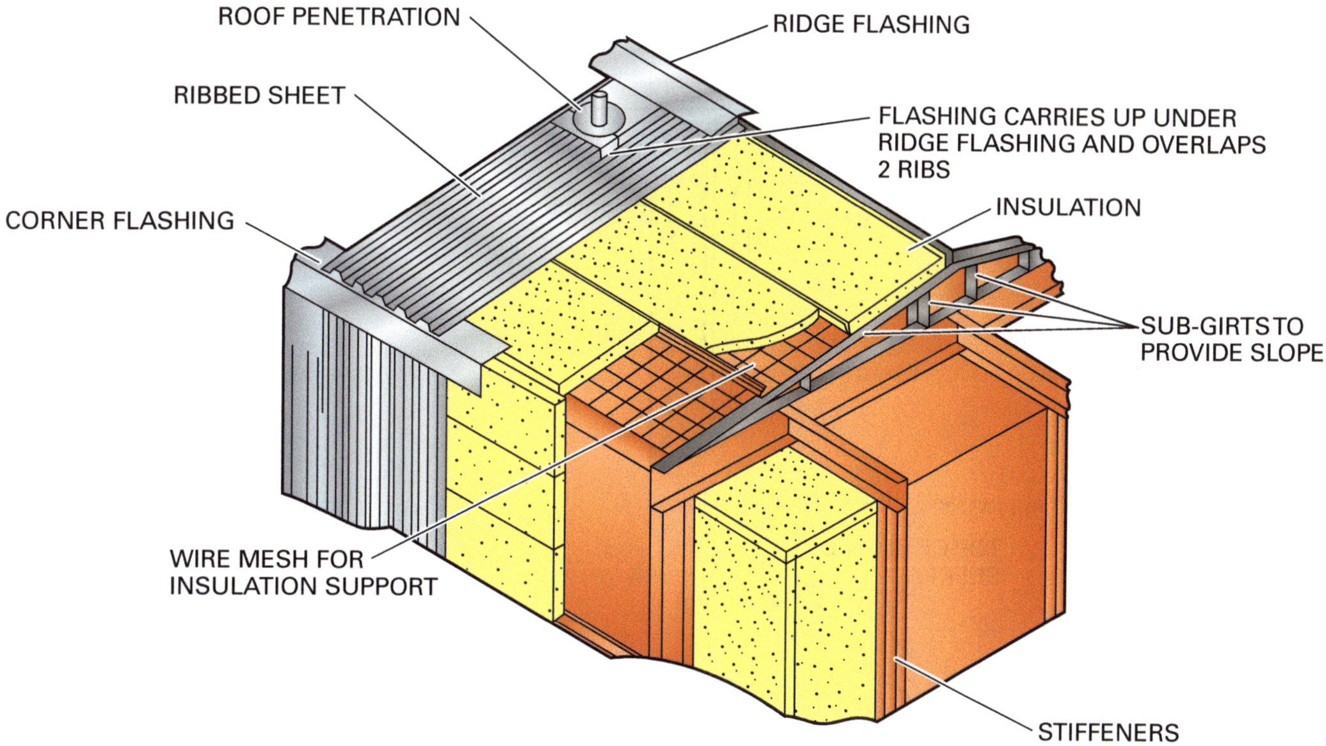

Figure 11 Section of a boiler flue structure showing insulation and lagging components.

Figure 12 Improperly installed lagging humped over structural supports.

Figure 13 An example of a lagging end-closure strip.

length to span the entire surface but if not, they should be installed to overlap in a manner that will properly shed water. Screws should always be installed only in the high ribs of sheets installed on top surfaces to avoid creating paths for draining water flowing in lower channels to enter the insulation via the screw holes. Flashing strips must also be installed to shed water.

Figure 14 shows how lagging can be installed on a flat roof to provide a drip lip. Flashing is installed to seal the end of the sidewall.

The flashing at the juncture of a sloped lagged surface and a vertical one is shown in *Figure 15*. Screws with neoprene rubber washers ensure a watertight securement.

Figure 16 shows flashing for offsets in a vertical surface and for overhangs. Note the crimped edges of the exposed flashing to form a drip lip and to keep it from bowing.

Flashing vertical inside corners prevents intrusion of water and weather at the joint in the corner (*Figure 17*). The flashing runs continuously up the corner until it is covered by a horizontal edge flashing. Upper sections of corner flashing overlap lower.

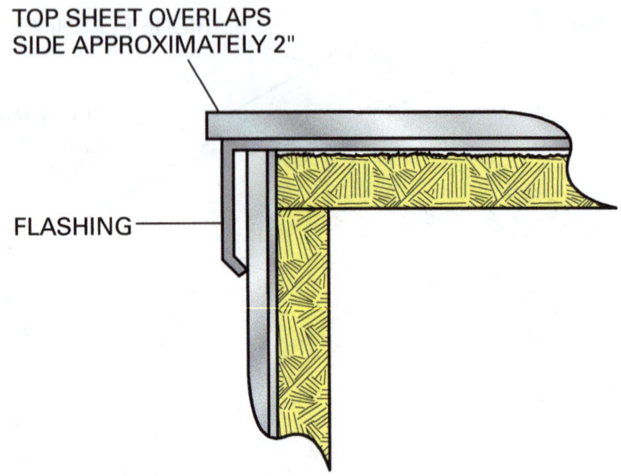

Figure 14 Flat surface overlapping vertical side and flashing.

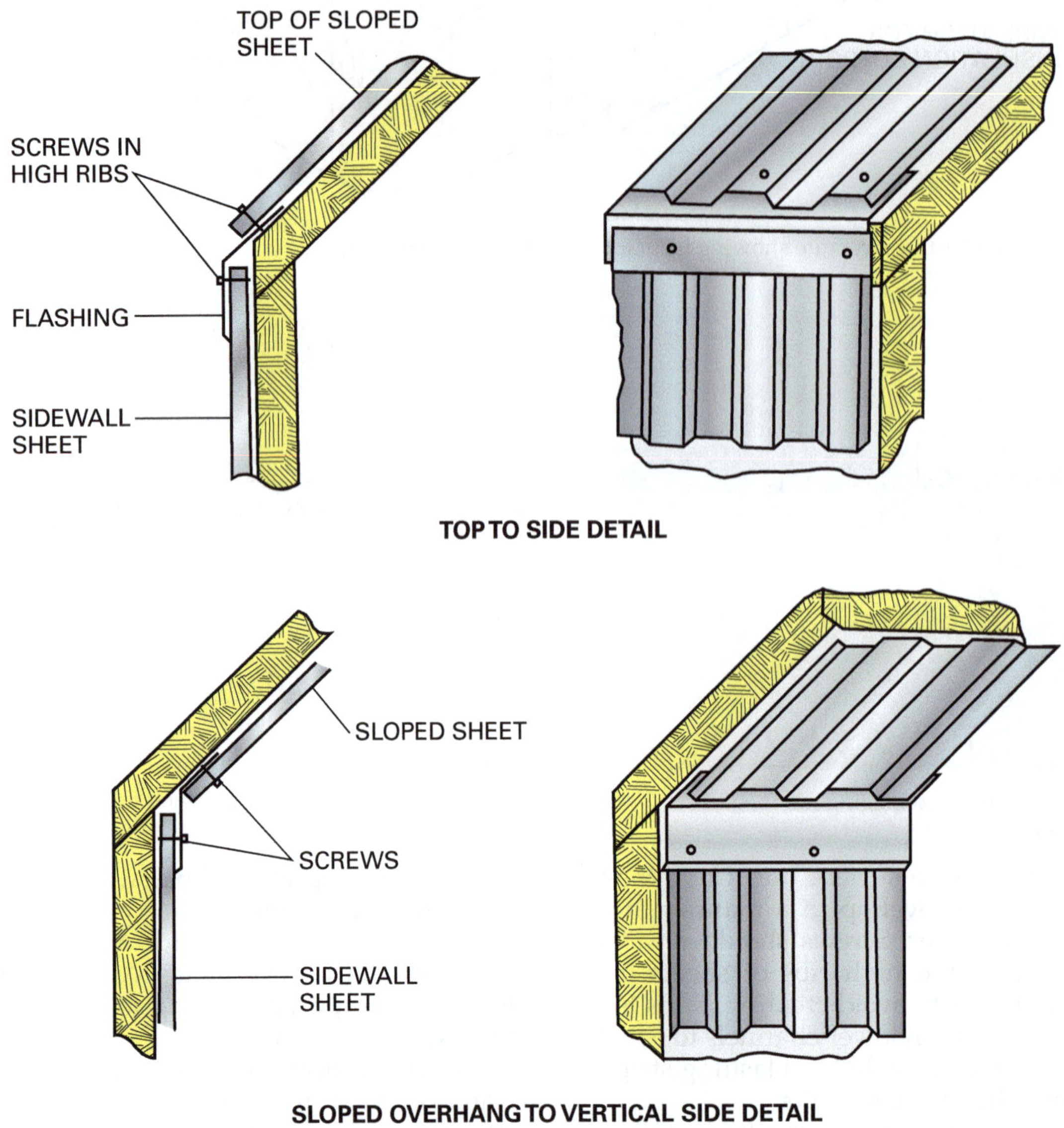

Figure 15 Flashing of a pitched lagged surface to a vertical surface (top); an overhanging lagged surface to a vertical surface (bottom).

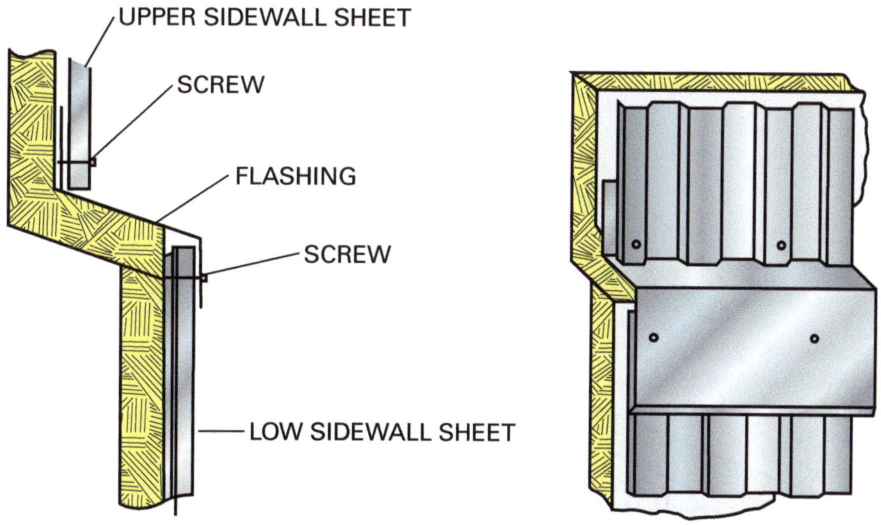

SIDEWALL OFFSET FLASHING DETAIL

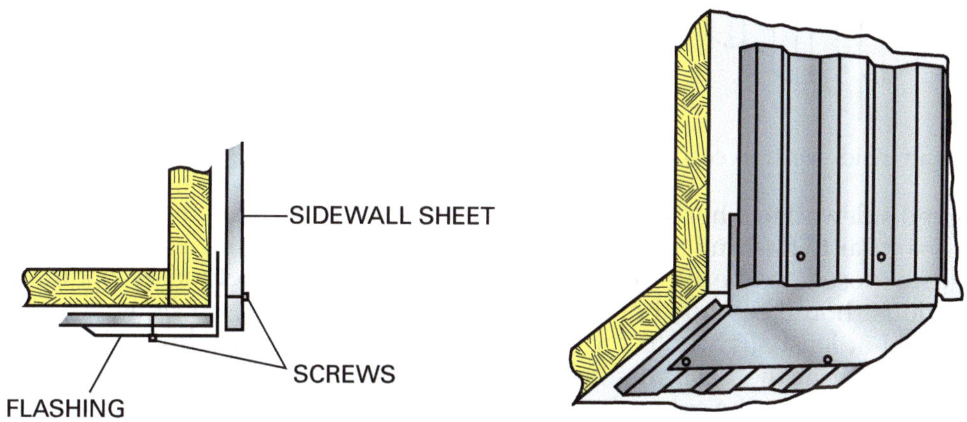

SIDEWALL TO BOTTOM DETAIL

Figure 16 Flashing sidewall offsets and overhangs.

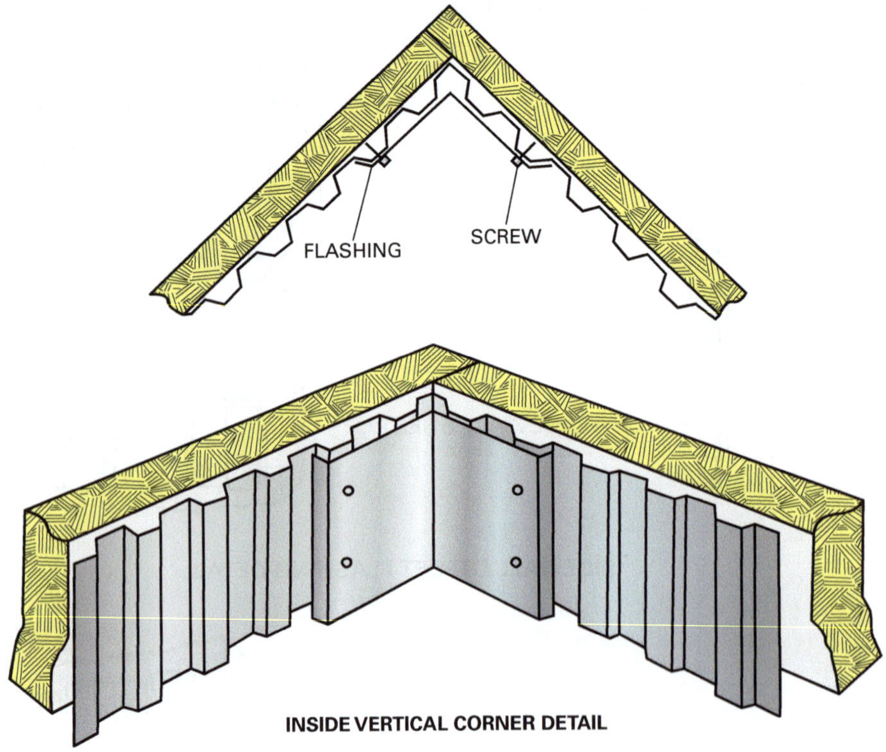

Figure 17 Flashing an inside vertical corner.

Flashing a lagged sidewall to a horizontal overhead (*Figure 18*) is similar to flashing an overhang. No upper surface exposed to the weather or to washdown operations should be truly horizontal, but should be sloped at least slightly to promote water drainage. Flashing at the vertical sidewall should direct water away from the sidewall to the sloped surface.

The approach to flashing of columns and vertical stiffeners that extend beyond the adjacent lagging surface will depend on the circumstances. One method shown in *Figure 19* requires bending and forming the sheet metal to closely fit the insulation on the column, and fastening the flashing to the sub-girts through the lagging sheets next to the column.

Expansion joints and supports are critical to effective lagging installations, especially for hot-service installations that experience large changes in temperature. Large power plant boiler enclosures can expand and contract several feet over the dimensions of the sidewalls. Lagging must be able to take up this motion without buckling or creating gaps. Certain types of lagging supports are naturally flexible (such as the knurled-stud and plate supports) and can move slightly as the underlying surface expands and contracts. However, engineered expansion joints need to be provided along long stretches of lagging.

One simple method is to join adjacent sections of lagging using an S-joint at the high point in the box ribbing as shown in *Figure 20*. This kind of joint can be folded in a shop using a machine similar to a Pittsburgh edger. The underlying sub-girts should be installed with a space of $^1/_8$" to $^1/_2$" (3 to 13 mm) between their ends to prevent buckling. Sub-girts should never be welded together at their ends in hot-service installations.

Flashing built-in expansion joints in flues and breechings requires bridging the expansion joint. The actual method to be used will usually be spelled out in engineering drawings. For smaller joints that may not be covered by drawings or specifications, *Figure 21* shows a method where flashing is field-fabricated to span the joint between lagged surfaces. Insulation pins are welded to the inside surface of the flashing. Insulation is impaled to the pins and secured with clips. The entire assembly is then secured to the lagging sub-girts on both sides of the joint. Many times the expansion joint will have a sheet metal cover, but no insulation. Fabric or rubberized expansion joints can overheat with insulation on them.

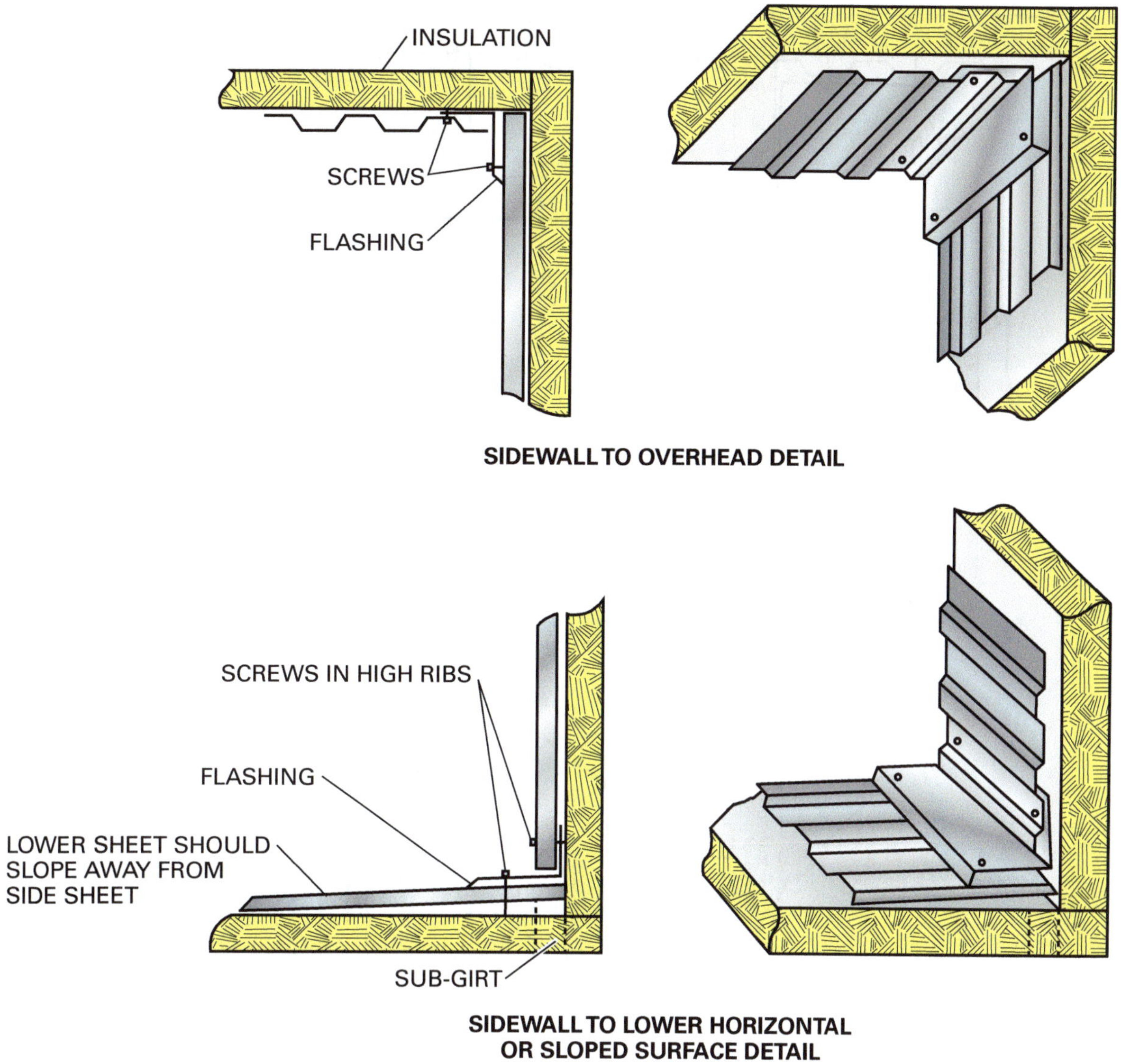

Figure 18 Flashing vertical lagged sidewalls to horizontal overheads and sloped top surfaces.

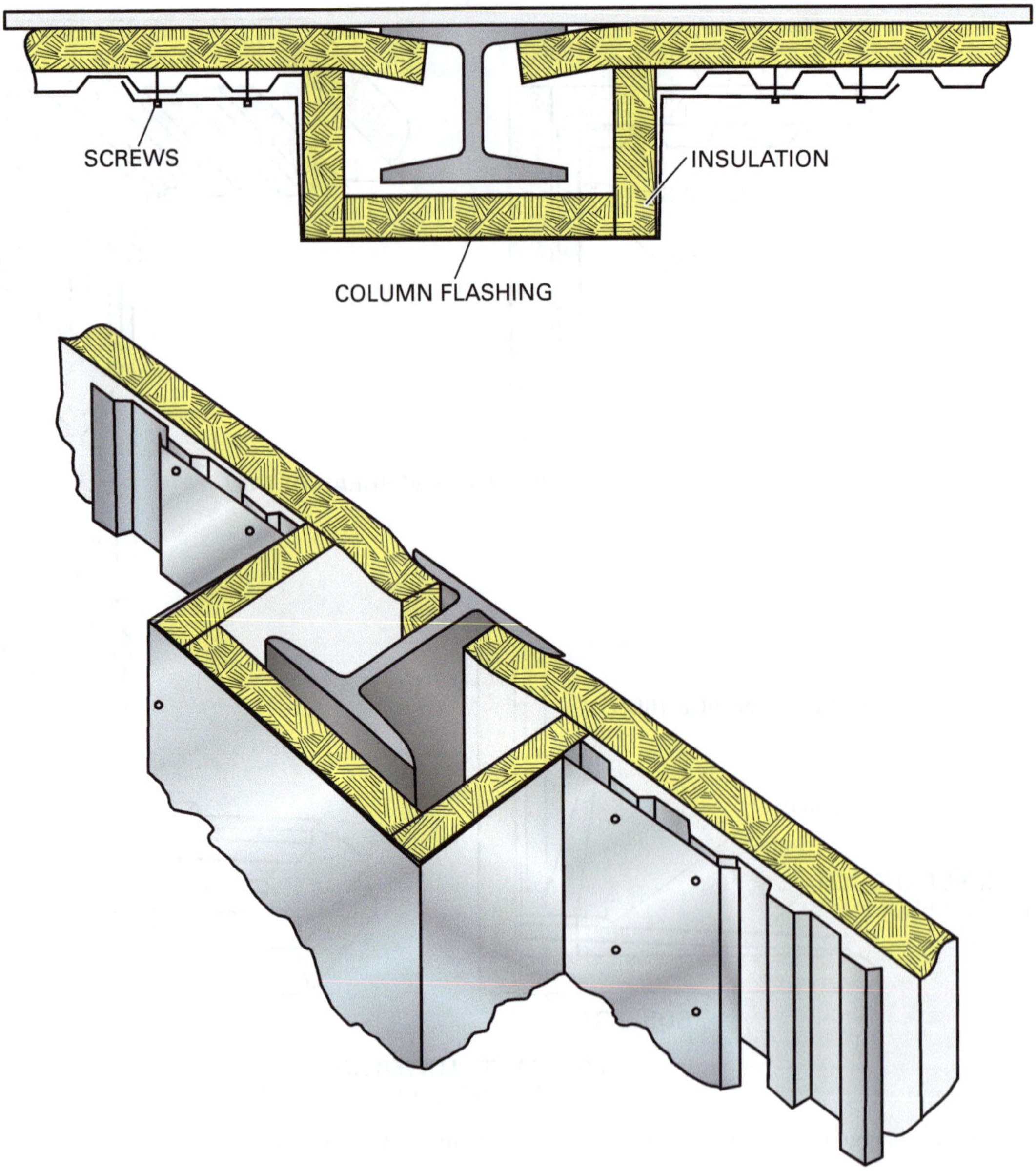

Figure 19 Flashing an insulated but unlagged column girder.

Figure 20 Cross section of an S-joint in lagging.

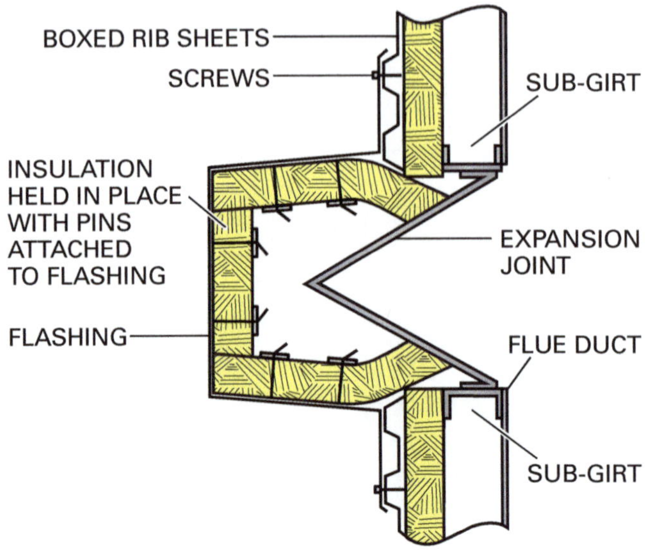

Figure 21 Flashing a flue duct expansion joint.

Metal Gauge System

Many workers often wonder where the mysterious sheet metal thickness units came from. There doesn't seem to be any relationship between the sheet metal gauge (Ga) number and the actual metal thickness.

Back in the 1700s in England when the Industrial Revolution was just getting started, steel rolling mills were set up and mill owners produced plate and sheet steel to meet customers' needs. It soon became apparent that some sort of thickness standards needed to be set. Many systems were suggested, but things remained pretty confusing until late in the 1800s.

American manufacturers first proposed a standard decimal gauge—numbers in thousandths of an inch. So a $\frac{1}{4}$" plate would have been 250-gauge thickness ($^{250}/_{1000}$"). Even later, in 1893, the US Congress established a system based on the weight of a cubic foot of wrought iron. This was done to help tax the metal and products. The weight of a square foot of sheet metal was proportional to its thickness compared to one foot.

Manufacturers didn't like this rule, mainly because they were making rolled steel, not wrought iron. Another thickness scale was proposed on uncoated mild steel. Others wanted to go back to a decimal gauge system. Eventually, the thickness scale based on weight won out and the dimensions were standardized as we see them today. This is why thicknesses for the same gauge of different metals are different—all metals have their own densities.

In the field, a worker can verify the thickness of the sheet metal on hand by using a simple thickness gauge, such as that shown in the photo. As metal gets thicker, its gauge number gets smaller. Many standards organizations are encouraging manufacturers and users to abandon the gauge system, but like many old systems of units, it continues to be used because it's familiar.

1.4.3 Pre-Insulated Lagging Panels

In the 1960s, the American power-generation industry introduced a new process to save insulation costs when erecting new power plant boilers. The insulation was combined to sheets of lagging and the materials were installed as a prefabricated unit to boiler sidewalls, boiler flues, and other high-temperature and low-temperature boiler island components. The high-temperature installations failed to perform and in more recent years, the pre-insulated panels have been (or should have been) limited to lower-temperature (less than 350°F or 175°C) systems.

Pre-insulated lagging panels may be fabricated in the shop or in the field (*Figure 22*). To build a pre-insulated panel, anchor pins are welded to the inside surface of a ribbed sheet on about 12" (30 cm) centers (6" or 15 cm from edges). The insulation is then impaled on the pins. The insulation is stopped one or two ribs short from one side and an equivalent distance from one end to provide an overlap when the panels are installed. Galvanized mesh or lightweight corrugated sheet metal is placed over the insulation and anchors to keep the insulation from settling. Insulation pin clips are installed and the pins are bent over in the direction of the ribs.

Pre-insulated panels may be installed directly to the stiffeners or framing of the structure being insulated/lagged. They may also be installed on a special subsystem of supports fabricated by angle iron. In either case, self-drilling screws can't be used for such materials. Holes must be drilled through the panel and the support, and heavy self-tapping bolts with neoprene washers are used to secure the panels.

Figure 22 Shop-fabrication of a pre-insulated lagging panel for a removable door access.

Installing prefabricated panels presents some challenges. For example, a single 4 × 8 panel of 0.040" (18 Ga) box ribbed aluminum lagging with 4" (10 cm) thick mineral wool at 8 lb/ft^3 (128 kg/m^3) attached weighs approximately 200 lb (91 kg). Lifting and handling this weight will require crane services and can be difficult in areas subject to wind.

Additional Resources

National Industrial and Commercial Insulation Standards Manual, Midwest Insulation Contractors Association (MICA).

National Insulation Association (NIA) website offers resources for products and training. **www.insulation.org**.

Mechanical Insulation Design Guide, National Institute of Building Sciences. Available at **www.wbdg.org/guides-specifications/mechanical-insulation-design-guide**.

The BRIL Book III—The Book of BRIL. Gary J. Bases. 2011. BRIL, Inc.

The BRIL Book V—Boiler Lagging. Gary J. Bases. 2015. BRIL, Inc.

Product Technical Guide for Self-Drilling Screws, **www.us.hilti.com**.

1.0.0 Section Review

1. In modern insulation usage, the term *lagging* refers to _____.
   a. blanket insulation with a fabric finish
   b. heavy metal sheathing over outside insulated surfaces
   c. piping system jacketing
   d. removable/reusable insulation used in steam plants

2. Which of the following statements is *not* true?
   a. Aluminum lagging metal is lighter per square foot or square meter than the same thickness of steel metal.
   b. Aluminum sheet metal thicknesses are given in inch or millimeter units.
   c. Corrugated sheet metal is not used as a lagging material.
   d. Ribbed lagging has the advantage of being rigid but can flex with the thermal expansion of the underlying structure.

3. Why is metal mesh often installed as a backing to insulation under lagging?
   a. To hold the insulation off of the insulated surface
   b. To hold the insulation flat against the lagging sheets
   c. To hold the insulation flat against the insulated surface
   d. To hold the insulation in place after the binder burns out

4. It is a general practice in the industry to install lagging _____.
   a. from top to bottom
   b. from outside corners to inside corners
   c. from right to left
   d. from left to right

Summary

Sheet-metal lagging work is a specialized part of the insulation trade. There are companies that focus on doing nothing but lagging, also known as flat work. The biggest challenge in sheet-metal lagging is to fabricate flashing that successfully keeps the insulation dry. The power generation industry, pollution control, and cement manufacturing are all industries that have a considerable need for this type of work.

Insulation workers may go their entire career and never be exposed to this particular type of installation. Nevertheless, a skilled mechanic should know how to work with and around sheet-metal lagging. The mechanic must acquire the skills and certifications to be a welder, scaffold builder, sheet-metal worker, and material distributor to successfully accomplish this type of installation.

Review Questions

1. Box-rib lagging is usually applied to which kind of surfaces?
   a. Small cylindrical vessels
   b. HVAC ducting
   c. Steam pipes
   d. Large flat surfaces

2. When installing lagging, flat sheet metal is normally reserved for _____.
   a. Pipe jacketing
   b. Duct sheathing
   c. Flashing
   d. Large flat surfaces

3. Which of the following statements is *true* about sheet metal used for lagging?
   a. A common form is stucco-embossed 0.040" (1.0 mm) box-ribbed aluminum.
   b. The thicknesses of all lagging sheet metals are only given in sheet metal gauge values.
   c. Mild sheet steel 0.10" (3 mm) or larger is preferred.
   d. The sheet metal used for flashing lagging is thinner and a different metal than the lagging.

4. Which of the following statements is *not* true for storing and handling galvanized steel sheet metal lagging?
   a. Avoid bending or creasing ribbed sheets.
   b. Stacks of sheets should be stored flat to avoid damage to edges.
   c. No weather protection is required for galvanized sheet since it is corrosion-resistant.
   d. Sheets should remain on their shipping skids until used.

5. Besides the burnout of fibrous insulation binders, what factor can cause the settling of such insulation in high-temperature power generation boilers?
   a. The vibrations caused by high-speed flue gases
   b. The thickness of the insulation
   c. The kind of insulation
   d. The spacing of the sub-girts

6. For any outside insulated industrial equipment or structures, which of the following statements is true?
   a. Equipment or structures without ribs or other supports would have to remain unlagged.
   b. Sub-girts must be installed on equipment or structures before insulating.
   c. Only those sections or surfaces of equipment and structures that have ribs or flanges can be lagged.
   d. Hot equipment and structures don't require lagging, because high temperatures keep the insulation dry.

7. Which of the following statements is *not* correct when installing lagging supports and attachments?
   a. Sub-girts may be attached with screws directly to boiler flue and equipment surfaces.
   b. Sub-girts must be welded to equipment and boiler flue surfaces.
   c. Sub-girts may be fastened with screws to boiler flue and equipment ribs and flanges.
   d. Sub-girts may be attached to the surface of insulation by insulation securements.

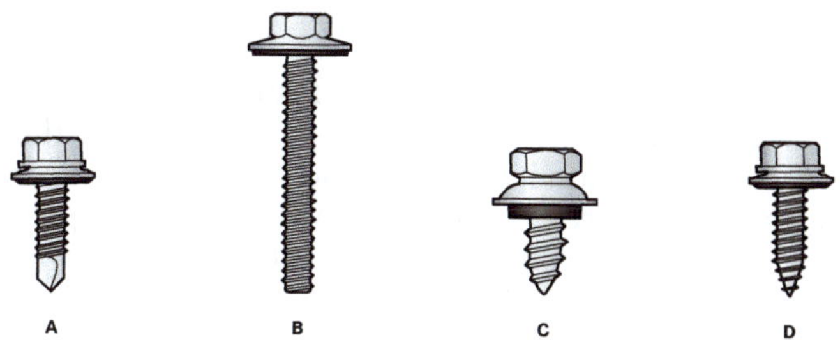

Figure RQ01

8. Which of the fasteners shown in *Figure RQ01* would you use for installing undrilled 0.040" (1.0 mm) aluminum lagging on 20-gauge galvanized steel sub-girts?
   a. A
   b. B
   c. C
   d. D

9. Which lagging accessory does *not* have the primary purpose of preventing the entry of water and wind into the lagging structure?
   a. Neoprene rubber screw washers
   b. Flashing
   c. Sub-girts
   d. End-closure strip

10. Flashing for lagging is always installed _____.
    a. on top of lagging edge seams installed edge to edge
    b. to direct water down-slope and to prevent its intrusion
    c. under lagging edge seams installed edge to edge
    d. to follow the peaks and valleys in the ribbed lagging

Trade Terms Introduced in This Module

Flashing: Narrow strips of sheet metal fabricated to protect joints in a covering as part of a weatherproofing system that may include caulking and neoprene washers for screw fasteners.

Galvanic corrosion: A form of corrosion (rusting) resulting from the creation of an electric current flowing between dissimilar metals in contact with each other.

Knurled: Having narrow milled ridges on a shaft or knob that are used to aid against slippage in a shaft hole or when gripped.

Lagging: With reference to insulation, heavy sheet metal sheathing over insulation, usually ribbed or corrugated, as weatherproofing and to protect insulation against industrial washdown operations.

Self-drilling: Describes a fastener (screw) that is equipped with a special drill-bit-like point that drills through the metalwork to be fastened and taps its own threaded hole. Compare with the term *self-tapping*.

Self-tapping: Describes a fastener equipped with a notched, tapered point that allows the screw to create its own threads in metals softer than the screw. These screws require a predrilled hole. Compare with the term *self-drilling*.

Sheet-metal brake: A heavily-constructed machine designed to create straight folds in sheet metal. The process of bending metal into a corner or fold is called *braking*.

Stiffeners: A piece of heavy formed steel, often with a T, H, I, or other cross section, welded to a flat surface to limit flexing. Also an angled ridge formed in sheet metal for the same purpose.

Sub-girts: In industrial construction, the term girt refers to a secondary support iron (T-, H-, or L-bar) attached to primary structural components and surfaces for stiffening, reinforcement, or to support other equipment. A subgirt is a similar bar attached to the underlying structure that supports an overlying cover or sheathing.

Additional Resources

This module is intended as a thorough resource for task training. The following reference works are suggested for further study.

National Industrial and Commercial Insulation Standards Manual, Midwest Insulation Contractors Association (MICA).

National Insulation Association (NIA) website offers resources for products and training. **www.insulation.org**

Mechanical Insulation Design Guide, National Institute of Building Sciences. Available at **www.wbdg.org/guides-specifications/mechanical-insulation-design-guide**

The BRIL Book III—The Book of BRIL. Gary J. Bases. 2011. BRIL, Inc.

The BRIL Book V—Boiler Lagging. Gary J. Bases. 2015. BRIL, Inc.

Product Technical Guide for Self-Drilling Screws, **www.us.hilti.com**

Figure Credits

Gary J. Bases - BRIL, Inc., Module opener, Figures 1, 3, 4, 7, 12, 13, 20, 22

SEALTITE Building Fasteners, Figure 10

The Eastwood Company, SA01

Section Review Answer Key

Answer	Section Reference	Objective
Section One		
1. b	1.0.0	1a
2. c	1.2.0	1b
3. d	1.3.1	1c
4. d	1.4.0	1d

NCCER CURRICULA — USER UPDATE

NCCER makes every effort to keep its textbooks up-to-date and free of technical errors. We appreciate your help in this process. If you find an error, a typographical mistake, or an inaccuracy in NCCER's curricula, please fill out this form (or a photocopy), or complete the online form at **www.nccer.org/olf**. Be sure to include the exact module ID number, page number, a detailed description, and your recommended correction. Your input will be brought to the attention of the Authoring Team. Thank you for your assistance.

Instructors – If you have an idea for improving this textbook, or have found that additional materials were necessary to teach this module effectively, please let us know so that we may present your suggestions to the Authoring Team.

NCCER Product Development and Revision

13614 Progress Blvd., Alachua, FL 32615

Email: curriculum@nccer.org

Online: www.nccer.org/olf

❑ Trainee Guide ❑ Lesson Plans ❑ Exam ❑ PowerPoints Other ____________

Craft / Level: Copyright Date:

Module ID Number / Title:

Section Number(s):

Description:

Recommended Correction:

Your Name:

Address:

Email: Phone:

NCCER CURRICULA — USER UPDATE

NCCER makes every effort to keep its textbooks up-to-date and free of technical errors. We appreciate your help in this process. If you find an error, a typographical error, or an inaccuracy in NCCER's curricula, please fill out this form (or a photocopy), or complete the online form at www.nccer.org/olf. Be sure to include the exact module ID number, page number, a detailed description, and your recommended correction. Your input will be brought to the attention of the Authoring Team. Thank you for your assistance.

[illegible]

NCCER Product Development and Revision

13614 Progress Blvd., Alachua, FL 32615

Email: curriculum@nccer.org

Online: www.nccer.org/olf

☐ Trainee Guide ☐ Lesson Plans ☐ Exam ☐ PowerPoints ☐ Other

[illegible]

Module Name:

[illegible]

Section Number(s):

Description:

Recommended Correction:

Your Name:

Email:

Jacketing Systems

Overview

Insulation materials are responsible for thermally isolating a mechanical system from its surroundings—either keeping heat in or out. However, if left unprotected, insulation would quickly become damaged and it would no longer perform its designed role. Not only that, insulation that quickly becomes useless must be replaced at great cost to the owner. To preserve the function and investment of insulation, it is standard practice to cover insulation with some form of jacketing as part of a complete insulation system. This module will emphasize the jackets integral to insulation, prefabricated PVC, and polymeric and laminate sheet and roll jacket materials.

Module 19313

Trainees with successful module completions may be eligible for credentialing through the NCCER Registry. To learn more, go to **www.nccer.org** or contact us at 1.888.622.3720. Our website has information on the latest product releases and training, as well as online versions of our *Cornerstone* magazine and Pearson's product catalog.

Your feedback is welcome. You may email your comments to **curriculum@nccer.org**, send general comments and inquiries to **info@nccer.org**, or fill in the User Update form at the back of this module.

This information is general in nature and intended for training purposes only. Actual performance of activities described in this manual requires compliance with all applicable operating, service, maintenance, and safety procedures under the direction of qualified personnel. References in this manual to patented or proprietary devices do not constitute a recommendation of their use.

19313 V2

19313
Jacketing Systems

Objective

When you have completed this module, you will be able to do the following:

1. Identify the purposes of insulation jacketing systems and the types of jacketing materials available.
   a. Describe the reasons for insulation jacketing and common applications for each.
   b. Identify important jacketing systems and materials, and describe their main properties.
   c. Review the methods of jacketing installation.

Performance Tasks

Under the supervision of your instructor, you should be able to do the following:

1. Install a protective jacket on insulated ductwork.
2. Install a vapor retarder jacket on insulated piping.
3. Install a laminated jacket on an insulated vessel sidewall.
4. Apply PVC jacketing to a straight run of piping using an adhesive and nylon bands.
5. Apply PVC jacket to a 90-degree elbow, a tee, and an end cap.

Trade Terms

Abrasion
Cathodic protection
Corrosion under insulation (CUI)
Emissivity
Fire-resistance rating
Infrared (IR)
Thermograph
Ultraviolet (UV)

Industry Recognized Credentials

If you are training through an NCCER-accredited sponsor, you may be eligible for credentials from NCCER's Registry. The ID number for this module is 19313. Note that this module may have been used in other NCCER curricula and may apply to other level completions. Contact NCCER's Registry at 888.622.3720 or go to **www.nccer.org** for more information.

Contents

Figures

SECTION ONE

1.0.0 INSULATION JACKETING SYSTEMS

Objective

Identify the purposes of insulation jacketing systems and the types of jacketing materials available.

a. Describe the reasons for insulation jacketing and common applications for each.
b. Identify important jacketing systems and materials, and describe their main properties.
c. Review the methods of jacketing installation.

Performance Tasks

1. Install a protective jacket on insulated ductwork.
2. Install a vapor retarder jacket on insulated piping.
3. Install a laminated jacket on an insulated vessel sidewall.
4. Apply PVC jacketing to a straight run of piping using an adhesive and nylon bands.
5. Apply PVC jacket to a 90-degree elbow, a tee, and an end cap.

Trade Terms

Abrasion: The roughening or wearing-away of a surface by rubbing or by the impact of particles. Sand and other materials can abrade surfaces as they move past them.

Cathodic protection: A method of preventing underground metal objects from corrosion aided by exposure to ground water. This may be done by either attaching a more easily-corrodible metal to the object or, in the case of underground pipes, supplying a low-voltage direct current through the metal that blocks the corrosive effect.

Corrosion under insulation (CUI): Corrosion occurring out of sight under pipe and equipment insulation due to the failure of the jacket, the vapor retarder boundary, or both, allowing water to penetrate the insulation and contact the metal surface.

Emissivity: A measure of how rapidly a material radiates heat away from its surface.

Fire-resistance rating: The time that a material by itself can resist the effects of a fire or continue to perform its function as measured by a standardized fire-resistance test.

Infrared (IR): A form of light not visible to humans that effectively transfers thermal energy from one object to any other that absorbs it.

Thermograph: An image created by detecting and recording infrared light. Colored thermographs show the intensity of infrared by colors, which indicate heat sources.

Ultraviolet (UV): A form of light invisible to humans that is more energetic than visible light, and can cause damaging chemical reactions when absorbed by skin or materials.

Understanding why and where insulation jacketing is required is the starting point for making informed decisions about the work required to successfully complete an insulation job. As you become more experienced, and take on more responsibilities, you will be required (and you will want) to take part in job bidding and planning. In this module, you will learn why jacketing is required, be introduced to typical jacketing materials that are currently being offered by suppliers and fabricators, and review the methods to installing these materials.

1.1.0 Purposes of Insulation Jacketing

Jacketing is considered an insulation accessory. It doesn't significantly contribute directly to the job of slowing the transfer of thermal, acoustic, or radiant energy through the boundary of the system. However, without jacketing, common environmental factors would quickly destroy the insulation and even damage the underlying mechanical system. Jacketing can also serve other purposes, like providing an easily-cleaned, hygienic surface, or a surface that protects against burns. It can also provide a nearly impenetrable boundary to water vapor or simply give a nice finished appearance to systems or equipment.

1.1.1 Protection from Weather and the Surroundings

Mechanical systems are often located outside without the protection of roofs or walls. Continual exposure to rain, wind, and sunlight will destroy unprotected insulation. Water is probably the most damaging. It is a much better conductor of heat than air. When it fills the air spaces in insulation, the insulating property is lost. Water can also chemically react with some forms of insulation or insulation accessories (such as kraft paper ASJ, steel fasteners, or wires), causing their deterioration. In cold or cryogenic systems, water penetration is almost immediately destructive. Water expands when it freezes. Just like frost heaves in the road that occur in northern climates, the expanding ice breaks up insulation. Obviously, melted snow entering insulation can freeze again with the same results (*Figure 1*).

Wind is another destructive factor in the environment. Even brief gusts can be strong enough to rip coverings off of structures. Most insulation materials are not very strong; they would be torn apart by continual exposure to typical winds, and that doesn't consider the effects of more severe storms.

Sunlight exposure over long periods of time is also very damaging. The **ultraviolet (UV)** component of sunlight that reaches the earth's surface causes chemical changes in most materials. Organic substances, such as insulation binders, and the polymeric foam and elastomeric insulations, are very sensitive to UV effects, even when they are coated with UV-screening compounds. These substances only slow but don't stop the damage.

In some situations, piping and equipment are periodically washed down to remove contaminants and dust that collect during the normal course of operations. These wash-downs usually involve high-pressure or high-velocity sprays (e.g., pressure washers). Without protection during these operations, insulation materials would be quickly torn to shreds or, at a minimum, soaked with water.

Figure 1 Inadequate wrap on these pipes would allow water to freeze within the insulation.

The protection of insulation from the environment isn't the only reason for jacketing. Water and chemicals that penetrate the insulation eventually get to the underlying metal. There, the problem of **corrosion under insulation (CUI)** occurs (*Figure 2*). One study in 2001 estimated that direct costs to all industries from corrosion in the United States alone exceeded $270 billion. Experts believe that the CUI costs-per-year are even higher today. The operational effects of CUI are especially hazardous because it can occur out of sight for years, and then the affected system can suddenly rupture without warning, causing secondary damage to other equipment, unplanned shutdowns, and even injury to people.

1.1.2 Impact and Abrasion Protection

In an industrial or commercial environment, there is always some kind of work going on. Architects and engineers try to design facilities so mechanical systems are compact and out of the way of normal activities, but this isn't always possible. In addition, normal maintenance work often requires climbing on and around insulated equipment, piping, and ducts.

Unprotected insulated systems near walkways would be eventually damaged by brief but frequent rubbing and contact with passing bodies. This is one form of **abrasion**. Pipes and equipment near points where maintenance occurs could receive damage due to the impact of tools, parts, and nuts and bolts dropped on them. Occasionally, the carelessly driven forklift may back into an insulated panel or snag a pipe with its tines. All of these kinds of occurrences can damage unprotected insulation (*Figure 3*).

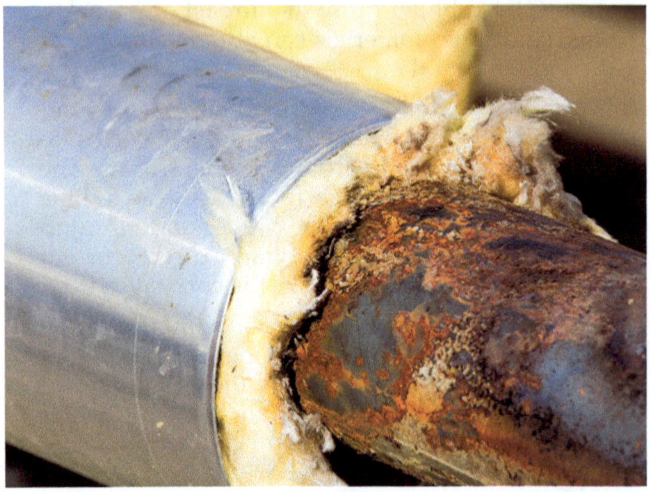

Figure 2 Corrosion under insulation (CUI) due to a failure of the insulation jacket.

Figure 3 Improperly maintained and protected insulation jacket resulted in the failure of this pipe insulation from weather, foot traffic, and abrasion.

1.1.3 Fire Resistance

Most high-temperature insulation materials such as calcium silicate/expanded perlite, mineral wool, and fiberglass are inherently resistant to flames because they are inorganic minerals and are chemically unreactive. However, the mineral fiber insulations usually have a binder that holds them together during installation. Direct exposure to high temperatures burns out the binder, and they would fall apart without something to hold them together.

However, fire-resistance is more of a concern for mechanical systems in structures built to a certain **fire-resistance rating**. Such a structure and its systems are designed to retard the spread of fire long enough for people to escape and for emergency response to put out the fire before catastrophic damage occurs. In these situations, systems coated with flammable vapor-retardant materials, for instance, would require a fire-resistant jacket in order to meet the facility fire-resistance rating.

1.1.4 Personal Safety

One important purpose of insulation is the protection of people from burns. In occupied spaces, high-temperature systems will have an appropriate amount of insulation to slow the movement of heat out of the system. But no insulation can completely isolate a system's heat from its environment. Heat eventually works its way to the surface of the insulation, where it collects until it can radiate away as **infrared (IR)** energy into the environment.

The rate of heat flow between a hot surface and the surroundings is determined by the thermal resistance of the insulating material between these points and the temperature difference. For a given insulation and thickness, the higher the system temperature is, the higher the insulation's outer surface temperature will be. Engineers specify a certain thickness of insulation as a trade-off between the heat-loss efficiency (and its cost) of a system, and the cost of insulation to prevent that heat loss. To lower the surface temperature, thicker insulation is needed.

Given that a jacket material is required, what properties should that material have to prevent burns? All materials have the ability to radiate heat. The rate of heat emission from a material's outer surface at a given temperature at its inner surface is called its **emissivity**. A material that has a high emissivity is able to get rid of its heat faster than a material with low emissivity. So a high-emissivity material will feel cooler to the touch and have a lower surface temperature than

Did You Know?

Thermographs

All objects at temperatures above absolute zero (–459.67°F, –273.15°C, or 0 K) emit some IR energy. Very hot objects emit or radiate a lot of IR. Infrared energy is a form of light energy, just like visible light, but humans are not equipped to see IR. However certain photographic films and other materials can detect IR.

Infrared imaging is used for many purposes. Probably the most familiar application is for night-vision goggles. Conservationists use aerial IR photographs to study the health of forests and the types of trees present. Medical use of IR images permits studying and diagnosing diseases in the body that either raise or lower body temperature from normal. Geologists can detect hot spots around a volcano that may indicate a potential for an eruption. Building inspectors use IR detectors to find insulation and structural sealant problems, where heat leaks into or out of walls and roofs. Industrial engineers can identify problems in piping system and equipment insulation. Thermographs can also identify poor electrical connections in panels and faults in components that generate hot spots.

Becoming familiar with thermography—the science of remotely measuring temperatures—could be a beneficial tool for mechanical insulators. Thermographs could be a useful tool for checking the quality of insulation work and securing contracts to repair or upgrade system insulation.

the low-emissivity material, which will hold back heat and become hotter. Therefore, it would make sense where burns are a concern to use a high-emissivity jacket, even if its presence will lower the overall thermal performance of the insulation system slightly.

Figure 4 shows a combination of the effects discussed here. The pipe jacket's average emissivity over its surface is nearly constant, but the **thermograph** shows some places where the insulation is either thin, missing, or holds leaking water (lighter spots), which is more conductive. At these spots, heat conducts more easily to the jacket and the jacket's temperature rises. More heat escapes at these points. Hot spots can be so intense that they can cause some jacket materials to melt.

Emissivity depends on the kind of material and its surface. Polished, shiny metals have a lower emissivities than ones that are corroded or are dull. Painted surfaces have higher emissivities than unpainted surfaces. Some organic materials, like PVC or fabrics, have higher emissivities than metals. So, if personnel protection is an important purpose of the insulating system, then choosing a high-emissivity jacket material could allow using a lesser thickness of insulation.

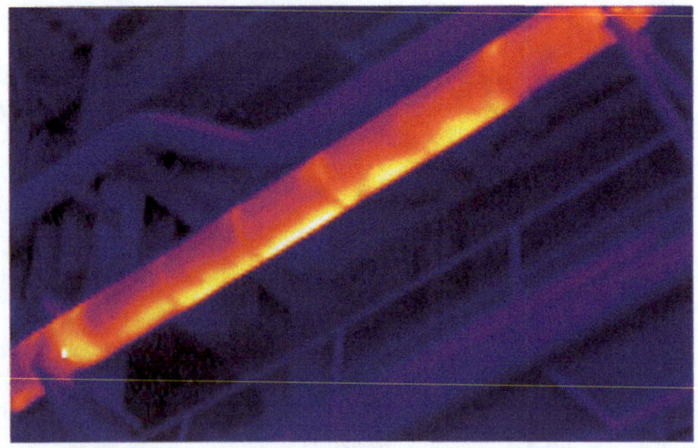

Figure 4 A thermograph of a hot pipe showing problems with the insulation under a jacket.

Figure 5 Several forms of jacketing used in an ice cream factory.

1.1.5 Food Processing/Preparation Hygiene

Commercial and industrial facilities that process and prepare foodstuffs must follow many health and sanitations standards. Regular surface washdowns with hot water and disinfectants are standard in such activities. Facility designs need to avoid flat horizontal surfaces and surfaces where water can pond or be retained. Insulated piping and equipment must be kept clean and sanitary, so they are required to have smooth impervious surfaces and tight joints at fittings and between jacketing sections.

Mechanical systems in food processing and preparation facilities are required to be spread out and accessible, allowing for inspection and cleaning, as well as ventilation to keep surfaces dry. The US Department of Agriculture (USDA) specifies that jacketing for food processing/preparation systems will not support bacterial or fungus growth, and will withstand not only high-pressure water spray, but steam cleaning as well. PVC and stainless steel are common jacket materials for these purposes (*Figure 5*), though PVC and plastic laminates may degrade with prolonged exposure to steam. They are also well-suited for serving the other functions discussed in this subsection.

1.1.6 Vapor Retarders

Vapor retarders are materials that are used to greatly limit the entry of water vapor into system insulation. This is mainly a problem for cold systems operating in temperature ranges below ambient to cryogenic. Water vapor next to cold surfaces condenses and can freeze. As water vapor in the air spaces next to the equipment or piping surface is removed in this way, more water vapor moves into the insulation to replace it, and the problem grows worse. As noted earlier, liquid water and ice can destroy insulation and contribute to CUI in cold systems. Water condensation or freezing in warmer systems are generally not a concern, since higher operating temperatures normally drive water vapor out of the insulation and condensation is not possible.

Cold-service systems must have tight, low-permeance coverings to minimize the entry of water vapor, though it can't be prevented. Since insulation materials alone are not capable of retarding vapor, other parts of the insulation system are responsible for that duty. These vapor retarder materials are usually a combination of low-permeance jackets, sheet or film membranes (*Figure 6*), tapes, mastics, and adhesives. Water vapor diffusion and vapor retarder systems are covered in detail in another module in the NCCER Mechanical Insulating curriculum.

1.1.7 Aesthetic Covers

In many commercial installations built on a warehouse theme (e.g., some big-box stores), the plumbing and HVAC ductwork is exposed to view high in the overhead. Facility owners desire an eye-pleasing covering to the mechanical systems that complements the building's décor. Insulation jacket may also serve as an attractive covering for such systems while also performing the other functions discussed in this section.

1.2.0 Insulation Jacketing Products

There are many insulation jacket products available for mechanical insulation systems today. These can generally be grouped into prefabricated covers, roll and sheet products, and mastics with or without fabric reinforcements. They all provide a durable, non-porous, protective cover for insulation. The properties and installation of rigid metal jackets are discussed thoroughly in other modules in this level, so only the organic, polymer, and laminate (two or more layers of different material bonded together) jacket materials will be covered below.

1.2.1 Prefabricated Covers

Prefabricated jacket covers for insulation come in a variety of forms. Probably the most common form is the all-service jacketing (ASJ) provided with fiberglass (*Figure 7*), mineral wool, and other forms of pipe insulation. The abbreviation ASJ has become a general term for any kind of jacketing used mainly as a vapor retarder, rather than an identity of a particular material. ASJ may be made of a foil-film laminate such as foil-scrim-kraft (FSK) and foil-scrim-polyethylene (FSP), resin-based composites, or other materials. These materials make effective jackets for service over a range of temperatures spanning 0°F to 850°F (–18°C to 454°C), depending on the insulation product.

Another common form of prefabricated jacket is split-sleeved cylindrical sections that fit over piping insulation. These covers are mainly PVC

Figure 6 Vapor retardant membrane applied to a straight pipe.

Figure 7 Fiberglass insulation faced with an ASJ supplied with an SSL.

(polyvinyl chloride) and are available in a range of thicknesses up to 40 mils (1 mm), depending on manufacturer. Pre-cut, pre-curled sections can be fit to insulation with outer diameters up to 30" (76 cm) or can be custom made. Many manufacturers offer a variety of jacket colors for system identification, and UV inhibitors for outside installations to avoid damage from sunlight. The company P.I.C Plastics is one source that supplies a wide range of prefabricated PVC covers, as well as sheet, roll, and accessory products for field fabrication. Split PVC pipe jackets may be obtained with or without self-sealing lap (SSL) systems. Standard jacket section lengths are either 36" or 48" (91 cm or 122 cm).

Prefab jacket covers are also available to cover standard piping system fittings and components. These include reducers and transitions, tees, and even standard valves, plus long- and short-radius elbows like those shown in *Figure 8*. The fitting cover products are made to match the associated pipe jacket materials in thickness and color. Manufacturers often supply joint tapes and adhesives that are compatible with their PVC jacket products. Because of its smooth surface, and adhesive-welded seams and joints, PVC is ideal for

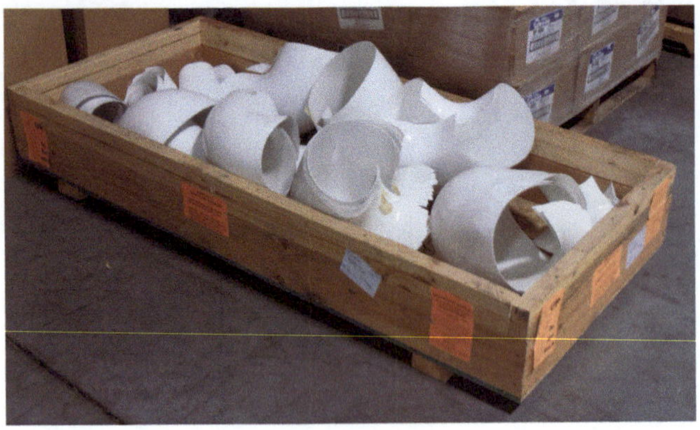

Figure 8 Custom-fitting PVC jackets.

installations involving food, pharmaceutical, and medical systems where maintaining excellent hygiene is important. Some products are designed to fit a range of pipe sizes. The ends of the covers are stepped at diameters that can be trimmed to accept the correct pipe sizes. This reduces costs by allowing a single item to fit pipes of various sizes and insulation thicknesses.

Vessel cover insulation jacket systems are also available in prefabricated panels and gores for vessel heads. Most of these products are sheet metal, but a few suppliers and fabricators offer custom prefabricated interlocking sidewall panels and vessel head gore systems (e.g., FTI's FibaClad™ system).

1.2.2 Roll and Sheet Products

Many jacketing products are offered in sheets and rolls. When applied, sheet products are mainly used as a finish or cladding on large flat surfaces. The thinner products are often mainly used as vapor retarders, or a combination of both jacket and vapor retarder, depending on their characteristics. Some roll materials are fiber-based or made from fabric, such as Lewco Specialty Products' silica and ceramic fiber textiles. These materials are commonly used either as covering materials for removable/reusable jackets or reinforcement for high temperature mastic/cement jacket and coatings. Rolls are most easily applied to piping and small-to-medium-sized equipment.

PVC products are mainly intended as a protective jacket. Common thicknesses are 20 mil and 30 mil (0.5 mm and 0.8 mm), and come in widths of 36" or 48" (91 cm or 122 cm). They can be cut in the field to custom dimensions to cover vessels as well as other flat or curved surfaces. A full range of accessories for installing and sealing this form of cladding is provided by the manufacturer. Vessel head jacket may be fabricated from the same materials. Some manufacturers can prefabricate PVC gores to cover standard-shaped vessel heads when given the required dimensions. PVC sheet and roll products have all the same advantages and limitations as prefabricated jacketing of the same materials.

Laminates such as FSK and FSP described earlier are supplied in rolls as a pipe and vessel wrap jacket. These materials are mainly used as vapor retarder jackets, but may serve as a finish jacket if not exposed to abuse from the environment. More advanced laminates may be constructed of PVC, aluminum foil, polymer/Mylar film or bitumen materials (*Figure 9*). These materials are flexible, resist denting, and have no special installation requirements different from other film materials. They may be obtained with or without an adhesive coating and come with tape and other accessories to aid installation. Recommended edge sealing may use either a solvent-based adhesive or heat welding.

PVDC (polyvinylidene chloride) is another laminate jacket best suited as a vapor retarder. It is constructed of a polymer film containing fibers arranged in two directions. One manufacturer's vapor retarder film comes in rolls 4 mils and 6 mils (0.1 mm and 0.15 mm) thick and 35.5" (90 cm) wide. Length of the rolls depends on the thickness of the film. Permeance values are 0.020 or less.

The number of roll and sheet insulation jacket products available is quite large. Those discussed here are just a few of the types offered by insulating product manufacturers. One last type of jacket that will round out the discussion is an aluminized-bitumen product. The combination of a metalized outer surface and a heavy rubberized bitumen adhesive layer makes for a rugged, weather- and UV-proof jacketing product suitable for exterior surfaces. It also has the advantage of being self-healing, easily sealing up slits and punctures in the jacket to maintain the weather barrier and vapor-retarder boundary. One product (*Figure 10*) is a jacket about 60 mils (1.5 mm) thick. It comes

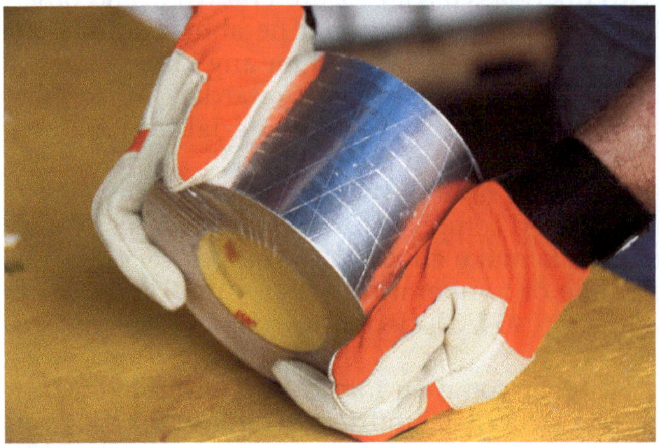

Figure 9 An example of an aluminized-bitumen laminate protective jacketing.

Figure 10 An aluminized laminated jacket for an exterior application.

in rolls 50' (15 m) long and several widths between of 4" and 35" (10 cm and 89 cm). In addition to its protective properties, its permeance rating is less than 0.01, making it an excellent vapor retarder. This product line also offers PTFE (polytetrafluoroethylene, commonly known as Teflon®) sliding saddles to protect the jacket at hangers and other kinds of pipe supports during pipe movement.

1.2.3 Mastics

Effective insulation jackets can be field-fabricated using a coat of mastic alone or reinforced with some form of fabric. Mastics (and insulating cements) are especially useful for providing a jacket cover for insulation in and around crowded and complex equipment shapes and piping system installations where prefabricated, roll, or sheet products would be difficult to shape and fit. Mastics can provide both an effective protective jacket as well as being a key part of a vapor retarder boundary. Insulating mastics and cements are covered in detail in other modules of the NCCER Mechanical Insulating curriculum, so they will not be addressed further here.

1.2.4 Removable/Reusable Flexible Insulation Jackets

The final type of jacket to be discussed here is removable and reusable covers. These are generally prefabricated, custom-fitted insulation sections sewn into and encased by an appropriate one-piece outer cover fabric or material. The covers' flaps and wraps are placed around the pipe or component to be insulated and secured with various devices such as hook-and-eye, hook-and-wire lacing, snaps, D-ring straps, or even hook-and-loop (e.g., Velcro™) fasteners (*Figure 11*). The outer fabric serves as the jacket for the flexible insulation cover. They are designed to be easily

Figure 11 Removable/reusable insulation covers.

removed for periodic inspection or maintenance of the insulated system and then reinstalled.

Though removable/reusable insulation jackets are initially more expensive to manufacture, over time they are far more economical than destroying and replacing permanent insulation and jacket. Their thermal performance can be made to be comparable to standard insulation as well. With the ready availability of computer-aided design and manufacturing (CAD/CAM) software and laser-scanning measuring instruments, many industries are moving to insulating large sections of their systems with custom-fitted removable/reusable insulation components. This kind of insulation, and its fabrication and installation, are described in NCCER Module 19314, "Removable and Reusable Flexible Insulation Covers," so these topics will not be considered further here.

1.3.0 Jacketing Installation Methods

Installing insulation jacket is a key step in the overall insulation work package. For small jobs, the mechanical insulator may install both the insulation and the jacket. For larger jobs, there may be separate crews for insulation and jacketing.

The methods for jacket installation vary based on the types discussed earlier in this module, but within each type, the approach is similar regardless of the materials involved. Since step-by-step jacket application is presented in other modules in the NCCER *Mechanical Insulating* curriculum, these methods will be presented here as a summary. In all cases, follow the manufacturer's instructions and installation guides for a given product. Study the product safety data sheets to understand and protect against any hazards associated with their use.

1.3.1 Installing Prefabricated Jacket

Prefabricated jacket may be applied separately from the insulation, or it may be an integral part of each insulation section. Pipe insulation will often be provided with an ASJ side-slit jacket. The insulation is applied to the pipe, and the SSL, if provided, seals the longitudinal seam. Insulation packages include matching butt strips to seal the circumferential seams.

Prefabricated jacket covers for pipe fittings are usually supplied in either one or two pieces. If the fitting is already insulated with a molded section of insulation, the jacket can be installed immediately. If applicable, trim off the stepped ends of the jacket to fit the diameter of the insulated pipe. Place the jacket cover around the fitting and secure it with an adhesive joint tape at its seams. PVC covers use a special adhesive to weld the edges of the cover together. Some manufacturers provide serrated stainless steel tacks to help secure the covers (*Figure 12*).

If the fitting is not insulated, then an approved fibrous or fabric insulation material is first wrapped around the fitting to fill up the space within the jacket cover (*Figure 13*). The cover is then installed as usual. Some specifications permit lining the fitting jacket cover with a thick mastic or insulating cement to fill in the air space, then pressing the cover into place around the fitting. Cut off excess filler material that sticks out and seal the joints and seams with a matching tape provided by the manufacturer.

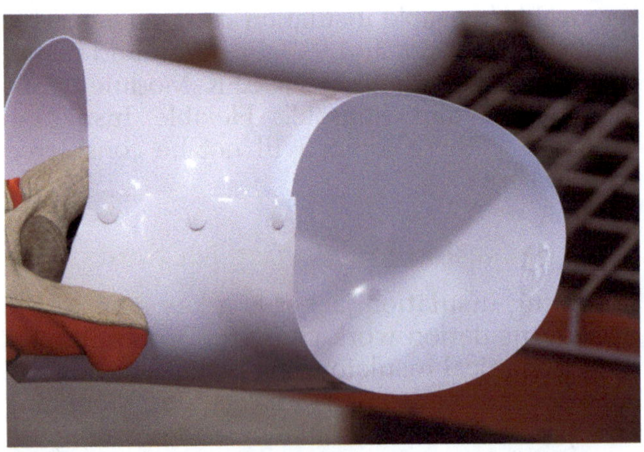

Figure 12 Serrated tacks securing a PVC elbow cover.

The following are some considerations for installing prefabricated, pre-curled, split-sleeved PVC jacket.

- Apply the straight-pipe jacket covers first, then fitting covers.
- Some specifications may require application of a mastic or adhesive under the jacket as part of the vapor retarder boundary.
- It is standard practice to stagger the joints of jacket system in relation to the underlying joints of insulation. Precut the jacket to achieve this and begin installation at a fitting.
- Install pipe jacket by applying the cover over the insulation and securing it using the SSL provided or gluing the lap with compatible adhesive.
- Jacket ends should lap adjacent sections by at least 1" (25 mm).
- The longitudinal lap and circumferential joints are sealed with a PVC welding cement (*Figure 14*), a matching adhesive tape, or both.

> **CAUTION**
> Always check the SDS for solvent-based adhesives and use the appropriate breathing and skin protection when applying such adhesives.

- Do not stretch PVC tape when installing it. It will retract and may expose the seam.
- If the jacket will be exposed to weather or wash-down operations, the longitudinal lap edge should be at the 3- or 9-o'clock positions and overlap the underside edge so water can't enter the seam.
- For interior installations not subject to water, it is a good practice to rotate the longitudinal lap so that it is not observable from occupied spaces.

1.3.2 Installing Roll and Sheet Jacket

Roll and sheet jacket is required when field-fabrication of the jacket system is required. There are two general approaches for installing these products. Sheet materials are generally measured and cut to size first, if required, and then installed. Roll materials remain on the roll as they are applied. They are wrapped around the surface and cut off after allowing an appropriate lap. Except for small-diameter pipes, most installations of

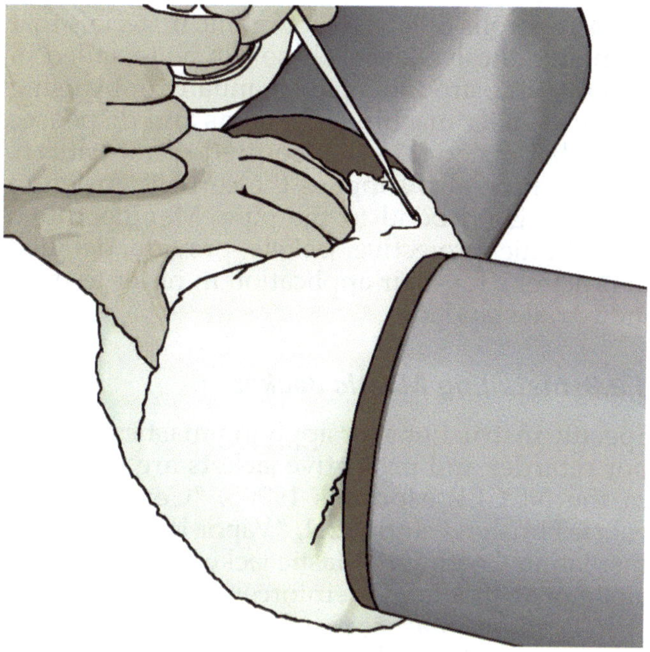

Figure 13 Insulating an elbow before installing the jacket.

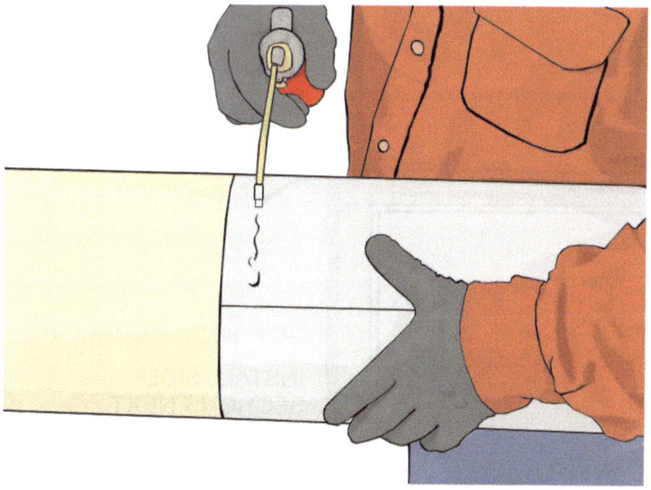

Figure 14 Application of a PVC welding cement to the circumferential joint.

sheet and roll jackets are best performed by two workers. This way, one person can be handling the roll while the other checks to make sure the material is being applied smoothly with no wrinkles (*Figure 15*). If the material has an adhesive coating with a protective covering, one person usually pulls the protective cover off as the other smoothly adheres the jacket to the insulation.

Each type of jacket has specific installation requirements. For example, metalized-asphaltic laminate weather-proof jacket must be applied over dry, clean, faced fiberglass or closed-cell polymer foam board so that its adhesive layer has a continuous surface to stick to. Most adhesive-backed roll materials are best applied to closed cell foam or faced insulation for good adhesion.

Figure 15 Checking for a smooth jacket installation.

The goal of such installations is a smooth, wrinkle-free surface, which takes practice to achieve.

The following are some considerations for installing an aluminized laminated roll jacket on ducts:

- Upper surfaces of ducts and equipment should be insulated so that they are sloped to permit water runoff.
- Round ducts should be jacketed so that the longitudinal seams are at the 3- or 9-o'clock positions with the lap edge on the upper side. The seams should be pressed with a roller to make a good seal (*Figure 16*).
- Ends of jacket sections should overlap by 4" (10 cm).
- When covering rectangular ducts, apply the bottom surface jacket first and wrap around to the sides by approximately 6" (15 cm). Then cover the sides, and last, the top (*Figure 17*).
- Alternatively, you can also jacket a duct by using a single piece over the sides and top as shown in *Figure 18*.

Large sections of PVC sheet jacket may be installed on insulated vessel sidewalls or equipment surfaces, especially in hygienic installations. Typically, the sheets are carefully measured and cut with electric sheet-stock shears for a clean edge. Any cutouts and holes for pipes or fittings are measured and cut. The sheet is test-fitted before installation. A contact adhesive is usually used to adhere the sheet material to the insulation surface. The adhesive is applied to the jacket by spraying or brushing according to the manufacturer's instructions. The jacket is positioned against the insulation surface and pressed into contact. On rigid or semi-rigid insulation such as foam glass or polymeric cellular foam, a roller may be used to ensure full-surface contact. Some manufacturers may provide custom edging

Figure 16 Rolling the lap seal of a laminate jacket.

accessories to join adjacent sections. Otherwise, edge seams are bonded using a PVC adhesive.

Some jacket materials form part of multi-layered anti-corrosion **cathodic protection** systems, or provide a protective cover for such systems. These installations are typically used for underground steel pipes like those found in the petroleum industry. The materials are often supplied as narrow rolls of fabrics or organic materials that form an adhesive layer. These can be installed in a spiral wrap around a pipe manually or by using a special roller machine that holds and dispenses the jacket material in the required spiral pattern. The machine also supplies the correct tension to ensure a good bond to the pipe. Manufacturers of cathodic protective jackets provide detailed instructions for their application in order to meet industrial standards.

1.3.3 Installing Mastic Jacket

Specific instructions for applying mastics as a vapor retarder and protective jackets are provided in the NCCER Modules 19208, "Cements and Fabric Finishes," and 19211, "Vapor Retarders and Insulation Coatings." Mastic jackets may be used alone or with a fabric reinforcement according to specifications. Mastics may also be included as an underlayment or filler to prefabricated, molded, sheet, or roll jacket.

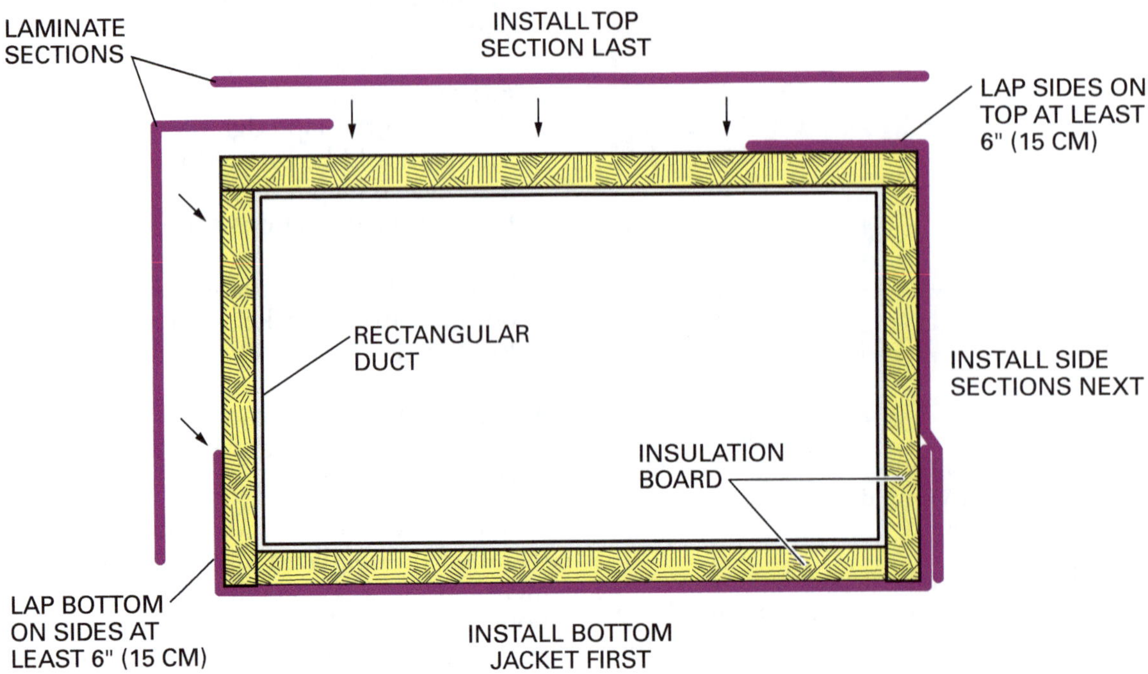

Figure 17 Roll or sheet jacket installation for a rectangular duct.

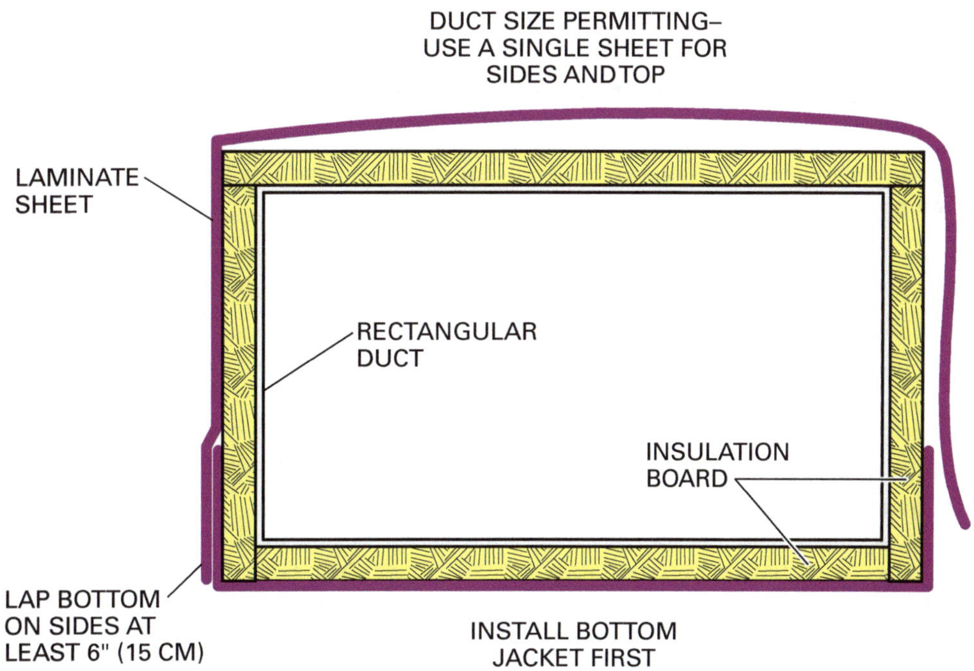

Figure 18 Jacketing a rectangular duct with a single piece covering the sides and top.

Additional Resources

National Industrial and Commercial Insulation Standards Manual, Midwest Insulation Contractors Association (MICA).

National Insulation Association (NIA) website offers resources for products and training. **www.insulation.org**.

Mechanical Insulation Design Guide, National Institute of Building Sciences. Available at **www.wbdg.org/guides-specifications/mechanical-insulation-design-guide**.

1.0.0 Section Review

1. What is a key purpose of insulation jacketing?
   a. Reduce the amount of insulation required
   b. Protect insulation against environmental factors
   c. Increase the overall thermal resistance value of insulation
   d. Permit insulation to breathe

2. A common material for various prefabricated and field-fabricated jacket applications is _____.
   a. CalSil
   b. foam glass
   c. PVC
   d. EPS

3. Of the listed items, which is *not* normally used to secure or seal prefabricated pipe fitting jacket covers?
   a. Glass fiber-reinforced tape
   b. Adhesive tape provided by the manufacturer
   c. Serrated stainless steel tacks
   d. Solvent-based adhesive

Summary

Mechanical system insulation is essential for thermal efficiency, economy, sound control, personnel protection, and other important reasons. But most insulating materials are fragile and can easily be damaged or destroyed by common environmental factors. This is why insulated mechanical piping systems and equipment are normally covered in a protective jacket material. Insulation jacketing serves many purposes—for environmental, impact, and abrasion protection, as vapor retardants, as personnel burn protection, and as a hygienic, easily-cleaned surface. Many jacketing materials can accomplish several of these functions.

Jacketing materials are commonly supplied in three forms—prefabricated, sheet and roll, and mastic-types. The application of each form has common aspects without regard to the materials in which they are supplied. The installation of materials within each form varies mainly in the details of adhesives and types of securements. As always, manufacturer's product data sheets, safety data sheets, and installation instructions must be studied and followed for each jacket product.

Review Questions

1. What environmental factor of those listed is probably the most damaging?
   a. Infrared
   b. Wind
   c. Temperature
   d. Water

2. What aspect of CUI increases its ability to damage piping and equipment?
   a. It happens quickly.
   b. It affects mainly stainless steel.
   c. It can occur unobserved.
   d. It affects all cryogenic systems.

3. Dropping a wrench causing a dent in an insulation cover during maintenance would be considered a(n) _____.
   a. nuisance
   b. abrasion
   c. scour
   d. impact

4. Which of the listed jackets would likely be the best selection for frequently steam-cleaned systems and piping installed in a food processing plant?
   a. FSK/FSP
   b. Stainless steel
   c. PVC
   d. Stucco-embossed, aluminum-bitumen laminate

5. Which of the listed items would *not* be considered an effective insulation jacket?
   a. Glass fabric secured with fiber tape.
   b. Prefabricated PVC covers
   c. Polymer mastic coat
   d. Foil-film laminate sheet

6. For pipe and equipment insulation jackets used outdoors, an important property is _____.
   a. density
   b. UV-inhibitor additives
   c. color
   d. flexibility

7. Which of the listed jacket materials/systems would be best suited for protecting the insulation of a group of crowded pipes and fittings on a vessel head?
   a. Field-fabricated PVC sheet or roll
   b. Prefabricated PVC covers
   c. Silica fiber removable insulation covers
   d. Field-fabricated mastic and fabric

8. With regards to installing insulation jacket, which of the following statements is true?
   a. Straight-pipe PVC jacket covers are installed after the prefabricated pipe fitting covers.
   b. Pipe insulation jacket section joints should align with the underlying insulation joints.
   c. Stretch PVC joint tape to make a neat, tight, seal.
   d. Longitudinal pipe jacket seams should be at the 3- or 9-o'clock positions.

9. Which of the listed materials would *not* provide a good surface for an adhesive-backed sheet or roll facing?
   a. FSK-faced fiberglass board
   b. Unfaced mineral wool blanket
   c. Closed-cell foam board
   d. Foam glass board

10. On what kind of mechanical systems would jacket for a cathodic protection insulation be installed?
    a. Underground steel piping and foundations
    b. Hygienic food processing systems
    c. Cryogenic fluid storage vessels
    d. Power plant steam piping

Trade Terms Introduced in This Module

Abrasion: The roughening or wearing-away of a surface by rubbing or by the impact of particles. Sand and other materials can abrade surfaces as they move past them.

Cathodic protection: A method of preventing underground metal objects from corrosion aided by exposure to ground water. This may be done by either attaching a more easily corrodible metal to the object or, in the case of underground pipes, supplying a low-voltage direct current through the metal that blocks the corrosive effect.

Corrosion under insulation (CUI): Corrosion occurring out of sight under pipe and equipment insulation due to the failure of the jacket, the vapor retarder boundary, or both, allowing water to penetrate the insulation and contact the metal surface.

Emissivity: A measure of how rapidly a material radiates heat away from its surface.

Fire-resistance rating: The time that a material by itself can resist the effects of a fire or continue to perform its function as measured by a standardized fire-resistance test.

Infrared (IR): A form of light not visible to humans that effectively transfers thermal energy from one object to any other that absorbs it.

Thermograph: An image created by detecting and recording infrared light. Colored thermographs show the intensity of infrared by colors, which indicate heat sources.

Ultraviolet (UV): A form of light invisible to humans that is more energetic than visible light, and can cause damaging chemical reactions when absorbed by skin or materials.

Additional Resources

This module presents thorough resources for task training. The following reference material is recommended for further study.

National Industrial and Commercial Insulation Standards Manual, Midwest Insulation Contractors Association (MICA).

National Insulation Association (NIA) website offers resources for products and training. **www.insulation.org**.

Mechanical Insulation Design Guide, National Institute of Building Sciences. Available at **www.wbdg.org/guides-specifications/mechanical-insulation-design-guide**.

Figure Credits

Coverflex Manufacturing, Inc., Figure 11

© iStock.com/Brian Brown, Module opener

© iStock.com/Vladimir Gorsky, Figure 1

© iStock.com/Jevtic, Figure 2

© iStock.com/jirodav, Figure 3

The Snell Group, Figure 4

© iStock.com/baranozdemir, Figure 5

Polyguard Products, Inc. – www.polyguard.com, Figures 6, 10

Photos courtesy of Owens Corning, Figure 7

Section Review Answer Key

Answer	Section Reference	Objective
Section One		
1. b	1.1.0	1a
2. c	1.2.1	1b
3. a	1.3.1	1c

NCCER CURRICULA — USER UPDATE

NCCER makes every effort to keep its textbooks up-to-date and free of technical errors. We appreciate your help in this process. If you find an error, a typographical mistake, or an inaccuracy in NCCER's curricula, please fill out this form (or a photocopy), or complete the online form at **www.nccer.org/olf**. Be sure to include the exact module ID number, page number, a detailed description, and your recommended correction. Your input will be brought to the attention of the Authoring Team. Thank you for your assistance.

Instructors – If you have an idea for improving this textbook, or have found that additional materials were necessary to teach this module effectively, please let us know so that we may present your suggestions to the Authoring Team.

NCCER Product Development and Revision
13614 Progress Blvd., Alachua, FL 32615
Email: curriculum@nccer.org
Online: www.nccer.org/olf

❑ Trainee Guide ❑ Lesson Plans ❑ Exam ❑ PowerPoints Other ____________

Craft / Level: Copyright Date:

Module ID Number / Title:

Section Number(s):

Description:

Recommended Correction:

Your Name:

Address:

Email: Phone:

Jacketing Fabrication— Piping and Fittings

Overview

Insulation jacketing has many purposes. It primarily serves to protect insulation against weather and the entry of liquid water into the insulation. These factors can destroy insulation or its ability to insulate. Water can also cause corrosion of the pipe or equipment under the insulation. Jacketing can be used as a part of the water vapor retarder system, to reduce the possibility of burns from insulated hot systems, or to provide a sanitary, easy-to-clean, and sightly finish to a mechanical system. This module describes how to fabricate and install metal jacket on pipes and fittings.

Module 19313

Trainees with successful module completions may be eligible for credentialing through the NCCER Registry. To learn more, go to **www.nccer.org** or contact us at 1.888.622.3720. Our website has information on the latest product releases and training, as well as online versions of our *Cornerstone* magazine and Pearson's product catalog.

Your feedback is welcome. You may email your comments to **curriculum@nccer.org**, send general comments and inquiries to **info@nccer.org**, or fill in the User Update form at the back of this module.

This information is general in nature and intended for training purposes only. Actual performance of activities described in this manual requires compliance with all applicable operating, service, maintenance, and safety procedures under the direction of qualified personnel. References in this manual to patented or proprietary devices do not constitute a recommendation of their use.

19310 V2

19310

Jacketing Fabrication – Pipe and Fittings

Objectives

When you have completed this module, you will be able to do the following:

1. Describe metal jacketing materials and general installation considerations.
   a. Describe common jacketing materials and their function.
   b. Describe general installation considerations.
   c. Describe how to measure and order jacketing materials.
   d. Describe various methods of cutting and forming jacketing materials.
2. Explain how to install jacketing on pipe and fittings.
   a. Explain how to install jacketing on horizontal pipe.
   b. Explain how to install jacketing on vertical pipe.
   c. Explain how to install jacketing around jacket penetrations.
   d. Explain how to install jacketing on pipe elbows.
   e. Explain how to install jacketing on pipe tees.
   f. Explain how to install jacketing on other common piping shapes.

Performance Tasks

Under the supervision of your instructor, you should be able to do the following:

1. Properly cut, roll, and apply aluminum jacket to a straight run of pipe.
2. Properly apply two-piece metal jacket to an insulated elbow.
3. Lay out a gored metal jacket for an insulated pipe elbow. Cut the gores, apply the proper beading and crimping, and install.
4. Properly lay out, cut, and apply metal jacket to an insulated piping tee.
5. Properly lay out, cut, and apply a flat metal end cap.
6. Properly lay out, cut, and apply a coned metal end cap.

Trade Terms

Batten
Bead
Bisector
Concentric
Crimp
Eccentric
Equal tee
Hemmed
Laser cutter
Lateral length
Offset compound shears
Pittsburgh seam
Plasma cutter
Pressed cover
Radial lines
Sector
Sweep elbow
Takeoff
Unequal tee
Wing seal
X-Y controller

Industry Recognized Credentials

If you are training through an NCCER-accredited sponsor, you may be eligible for credentials from NCCER's Registry. The ID number for this module is 19310. Note that this module may have been used in other NCCER curricula and may apply to other level completions. Contact NCCER's Registry at 888.622.3720 or go to **www.nccer.org** for more information.

Contents

Figures and Tables

SECTION ONE

1.0.0 METAL JACKETING

Objective

Describe metal jacketing materials and general installation considerations.

a. Describe common jacketing materials and their function.
b. Describe general installation considerations.
c. Describe how to measure and order jacketing materials.
d. Describe various methods of cutting and forming jacketing materials.

Trade Terms

Bead: A raised ridge or groove along an edge of sheet metal.

Crimp: A treatment of an edge of sheet metal to improve its rigidity, consisting of either many peaks and valleys along the edge, like gear teeth, or a single shallow fold parallel to the edge.

Eccentric: The condition where two circles, circular objects, or rotating objects do not have the same center point or center of rotation.

Hemmed: The folded-over edge of a sheet of material, as in a fabric hem. A hemmed edge is smooth and eliminates a safety hazard.

Laser cutter: A device that uses a form of high-energy light, usually of a single color and produced in a tight, intense beam, to cut materials. Laser cutters can be designed for extremely precise shape fabrication.

Pittsburgh seam: A seam between two sheet-metal edges consisting of a right-angle flanged male tab on one edge and a folded female pocket on the other edge. The flange is inserted into the pocket and the side extension of the female pocket is folded over to lock in the male flange. The various folds and closures can be manually- or machine-formed in the sheet-metal edges.

Plasma cutter: A device that uses a high-velocity, high-temperature jet of ionized air or other gas emitted from a nozzle to cut and shape materials.

Sweep elbow: A type of pipe elbow that has a longer radius than the standard short- or long-radius elbows.

Takeoff: The process of surveying, measuring, itemizing, and counting all materials and equipment needed for a construction project, as indicated by the drawings.

Wing seal: A type of metal banding seal or buckle equipped with two tabs or wings that are bent over the folded end of the band to secure it.

X-Y controller: A computer-controlled device that precisely positions a machine or tool in the front-to-back and side-to-side directions in a plane.

Insulated mechanical systems and equipment are usually provided with some type of jacketing, cladding, or lagging, depending on where they are located and the environmental factors affecting them. Insulation jacketing is commonly made from rigid, relatively thin sheet metals, plastics (or other polymer materials), laminated and composite materials, and mastics. The fabrication and installation of jacket made from non-rigid metal and nonmetal materials are described in other modules from NCCER's *Mechanical Insulating* curriculum. Rigid metal jacketing comes in a variety of alloys and forms. Considerations for installation and general fabrication methods are applicable to all materials.

1.1.0 Metal Jacketing Materials

Metal jacketing material can be mild steel, stainless steel, galvanized steel, or epoxy-coated aluminum or steel. Each of these materials is used for specific purposes and their use depends on the individual job specifications.

1.1.1 Sheet-Metal Properties

These materials come in many different thicknesses. The most commonly used material for piping is 0.016" (0.4 mm) aluminum. However, individual specifications may call for heavier thicknesses in areas that are subject to natural or maintenance damage, or lighter thicknesses for areas that do not present a mechanical hazard. A thickness of 0.040" (1 mm) is about the heaviest aluminum jacket used over pipe insulation.

Stainless steel jacket, because it is stronger than aluminum, is supplied in thicknesses from 0.010" to 0.032" (0.3–0.8 mm). Required jacket thickness is determined mainly by the diameter of the insulated pipe or vessel, and whether the insulation is rigid or non-rigid. Larger insulated diameters and non-rigid insulation require thicker jacket.

All types of metal jacketing used over insulation should contain a moisture barrier. The moisture barrier may be an acrylic paint coating, polyethylene and kraft paper sheet, or epoxy resin, polymer, or polyethylene coatings applied directly to the metal. The purpose of this barrier is to prevent galvanic corrosion from occurring. This kind of corrosion is common when the jacket and underlying metals are different and is made worse by direct contact with moisture.

There are many types of finishes that can be used for jacketing materials, as shown in *Figure 1*. Metal jacketing can be obtained smooth, stucco-embossed, corrugated, or epoxy finished. Metal jacketing can also be painted or coated in different color schemes. Corrugated materials provide extra rigidity but are harder to install than other types and are generally more expensive. On the other hand, corrugated jacket reduces glaring sunlight reflections, minimizes the appearance of dents, and allows for more expansion and contraction than other forms of metal jacket. Smooth, shiny surfaces can reflect sunlight glare and make dents from hail more obvious.

Jacket sheet metals are supplied in various amounts and sizes. Metals may be sold in 50' to 300' (15–90 m) rolls depending on the type of metal and distributor. The sheet-metal rolls are generally 36" or 48" (91 cm or 122 cm) wide and packed in cardboard boxes. Most suppliers will also cut the materials to the required sizes as coils or sheets of metal. The same materials used on piping are available for all of the different types of fittings. Fitting jacket covers can be ordered from the manufacturer or distributor, or field-fabricated, as needed.

1.1.2 Jacket Accessories

Metal jacketing is installed with accessory materials. All jacketing materials can be secured with banding, which may be obtained in galvanized steel, aluminum, or stainless steel. Banding width varies from 3⁄8–3⁄4" (10–19 mm) and the thicknesses run between 0.015" and 0.020" (0.4–0.5 mm). In general, large-width banding is required for large-diameter pipe and vessels. The use of aluminum banding has some restrictions, including use only in noncorrosive environments and on insulation diameters less than 8" (20 cm).

CAUTION

Some specifications may require color-coded banding. For example, older facilities may signify asbestos-free insulation under metal jacket with a specific band color to distinguish it from original asbestos insulation. Any repairs involving asbestos replacement and new work in such a facility requires the use of these jacket band colors. Banding colors may also signify system function, or the type of jacket/banding material.

An insulation jacket band is secured with a **wing seal**. The wing seal does not require a special crimping tool to secure it. The seal is applied while using a band-tensioning tool, which may be of the pistol-grip or ratchet type. The ratchet tensioner can apply more force to the band. Seals are sized to the banding to be secured. Seal metals should match the banding metals to avoid galvanic corrosion.

Metal jacket can be secured with many different types of fasteners. However, jacket installed on systems operating at below-ambient temperatures and equipped with a vapor retarder boundary must not be secured with any kind of penetrating fastener (screws, staples, tacks, or rivets). These can puncture the vapor retarder, allowing water vapor to enter the insulation. Screws can be galvanized, coated, or stainless. They may be hex-head or pan-head, with or without neoprene washers, and can be self-tapping or self-drilling. Aluminum blind rivets are sometimes specified for jacket securement. Rivets are not in widespread use.

Jacketing is normally sealed at its seams during installation with some form of caulking prior to securement. For below-ambient service, jacketing is sealed with an approved vapor retarder

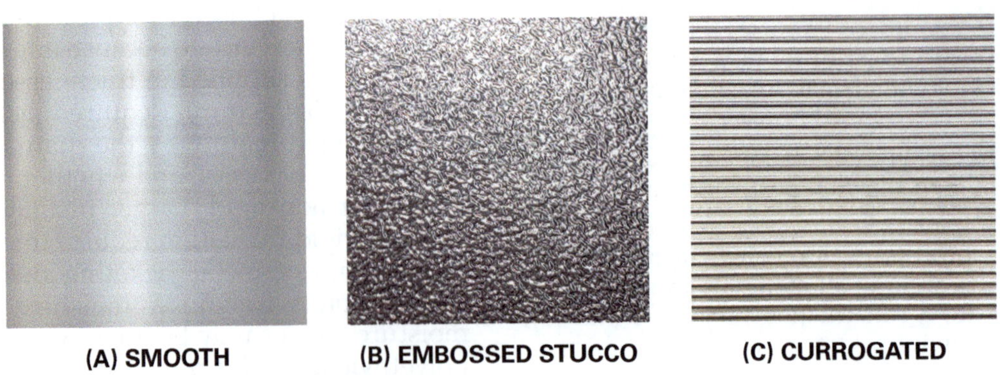

Figure 1 Sheet-metal jacket materials.

mastic or sealant that is part of the overall vapor retarder boundary. Sealants and caulking will be identified in construction drawings or in other applicable specifications.

1.2.0 Jacket Installation Considerations

Jacket installation is generally one of the last steps an insulator will complete in an insulating job. Before beginning jacketing work, there are a number of things to be aware of, including the following:

> **WARNING!**
> Stainless steel is very difficult to work with and represents a severe cutting hazard. Cut-resistant gloves and Kevlar® sleeves should be worn when handling this material.

- Jacket may be field- or shop-cut, or can be obtained from the manufacturer or distributor precut to fit all pipe sizes and thickness of insulation. If prefabricated jacket is not available, lengths of jacket are cut from a roll of sheet metal.
- When jacket is installed on large-diameter pipe insulation, the natural curvature of rolled jacketing helps the jacket to fit around the pipe insulation. If the pipe insulation is around 12" (30 cm) or less in diameter, the jacket should be rolled with a sheet-metal roller for proper fit.
- The edge overlaps on jacketing materials are determined by the job specifications. However, it is a standard practice for longitudinal laps to allow for at least a 2" (5 cm) overlap for insulation diameters less than 16" (41 cm), and at least a 3" (8 cm) overlap for larger diameters, unless specifications dictate otherwise. Butt laps on adjacent straight pipe and fittings jacket should be at least 2" (5 cm).
- All longitudinal laps should be arranged to shed water and positioned away from casual view if possible. Seams should be normally positioned below the horizontal plane through the pipe (the 3 to 4 o'clock or 8 to 9 o'clock positions).
- Longitudinal edges may be **hemmed** (folded) to eliminate a cutting hazard.
- Corrugated jacket is installed with the corrugations running parallel to the pipe.
- Pipe jacket should run continuously through pipe hangers, saddles, and similar horizontal weight-bearing pipe supports. Rigid inserts should be provided within the insulation to prevent jacket from compressing the insulation at the supports.
- Many contractors have shops where jackets are precut, rolled, and fabricated for jobs. Some large jobs may have an on-site shop. The manufacturer, distributor, and some large contractors have automatic machines that cut, roll, and fabricate the jackets as required. Jacketing may also be hand-cut for custom fitting and for small jobs.
- For securement, bands should be applied directly over all circumferential joints. Specifications will determine band spacing, but a maximum of 12" (30 cm) is standard.
- For crimped and mitered covers on short- or long-radius pipe elbows, a band is installed between each **crimp** or on gores between each edge **bead**. For fabricated jacket on a longer **sweep elbow**, each gore segment may have two bands. The end bands are installed directly over the overlapping sections with the adjoining straight pipe jacket.
- Screws and rivets may be used on hot systems or those without a vapor retarder, according to specifications. They must *not* be used to secure cold-system jacket where a vapor retarder is present.
- If rivets are permitted, ensure the pilot holes are drilled to match the rivets' manufacturer instructions. Too-large holes will result in the rivets pulling out.
- When using screws, they should be spaced evenly at a maximum of 6" (15 cm) apart on the longitudinal laps of straight pipe jacket. For elbow jacket longitudinal joints, if screws are needed, they should be spaced a maximum of 3" (8 cm) apart.
- If a seam sealant is required at longitudinal and butt edges, it must be applied between the metal and lap before securing the jacket. Do not apply a bead of sealant along edges after the jacket is secured.
- S-clips for supporting vertical pipe jacketing should be folded into three sections to create a butt seam overlap of 2" to 3" (5–8 cm). These clips support the upper jacket sections.
- When precut or prefabricated jackets are used, the quantity is estimated and the material is ordered by pipe size, insulation thickness, and total pieces required. The next section describes how this is done.

1.3.0 Estimating and Ordering Jacket Materials

As an insulating contractor, you will need to estimate the amount of materials required to complete the job, but not have an excessive amount of waste or materials that can't be returned to the distributor. The first requirement is fairly accurate knowledge of the dimensions of the system(s) to be worked on. This will involve studying as-built drawings, taking on-site measurements, or both, and writing up your findings in what the construction trades call a **takeoff**. (The takeoff process is described in some detail in NCCER Module 19309, "Construction Drawings and Specifications.") Specifications found in plans inform you of the kinds of materials to use, the size of pipes, and the thickness and types of insulation.

The jacket estimation process that follows is just a lot of simple arithmetic combined with the ability to visualize the jacket installation. For simplicity, the following example involves only straight pipe, so no fittings are included in the calculations. Assume the following job requirements:

- 120' (37 m) of nominal 6" (DN 150) straight steel pipe
- Insulation is $1\frac{1}{2}$" (3.8 cm) thick
- Specified jacket material is 0.016" (0.4 mm) aluminum jacket with a moisture barrier
- Jacket is to be shop-fabricated from roll sheet metal 36" (91 cm) wide
- Longitudinal laps are 2" (5 cm); butt laps are 2" (5 cm)

The following steps are an example for estimating the amount of material (for clarity, only Imperial units of measurement will be used in this example):

NOTE

It is a good practice when estimating or making other calculations in mechanical insulating work to round up your answers to avoid coming up short with materials or fabricated dimensions. It's always easier to return excess materials or trim something that's too large. With experience, you will learn what kind of rounding is appropriate in each case.

Step 1 The ASTM-specified diameter of 6" (DN 150) steel pipe is $6\frac{5}{8}$", or 6.63". Add twice the thickness of insulation ($2 \times 1\frac{1}{2}$" = 3") to the pipe diameter to determine the insulated diameter of 9.63" (since 6.63" + 3" = 9.63").

Step 2 Find the insulated circumference (C_I) of the insulated pipe. Since the formula for determining circumference is $C = \pi D$, the insulated circumference is calculated as follows:

C_I = insulated diameter $\times$ π
C_I = 9.63" $\times$ 3.14
C_I = 30.24"

Step 3 Find the number of 36" sections required to cover the pipe:

- Since a 2" (5 cm) butt overlap is required, the exposed length of each section is only 34" (86.4 cm).
- Convert the length of the pipe to inches by multiplying it by 12 (since there are 12 inches in a foot):

120' = 1,440"

- Divide the pipe length by the exposed length of jacket per section:

1,440" ÷ 34" per section = 42.4
(round up to 43 sections)

Step 4 Find the total length of roll jacket required. First, add the 2" (5 cm) lap to the insulation circumference to find the cut width of a section of jacket:

cut width = 30.24" + 2" = 32.24"

Step 5 Practically speaking, the sections should be cut to $32\frac{1}{4}$" (82 cm) wide. Multiply the cut width by the number of sections:

Length = 32.25" $\times$ 43 sections
Length = 1,387" *or* 116'

Therefore, to jacket the example pipe requires 43 sections of 36" (91 cm) roll jacket cut to $32\frac{1}{4}$" (82 cm) widths, for a total jacket length of 116' (35.4 m). It is always a good practice to add 5–10% extra sections to the estimate for mistakes and waste.

NOTE

When ordering jacketing from the shop, list the pipe size, the thickness of insulation, and the number of pieces required, or order the number of pieces cut to specific widths. Always allow for overlaps, both longitudinal and circumferential.

1.4.0 Cutting and Forming Metal Jacket

Prefabricated, pre-curled jacket obtained from a distributor is ideal for pipe jacketing jobs. However, if the systems to be worked on vary significantly in pipe sizes and insulation thicknesses, it generally is more practical to make on-site estimates of material requirements and submit them to the sheet-metal shop for local fabrication. Depending on the work force organization, you may be involved in the requisitioning and even the cutting of the jacket for your job. There are several approaches introduced in this section for cutting and preparing sheet metal for insulation jacketing.

1.4.1 Cutting Sheet-Metal Jacket

A variety of equipment and tools are available for cutting jacket. Some of these include the following:

- *Automatic cut-and-roll machine* – Jacket roll is placed on the machine (*Figure 2*) and the machine is set for the length and rolled diameter of the finished pieces. The number of pieces is entered in the control unit and the jacket is automatically cut and rolled. These machines are the easiest and most efficient method for cutting large numbers of identical jacket sections.
- *Sheet-metal shop shears* – There are three types of shop shears used to cut jacket metal. One is a drop-shear, where the angled punch blade drops vertically against the stationary shearing die blade. These are heavy, floor-mounted pieces of equipment. The drop blade may be operated by a foot pedal or hydraulically. A second type is called a guillotine shear. In these cutters, one end of a heavy blade is hinged and the blade cuts by rotating around the hinge against a stationary cutter blade (*Figure 3*). Both manual and hydraulic versions

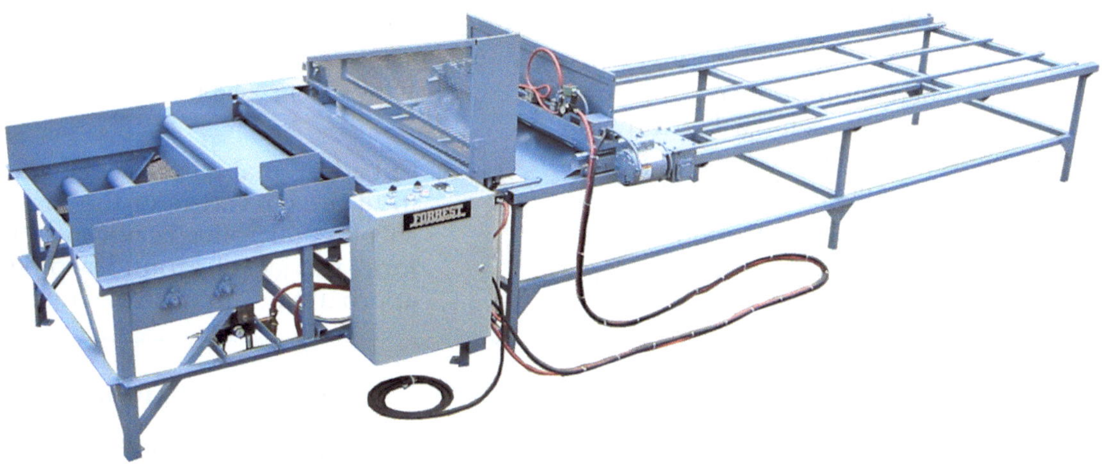

Figure 2 Automatic sheet-metal cut and roll machine.

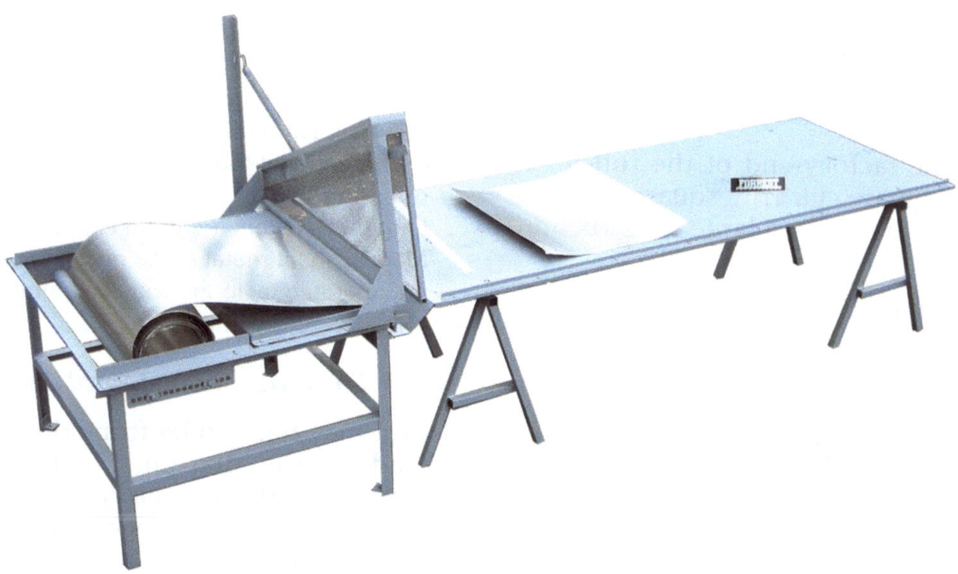

Figure 3 Manual sheet-metal shear and cutting table.

are available. Some hand-operated versions set on a table. Hand operated shears are suitable only for light aluminum sheet metal. The third type is a wheel cutting shear. For shop fabrication, a guide bar mounts a carriage on which a heavy, sharp-edged wheel bears against a cutting edge. The carriage is pulled by hand across the material and the wheel presses the metal against the blade, cutting it. A ruler should be installed on tables to aid measuring the cut widths of jacket.

- *Energy cutters* – These machines are high-end sheet-metal cutters that might be seen in some fabricator facilities. They are equipped with either an industrial laser cutter or a plasma cutter. The X-Y controller guides the laser beam or plasma jet to specific X-Y coordinates to vaporize the metal to the desired shape (*Figure 4*). The cutters are normally guided by computer programs and are capable of cutting out complex shapes (e.g., elbow gores and eccentric reducer covers) as well as straight cuts very precisely.

For small jobs, or when appropriate equipment is not readily available, sheet-metal jacket can be cut using basic hand tools. The following steps present one approach to hand-cutting jacket:

Step 1 Select a clear, clean area of the floor.

Step 2 Open the carton, remove the roll of metal, and place it on the floor.

CAUTION

To protect the floor, place a piece of cardboard or wood on the floor. It will also provide a cushioned cutting surface and minimize the dulling of the knife blade when cutting.

Step 3 Remove the tape from the roll of jacket, being careful not to allow the roll to uncurl.

Step 4 Check the factory end of the roll for a straight, square cut. Trim square if necessary with the aid of a large construction T-square.

Step 5 Using a tape measure, mark each end of the jacket (moisture barrier side up, if present) for the size of the piece to be cut.

Step 6 Place a straight edge between marks. Use a razor knife and cut along the straight edge. Place the knife on the jacket and pull toward the edge, pressing firmly. The point of the knife will score the jacket.

Figure 4 Laser sheet-metal cutting machine.

Step 7 Hold the end of the jacket and flex at the knife score. The jacket should break clean.

NOTE

Handheld electric sheet-metal shears can also be used to quickly cut sheet metal. Use a heavy straightedge as a guide and allow for the blade standoff before cutting.

Step 8 Cut additional pieces as required and roll to fit the curvature of the insulation. If a sheet-metal roller is not available, hand-form the jacket around a pipe to fit the approximate curvature of insulation.

Did You Know?

Sheet-Metal Brakes

Many workers believe the term sheet-metal *brake* is spelled "break," but this isn't so. The word for the machine comes down to the present from a combination of an Old Dutch word for crushing and a French word for a lever or handle. Together they came into the English language referring to a long-handled tool or machine for controlling or performing an action. Old car and wagon handbrake levers are good examples of this usage. Manual sheet metal brake machines were and still are operated by long lever-like handles.

1.4.2 Forming Sheet-Metal Jacket

Jacket may need to be formed or otherwise prepared for installation. Most common is pre-curling straight-pipe jacket to make it easier to fit and secure. Some specifications may require hemming or other forms of edge preparation. Mitered-elbow jacket covers as well as rounded vessel head jacket gores require beading and/or crimping to ensure a secure and professional looking installation. Some jacket preparation and fabrication methods include the following:

- *Sheet-metal rollers* – After the jacket sections are cut, the pieces are moved to the roller table and run through a sheet-metal roller (*Figure 5*), if required. Smaller-diameter pipe will require the jacket to be rolled to obtain the proper curvature to fit the insulation. Jacket must be fed through the roller straight in order to avoid damage to the outer edges. After rolling, the sections are then stacked together and marked as to size and thickness. Rolling machines may be either manually- or motor-operated.
- *Beading and crimping* – Some jacket fabrication methods require a means of reinforcing the connections between adjacent sections. To ensure a close, friction fit, each section receives a crimped edge or a bead, depending on the kind of seam. Special machines, called *beaders, crimpers,* or *beader-crimpers* if they can perform both functions (*Figure 6*), are required to form neat, effective seams. The widths and depths of the beads and crimps are selected based on the thickness of the sheet metal and the type of installation. Beading and crimping rollers come in pairs to make the required bead profiles. Refer to the jacket specifications for this information.
- *Bending* – Jacket may need to be fabricated with folded corners of various angles at the edges. Neat, crisp corners are formed with a standard shop machine called a *sheet-metal brake*. Large motor- or hydraulic-operated brakes may be used for bending long sections of ductwork and rectangular jacket. Some sheet-metal fabricators use brakes for hemming jacket edges. Smaller, hand-operated brakes (*Figure 7*) may be used for fittings and their covers.
- *Hemming* – Some specifications may require finishing one or more exposed edges of jacketing in a way that reduces or eliminates the cutting hazard of raw-edged sheet-metal jacketing. Bending over the edge presents a smooth seam that minimizes the hazard to people rubbing up against the jacketing while walking by or when performing maintenance. Hemming sheet metal is similar to creating a hem in a piece of cloth fabric. The edge is folded under the exposed surface. Sheet metal is stiff enough that no other securement for the hem is needed. Hemming can be formed by manual tools or

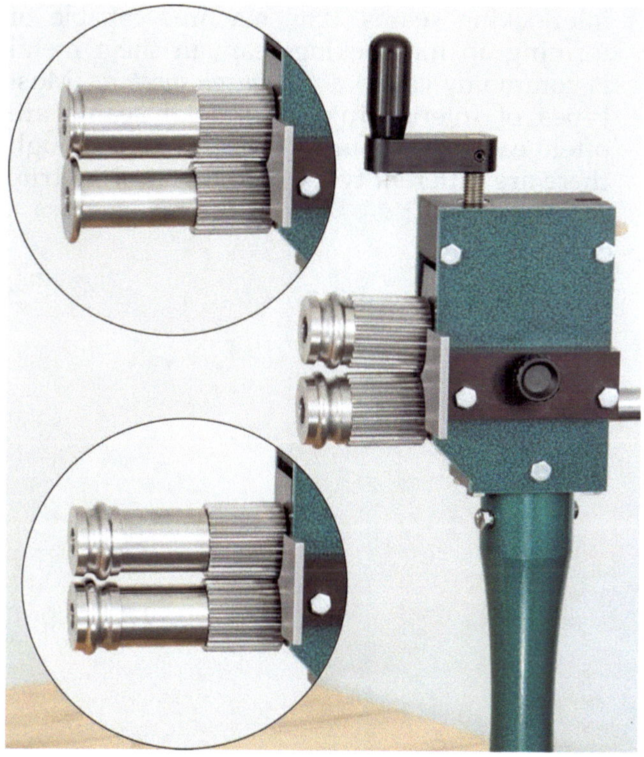

Figure 6 Sheet-metal beader-crimper.

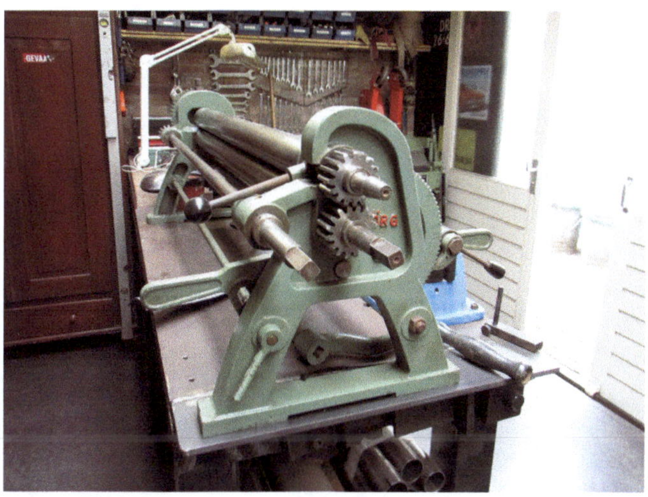

Figure 5 Manual sheet-metal roller.

Figure 7 Hand-operated sheet-metal brake for creating folds and corners in sheet metal.

large floor-mounted shop machines (*Figure 8*), as well as sheet-metal brakes. Hemming requires at least two steps to create the folded edge.

- **Pittsburgh seam** – While pieces of sheet metal can be assembled by using screws or rivets in overlaps at seams and edges, they can also be securely assembled at specially folded, interlocking joints. No fasteners are required, and the results look clean and are very strong. Such seams can be fabricated using manual sheet-metal brakes, but the process is laborious. However, special machines are available that quickly and uniformly fabricate the necessary bends, pockets, and tabs required to form interlocking seams. Any machine capable of forming an interlocking seam in sheet metal is commonly called a *Pittsburgh machine*. Most types of interlocking sheet-metal seams are often called Pittsburgh seams, even though there are different types of seams besides true Pittsburgh seams. Interlocking seams may be used in various jacket applications where sheet metal has to be assembled in complex shapes.

Figure 8 Shop sheet-metal hemming machine making the first crimp.

Additional Resources

Mechanical Insulation Design Guide, National Institute of Building Sciences. Available at **www.wbdg.org/guides-specifications/mechanical-insulation-design-guide.**

National Industrial and Commercial Insulation Standards Manual, Midwest Insulation Contractors Association (MICA).

National Insulation Association (NIA) website offers resources for products and training. **www.insulation.org.**

Installation Guidelines for ITW Insulation Systems Metal Jacketing Used on Insulated Pipe, ITW Insulation Systems. 2016. Available at **www.itwinsulation.com**

1.0.0 Section Review

1. Which of the following factors is *not* normally considered when selecting the weight of stainless steel jacket?
   a. Total insulation diameter
   b. Type of sheet metal
   c. Rigidity of the insulation
   d. Service temperature

2. As a general trade practice, longitudinal seam overlaps for jacket installed on insulated diameters of 16" (41 cm) and greater should be at least _____.
   a. 4" (10 cm)
   b. 3" (8 cm)
   c. 2" (5 cm)
   d. 1" (3 cm)

3. Which kind of information provided here would you *not* require when developing a quantity estimate for a jacketing job?
   a. The kinds of jacket materials to use
   b. The sizes and lengths of system piping, and the kinds and numbers of fittings
   c. The thicknesses of insulation required
   d. The materials from which the piping system is constructed

4. Using which tool is the easiest and most efficient for cutting large numbers of identical insulation jacket sections?
   a. Hand-cutting with manual shears
   b. Guillotine shears
   c. Plasma cutter
   d. Automatic cut-and-roll machine

SECTION TWO

2.0.0 INSTALLING JACKET ON PIPE AND FITTINGS

Objective

Explain how to install jacketing on pipe and fittings.

a. Explain how to install jacketing on horizontal pipe.
b. Explain how to install jacketing on vertical pipe.
c. Explain how to install jacketing around jacket penetrations.
d. Explain how to install jacketing on pipe elbows.
e. Explain how to install jacketing on pipe tees.
f. Explain how to install jacketing on other common piping shapes.

Performance Tasks

Under the supervision of your instructor, you should be able to do the following:

1. Properly cut, roll, and apply aluminum jacket to a straight run of pipe.
2. Properly apply two-piece metal jacket to an insulated elbow.
3. Lay out a gored metal jacket for an insulated pipe elbow. Cut the gores, apply the proper beading and crimping, and install.
4. Properly lay out, cut, and apply metal jacket to an insulated piping tee.
5. Properly lay out, cut, and apply a flat metal end cap.
6. Properly lay out, cut, and apply a coned metal end cap.

Trade Terms

Batten: In layout work, a long, thin, flexible strip of wood or other material that can evenly bend without kinks. Used for creating smooth, sweeping single- and compound-curves.

Bisector: In geometry, a line that divides an angle in half or bisects it.

Concentric: The condition where two or more circles or circular objects have the same center point or lie on the same axis.

Equal tee: Identifies a piping tee where the main and branch pipes have the same diameters.

Lateral length: For a conical fitting, the top-to-bottom length of the fitting as measured along its side (or lateral) surface.

Offset compound shears: A type of hand shears with jaws designed to permit cutting small-radius curves in sheet metal without crimping the material and keeping the worker's hand clear of the cut. Most manufacturers supply color-coded tools to identify the designed direction of the cut: green is clockwise cutting; red is counterclockwise cutting; yellow is straight cutting. Also called *aviation snips*.

Pressed cover: A type of preformed insulation jacket cover for elbows and tees available from many insulation distributors. The cover is made by shaping sheet metal in a form press and can be identified by the crimped fingers in the throat of the cover.

Radial lines: In geometry, straight lines drawn outward from the center of a circle, like a radius.

Sector: In geometry, an area contained inside a circle's perimeter between two radial lines.

Unequal tee: A type of piping tee where the main and branch pipes are of unequal size. The branch pipe may have either a smaller or larger diameter than the main pipe.

Being able to fabricate and install jacketing on pipes and fittings is an essential skill for mechanical insulators. In this section, you will be introduced to the more common kinds of insulation jacket installations for piping systems.

2.1.0 Installing Jacket on Horizontal Pipe

A typical horizontal piping application would be two insulated chilled water pipes located on the roof connected to air handling units. The pipes pass to the roof area through a sleeved penetration in the mechanical room wall on the roof. These pipes run level, approximately 6" (15 cm) apart and sit on common supports.

This example assumes the insulation is already in place. However, in practice, if the pipe supports or hangers must be adjusted for installation of insulation or jacket, the insulation and jacket should be installed at the same time to avoid adjusting hangers twice. It is not uncommon that piping runs, both horizontal and vertical, may include attachments, instrumentation wells, welded bosses, and support brackets clamped directly to the pipe. For situations in which fittings and supports are attached to the pipe and penetrate the insulation and jacket, the jacket must be cut and fitted around the penetrating object. Section 2.3.0 discusses the approach to fitting jacket around penetrations. For the following example, it is assumed no penetrations will be encountered:

Step 1 Move cut and rolled pieces of jacket to the roof where the jacket is to be installed. Move bands, tie wire, and tools to the same location. Place the jacket and wire where they may be reached from the pipe.

Step 2 Use appropriate personal protective equipment (PPE), as required.

Step 3 Make sure to start pipe jacketing at the lowest point of the run if it is not level. Place a piece of jacket around one pipe at the room wall with the longitudinal lap facing the other pipe run. Snug the jacket to the insulation with the lap at or slightly below the 3 o'clock or 9 o'clock position. Push the jacket into the room wall sleeve approximately 2" to 3" (5–8 cm). Place the tie wire snugly around the jacket. Readjust the jacket, if needed, and tighten the wire where it crosses the longitudinal lap.

Step 4 Proceed with the same application on the second pipe, again placing the lap on the inside (facing the other pipe).

Step 5 Loosely position sections of the jacket on the pipes to the first pipe support. Provide a gap at the supports and continue placing loose sections of the jacket to the chiller unit. This method allows the jacket to be applied with more ease rather than removing each piece of jacket from the box as each piece is attached.

Step 6 Return to the mechanical room wall and move the second section of jacket into a permanent position overlapping the first jacket section 2" to 3" (5–8 cm). Be sure the longitudinal seams are even and straight, and the jacket is fitted to the insulation. Place wire snugly around jacket. Readjust the jacket, if necessary, and tighten the wire. Continue placing the jacket sections in their permanent positions until the jacket meets the first pipe support.

Step 7 Loosen the pipe support to allow installation of the jacket. If rigid insulation inserts are present, be sure they are not disturbed while installing the jacket. Place bands on section passing through the pipe support as required by specifications. Restore the pipe supports according to the system's specifications to complete the installation before moving to the next section. Never loosen more than one pipe support in a pipe run at a time.

NOTE

If saddles are required at the support, install them after jacketing has been installed and prior to restoring the piping weight carried by the support—jacket first, then saddle.

Step 8 Continue installing the jacket on the pipe and through supports until the pipe runs are completed. Pay close attention to overlapping each succeeding piece at least 2" (5 cm). Be sure all longitudinal laps are even and in a straight line as in *Figure 9.*

Step 9 Return to the mechanical room wall and start placing bands. Bands should be placed no more than 12" (30 cm) apart. Wing seals should be placed on the jacket lap in a straight line. A band should be placed on the circumferential overlap of each piece of jacket. A 36" (91 cm) section should have four bands evenly spaced. When bands are in place on a piece of jacket, cut the temporary wires from the jacket. Place cut wires and band cuttings in a waste receptacle for proper disposal.

CAUTION

No unjacketed insulation exposed to weather should be permitted to remain after quitting time. All permanent banding on exposed jacket should be installed before the end of work. If this is not possible, cover exposed insulation with tarps or plastic sheeting and tape in place.

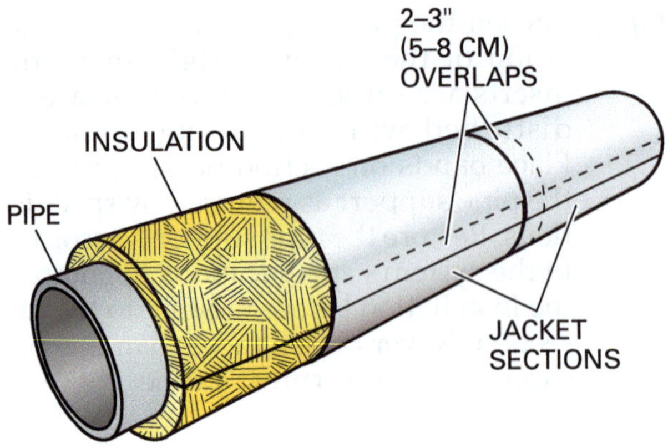

Figure 9 Jacket placement on horizontal pipe.

The following are important guidelines/considerations to emphasize for jacket installations on horizontal (or nearly horizontal) and vertical pipe runs:

- Start jacket installation at the lowest point of the pipe run. This ensures jacket seams will shed water falling on the pipe.
- Ensure jacket fits tightly to underlying insulation.
- Do not overtighten the bands and crimp the jacket.

2.2.0 Installing Jacket on Vertical Pipe

The approach to installing jacket on a vertical pipe is somewhat similar to working on horizontal pipe runs, but there are some key differences as well. Most importantly is the need to physically support the upper sections of jacket and working around pipe supports, which normally cannot be loosened to pass the insulation or jacket through.

2.2.1 Riser Jacket Installation

A typical vertical piping installation consists of chilled water pipe risers from a package chiller up the wall to air handling units on the roof. There are wall-mounted brackets attached to the pipe up the wall that will create penetrations in the jacket. These will require that the jacket be cut to fit around the brackets, which is discussed in Section 2.3.0. The jacket and bands are prefabricated. Wire is used to hold the jacket in place until bands are installed. Scaffolding is in place or a lift is available allowing access to the entire riser. The pipe insulation is installed and metal jacket covers are installed at the lower elbows.

When jacketing is installed on vertical pipe, provisions must be made to prevent the jacket and bands from sliding down the insulation. Vibration and changes in temperature over a period of time may cause the jackets to slip. If job specifications permit, a screw may be installed at each overlap and under each band. However, in chilled water system piping equipped with a vapor retarder, no screws or other penetrating fasteners should be used.

The preferred method to support metal jacket is to install a minimum of two S-clips at each jacket circumferential lap for smaller pipe. The number of clips increases with pipe size in order to properly support the next section of jacket. The clips are usually field-fabricated from a piece of band material about 4" to 6" (10–15 cm) long. The strips are bent into three segments of a flattened S-shape (as shown in *Figure 10*) so that there are 2" to 3" (5–8 cm) between the ends of the clip. This dimension sets the jacket butt-seam overlap. It is important that S-clip lengths are identical or nearly so for a given butt joint so they all bear the same amount of weight. The inner hook of the clip is hung on the top edge of the lower jacket section and the next section is placed in the outer hook.

> **NOTE**
> Experienced insulation mechanics can make an S-clip jig from a piece of wood and two nails placed at the required overlap spacing.

To install jacket on vertical pipe, use the following steps:

Step 1 Use appropriate PPE, as required. For elevated work, this may include fall-protection harnesses and lanyards.

Step 2 Determine the number of pieces of jacket, bands, and clips required to completely jacket the risers.

Step 3 Place jacket, bands, clips, a small roll of wire, and tools on the scaffold or in the lift.

Step 4 Starting at the bottom of the pipe run, place a section of jacket over the pipe insulation with the longitudinal seam and lap on the far side of the pipe (the least visible side that faces another pipe or a wall). If the pipe run starts at an elbow, ensure the elbow jacket is in place first.

Step 5 Place a loop of wire loosely around the jacket section.

Step 6 Slide and adjust jacket to overlap the elbow jacket a minimum of 1" (3 cm) or to a point on the elbow jacket where the turn starts.

> **NOTE**
> The position of the jacket's longitudinal lap should be selected in relation to the pipe's wall supports so that the jacket penetration cutouts for the supports do not fall within the overlap, if possible.

Step 7 Tighten the wire. Place the second wire on the jacket if required for a snug fit.

Step 8 Place a number of S-clips on the top lip of the jacket section appropriate for the pipe size.Step 9 Place another section of jacket in the clips (*Figure 11*) with the longitudinal seam in line with the first section's seam.

Step 10 Continue installing the jackets on both risers until the installation is complete at the scaffold level, if applicable, or until a pipe support is encountered. Jacket installation around pipe supports and other penetrations is discussed in Section 2.3.0. Install bands on the jacket to complete the installation. Remove all temporary wires. Place wires and band cuttings in a waste receptacle.

Step 11 Move the jacket materials to the next scaffold level, if necessary, and continue the installation up through the last whole section of the jacket before the top elbow or roof penetration.

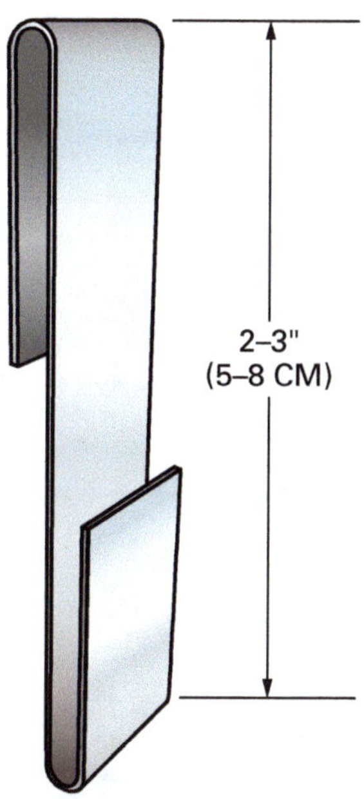

Figure 10 S-clip for supporting vertical pipe jacket sections.

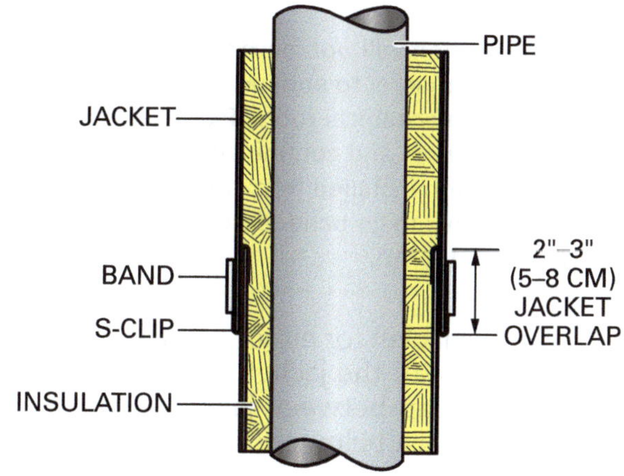

Figure 11 Vertical pipe jacket section support.

2.2.2 Finishing a Riser Jacket Installation

The riser may end at an elbow inside the mechanical room or continue on through the roof to the rooftop piping. If the riser ends at an elbow inside the mechanical room, complete the piping run jacket installation as follows.

Step 1 Temporarily install the elbow cover with wire. Measure and cut the last section of the jacket to overlap the elbow by at least 1" (3 cm).

Step 2 Remove the elbow cover. Position the jacket over the insulation. Verify the butt overlap with the next lower section is correct and that the longitudinal seam aligns with the lower section's seam.

Step 3 Permanently band the straight-pipe jacket section in place. Then install the elbow cover, ensuring the minimum 1" (3 cm) of overlap exists a shown in *Figure 12*. Band the elbow cover in place as previously described.

If the riser penetrates the roof, and no transition sleeve is provided, then the following steps will be necessary to finish the riser jacketing if specifications require it be jacketed its entire length.

Step 1 If a full section of jacket is long enough to reach through the roof penetration, the jacket installation is continued as before. With the help of an assistant on the roof, measure to verify a full section can be installed before encountering a fitting or a pipe support.

> **NOTE**
>
> If a full section will *not* reach to roof surface, cut the riser jacket to end just below the roof inside the mechanical room. Place S-clips on the top section, and continue the jacket with a jacket section installed from the roof. The lower end will have to be banded from inside the mechanical room.

Step 2 Cut the jacket for either a bracket or an elbow. Install the jacket, placing a band at the overlap between jacket sections and another just below the roof.

Step 3 On the roof, continue the riser jacket to the elbow. Measure and cut the last riser jacket section at an elbow. Install the jacket section and band it in place.Jacketing through a roof penetration may require the installation of flashing and other weatherproofing at the opening (*Figure 13*). The riser jacket will need to be installed with the flashing so that it properly sheds water. The roof penetration also may require a filler material after the jacket is installed. Make sure the roofing contractor is involved in flashing the pipe so as to not jeopardize the roof warranty.

Step 4 After the riser installation is finished, check the risers completely for proper jacket alignment and spacing of the bands before removing the scaffolding, if applicable.

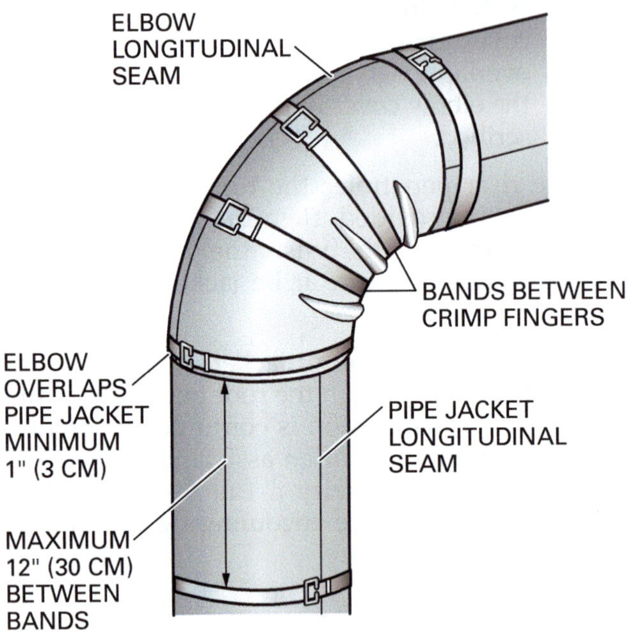

Figure 12 Finishing riser jacket at an elbow.

Figure 13 Pipe flashing of a roof penetration.

2.3.0 Pipe Jacket Penetrations

Vertical pipes and risers are required to be supported at regular intervals by clamp-on brackets. Horizontal pipes may also be supported by clamp-on brackets rather than saddles or hangers. All piping runs may include various kinds of insulated and uninsulated fittings that will penetrate the jacket, especially instrumentation wells (*Figure 14*).

When a pipe bracket (or other pipe fixture that requires penetration of the jacket) is encountered, it is a common practice to end the affected jacket section at or just above the bracket to minimize the cutting and fitting required. The following steps describe a method for applying jacket

Figure 14 An instrumentation junction box penetrating pipe jacket.

around fixed pipe brackets. This method is also applicable to other kinds of small penetrations. (Larger insulated connections to pipes, such as tees will be addressed later in this module.) Refer to *Figure 15* as you read through the steps:

> **WARNING!** Never loosen the pipe brackets or supports for vertical piping runs when installing insulation or jacketing. Doing so could compromise the support for the entire pipe run, possibly leading to system damage and personal injury.

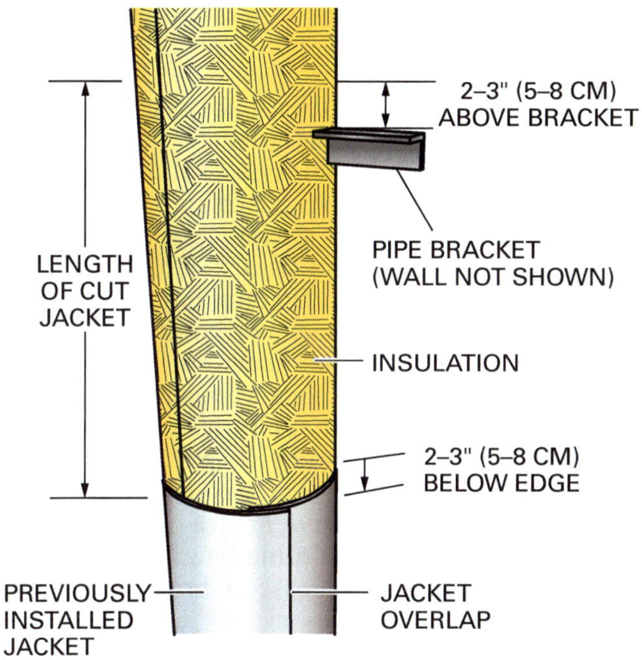

Figure 15 Measuring jacket at a vertical pipe bracket.

Step 1 Find the longitudinal position of the cutout at the pipe bracket by measuring the distance from the end of the last-installed jacket section to 2" (5 cm) above the pipe bracket. Add 2" to this length for the overlap with the lower jacket. As an example, use a measurement of 20" (51 cm) for this distance.

Step 2 Lay the section of jacket (outside of the jacket up) flat on a work surface. Mark the length of jacket at 20" (51 cm) and cut the jacket off at the mark using the score-and-break method or sheet-metal shears.

Step 3 Temporarily place the cut section of jacket around the pipe below the pipe bracket. Arrange the lap to match the section below the one you are working on and align the longitudinal seams. Make sure the jacket wraps the insulation tightly.

Step 4 Mark the points on the jacket where its top edge touches the outer edges of the pipe bracket. This is the width of the slot to be cut. If the bracket is an angle iron, U-channel, or has some other cross-sectional shape, sketch the shape onto the jacket to match the bracket as closely as possible. *Figure 16* shows how the jacket is notched and fit to the bracket.

Step 5 Make the cutout with sheet-metal shears. Flatten any kinks in the cutout edges, then place the jacket section on the pipe below bracket. Slip the jacket up around the pipe bracket. Check for proper fit and overlaps, and trim if necessary.

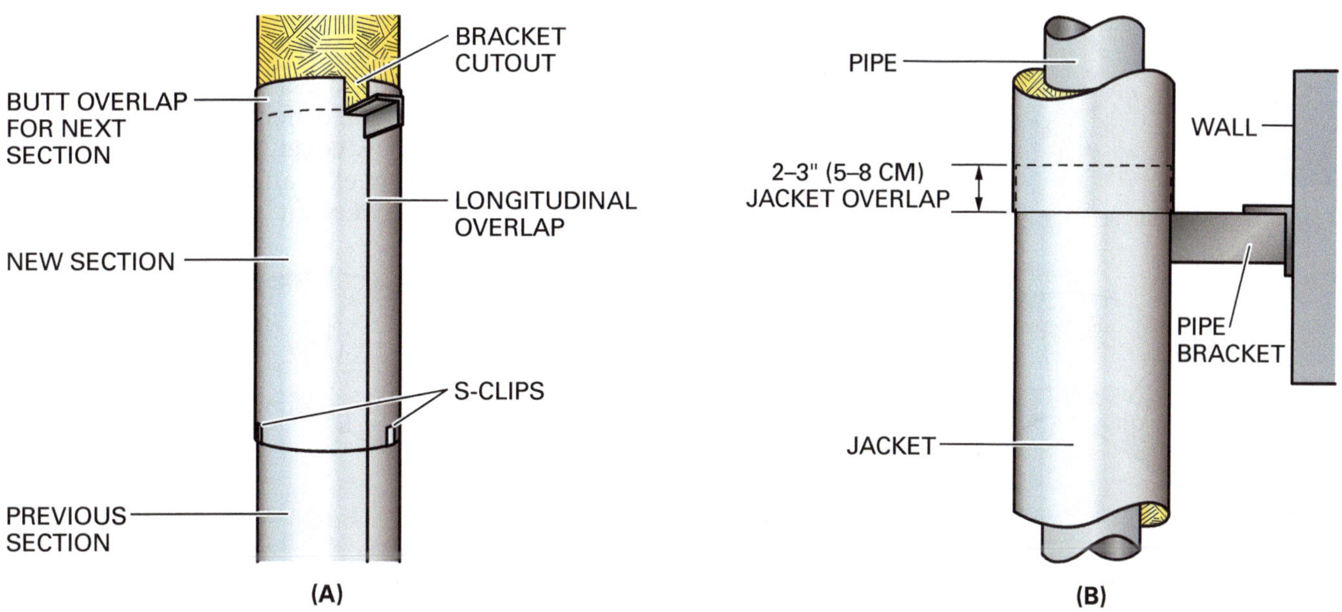

Figure 16 Shaping and fitting jacket to an angle-iron pipe bracket.

NOTE

All surfaces of the pipe insulation around the pipe bracket should be covered with jacket. The jacket should overlap the next lower section by at least 2" (5 cm).

Step 6 When the jacket fits properly, permanently secure the jacket except for the top edge.

Step 7 Place the next section of jacket above pipe bracket with bottom edge of jacket in contact with top of the pipe bracket. Continue placing jacket sections, clips, and bands to the top of risers.

An alternative approach for a bracket penetration is to cut the lower jacket off even with the top of the bracket and notch the upper section of jacket to provide the minimum 2" (5 cm) overlap (*Figure 17*). The former approach provides a cleaner appearance. Note the accurate jacket cutout around the angle-iron bracket.

Figure 17 Installed vertical pipe jacket at a bracket.

There may be cases where penetrations along either horizontal or vertical pipe runs do not line up nicely like a series of pipe brackets on a wall. There may be sensor wells, stub pipes left over from system testing, etc. These situations may require very short jacket sections notched at their upper edges as described above, or other acceptable approaches, such as cutting holes for short bosses. You will have to use your judgment to deal with these cases, following the principles given in this module. In every case, direct measurements on the affected piece of jacket are required for a neat and accurate installation.

2.4.0 Pipe Elbow Jacket

Distributers and fabricators supply jackets in the form of a two-pieced pressed cover or as mitered segments in metal thicknesses of the most common pipe jacketing. Aluminum and steel elbow covers are available by various suppliers to fit nominal pipe sizes up to 12" (DN 300), and insulation thicknesses up to 4" (10 cm). Larger sizes for some products can be special ordered. All distributors provide tables to help customers determine the right size covers for each pipe size and insulation thickness. Note that elbow covers specifically made for 90°, 45°, and sweep elbows are not interchangeable.

The pressed types (*Figure 18*) are constructed in two halves with an overlapping longitudinal seam. There are two types of pressed elbow covers. The standard elbow cover fits closely to the contour of the insulated elbow, with a specified overlap at the seam between cover halves. These pressed covers have finger folds in the throat of the cover. The standard covers are marked with the size and thickness they fit, such as 3x2 LR (3" long-radius elbow with 2" insulation). They may fit more than one pipe size and thickness and sometimes they are marked with several different sizes.

The universal elbow cover (also called a *humped elbow*) has a larger heel than a typical insulated elbow. Universal covers fit more combinations

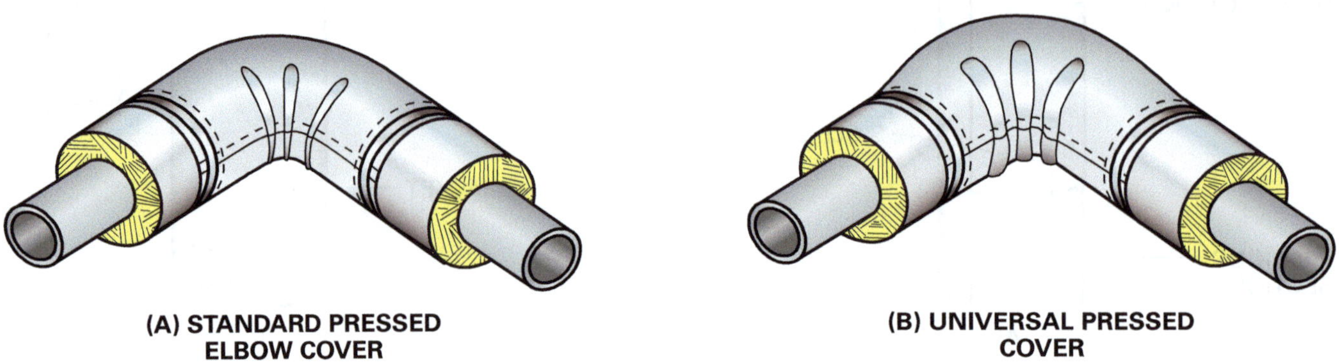

Figure 18 Prefabricated pressed covers

of pipe sizes and insulation thicknesses than the standard type. They are marked with a number (such as 5), which will fit several sizes and/or thicknesses of insulation according to a table of pipe sizes and insulation thicknesses (see *Table 1*).

> **NOTE**
>
> Universal covers will generally fit both short- and long-radius elbows of a nominal pipe size, but standard covers will not. When ordering elbow covers, always list the pipe size, insulation thickness, and whether the elbows are of short- or long-radius.

Aluminum elbow covers are typically made from 0.024" (0.6 mm) thickness sheet and coated on the inside with a moisture barrier material. Elbow covers may be attached with screws, bands, or rivets. Screws or rivets should not be used if the elbow being covered is on a low-temperature pipe because they might break the vapor barrier on the insulated elbow.

Prefabricated and field-fabricated mitered elbow covers are generally used for larger pipe sizes for which pressed elbow covers aren't available, as well as for sweep elbows. Their main advantage is that they can be custom-made to fit any sized elbow and any thickness insulation. They are constructed similar to mitered duct elbows, with overlapping gore segments. Typically, the heel surface of each gore has a bead on one edge and a crimp on the other as shown in *Figure 19*.

Prefabricated gore covers are available from jacket manufacturers and distributors in aluminum and stainless steel. They are also available for any radius elbow from short-radius to sweeps. They are ordered by pipe size and insulation thickness. If oversized insulation is used, the outer diameter of the insulation must be stated. The radius must be stated as long-radius or short-radius for a nominal pipe size, and the bend dimensions must be provided for sweep elbows.

Figure 19 Prefabricated mitered-elbow jacket covers.

Table 1 Example Universal Fitting Jacket Table

NPS	Insulation Thickness (Inches)								
	1	1½	2	2½	3	3½	4	4½	5
½	3	4	5	6	7	8	9	10	11
1¾	3	4	5	6	7	8	9	10	11
1	3½	4½	5½	6	7	8	9	10	11
1¼	3½	5	5½	6	7	8	9	10	11
1½	4	5	6	7	8	9	10	11	12
2	4½	5½	6	7	8	9	10	11	12
2½	5	6	7	8	9	10	11	12	
3	5½	6	7	8	9	10	11	12	
3½	6	7	8	9	10	11	12		
4	6	7	8	9	10	11	12		
4½	7	8	9	10	11	12			
5	7	8	9	10	11	12			
6	8	9	10	11	12				
7	9	10	11	12					
8	10	11	12						
9	11	12							
10	12								

Fitting cover number listed fits both long- and short-radius elbows.

2.4.1 Installing Pressed Metal Elbow Covers

If pipe being jacketed is outside or exposed to liquid water, pay close attention to the overlap of the elbow covers to ensure they are installed to shed water. For example, if an elbow is at the top of a pipe riser, the pipe riser should be jacketed first and the elbow cover should overlap the pipe jacket. If the elbow is at the bottom of the riser, the elbow cover should be installed first. The pipe riser jacket should overlap the elbow cover. If the pipe being jacketed is indoors and does not otherwise have to be waterproof, it may be more practical to install elbow covers before installing pipe jackets. Refer to *Figure 20* as you read through the following steps for installing elbow covers:

Step 1 Determine the best arrangement for the longitudinal overlap. If the pipe is near a wall and the bend is parallel to the wall, the front piece should overlap the back piece. In any case, place one half of the elbow cover on the elbow and check for fit. Then place the second half on the elbow and notice the amount of overlap at the longitudinal seam.

Step 2 Place two or three temporary tie wires around elbow cover and tighten as close to the throat as possible to avoid misalignment. Do not crimp the elbow cover. Check for fit and adjust if necessary.

Step 3 On hot systems, as well as systems without a vapor retarder boundary, screws should be used to secure the heel seams at the butt ends where they overlap the adjacent pipe jacket. If the seam gaps, place screws at about 3" (7 cm) centers as needed.

Step 4 Tighten screws snugly, but do not strip out their holes by overtightening.

Step 5 Install screws in the throat seam in the same way as in Steps 3 and 4.

Step 6 Remove temporary tie wires.

For cold-service installations and those with a vapor retarder, screws and rivets may not be used because of the likelihood of puncturing the vapor retarder. The process for installing a pressed elbow cover on a cold-service system is the same as just described, except that a vapor retarder sealant is applied between the laps before the jacket is assembled (if required), and bands are used to secure the cover halves. Place a band between each pressed finger in the throat of the elbow and a band over each butt seam with adjacent straight jacket sections.

2.4.2 Fabricating and Installing Mitered-Elbow Jacket

Mitered-elbow jacket gores may be field-fabricated or shop-fabricated from roll jacketing. For a quality job, a sheet-metal beader is necessary to form the interlocking laps. The general approach to fabricating mitered elbow covers from scratch begins with determining the number of gores to be made. The more gores, the closer the fit of the jacket to the insulation, but also the more cutting and installation work required.

The number of gores is usually even—four, six, or eight are most common. The allowable space between jacket and insulation may be slightly larger for large-radius bends and should be smaller for smaller elbows. An elbow jacket with a larger number of gores will also mate up better with the adjoining straight-pipe jacket. Adjust the number of gores as necessary. *Figure 21* shows how the number of gores affects the shape of the elbow cover and how it fitsTo find the width of an elbow gore, measure the distance between

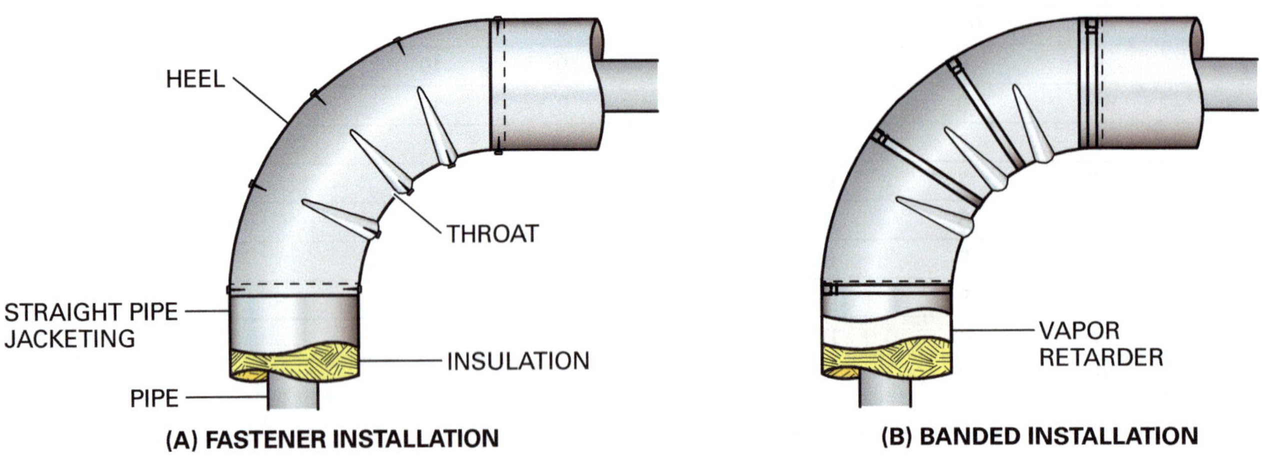

Figure 20 Securing pressed elbow jacket covers.

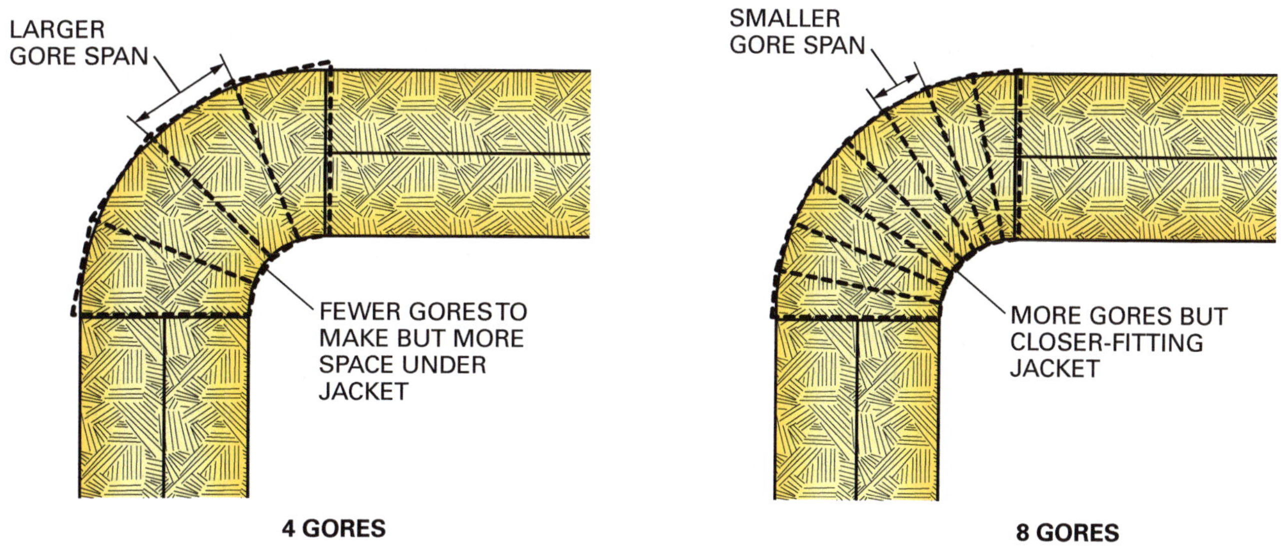

Figure 21 Number of gores affects jacket fit around the insulation.

the ends of the straight pipe insulation sections along the elbow heel. This should correspond to the weld-to-weld distance for standard welded pipe elbows. Add at least 2" to 4" (5–10 cm) to this measurement for butt seam overlap and then divide the total by the number of gores selected. This value will be one of the important dimensions for creating an elbow gore.

The following steps summarize the general approach to laying out and fabricating a gore for a 90° elbow insulation jacket. This approach produces a one-piece gore with a seam along one side of the elbow:

> **NOTE**
> For a cleaner-looking installation, plan to place the lap seam on the opposite side from the one that is visible.

> **NOTE**
> Detailed instructions for laying out patterns and developing two-dimensional true-shape plans of three-dimensional objects are provided in the NCCER Module 19212, "Trade Math and Layout". It is strongly recommended that this information be reviewed along with the instructions in this module.

Step 1 If not already obtained, measure the distance between the ends of adjacent straight-pipe insulation sections or welds along the heel and throat of the elbow. Divide the heel and throat measurements by the number of gores to find the heel and throat widths for each gore.

Step 2 Find the circumference of the insulated pipe. Take several circumference measurements at different locations along the elbow and average their values.

Step 3 Since multiple identical gores will be required, you should create a template to ensure the gores are the same shape. Obtain stiff, heavy cardstock or a scrap of flat sheet metal for the template. Refer to *Figure 22* for the remaining steps.

Step 4 Draw a horizontal center line on the template stock equal to the insulated pipe circumference (labeled $C_{insulation}$) of the elbow measured in Step 1, plus 4" (10 cm). The added 4" are required for the jacket seam overlaps between the upper and lower halves.

Step 5 Starting at the left end of the center line, measure in 2" (5 cm) and make a point labeled "A." From A, divide the center line to the right of A into four equal segments. The length of each segment is $\frac{1}{4}$ the circumference of the insulated pipe measured in Step 3, labeled "$\frac{1}{4}$C." Label these points "B," "C," "D," and "E," as shown in*Figure 22*. Measure 2" past E on the center line and make one last point. This is the endpoint of the pattern.

Step 6 Starting at the left end of the center line, draw vertical lines square to the center line at points A through E, and at the end points. Use a protractor or a contractor's square for accuracy.

Step 7 Divide the heel and throat lengths calculated in Step 1 in half for the following steps.

Step 8 Set your divider to the half-value for the throat measurement ($^1/_2T$) found in Step 7. Place one divider point at point B on the center line and prick points above and below the center line on line B.

Step 9 Set your divider to the half-value for the heel measurement ($^1/_2H$) found in Step 7. From point D, make prick marks on line D above and below thecenter line.

Step 10 Place a straightedge on the upper marks made on lines B and D (indicated by the dashed lines in *Figure 22*). Make a mark on line C where the straightedge crosses. Repeat using the lower marks on lines B and D.

Step 11 Set your divider to the distance of the mark on line C above the center line and transfer this height to lines A and E. Repeat for the lower marks on these lines.

Step 12 Make light punch marks (in the sheet metal) at the upper and lower marks on lines C and D. The marks on line C will aide in positioning the crimper-beader machine. The marks on line D will aid in aligning the heel sections of the gore during application.

Step 13 Set your divider between about $^1/_2$" and $^3/_4$" (13 and 19 mm) that matches the width of sheet metal needed to form an edge bead. Add this distance to each of the marks made on lines A through E above and below the center line.

NOTE

For a more accurate estimate of the metal required for a given bead, you may want to do a trial bead on a scrap piece of jacket sheet metal. Scribe parallel lines along a straight edge at $^1/_{16}$" (1.5 mm) intervals out to an inch or more from the edge. Then run the scrap through the beader to be used to find exactly how much metal is required to form a bead or crimp with that machine. Use this value for Step 13.

Step 14 Draw smooth, S-shaped curves through the sets of points marked in Step 11 above and below the center line as shown in *Figure 22*. Extend the curves to the ends of the template. A long flexible batten of thin wood or drafting curve may be useful for creating these curves. Cut out the template, notching the edges at line C.

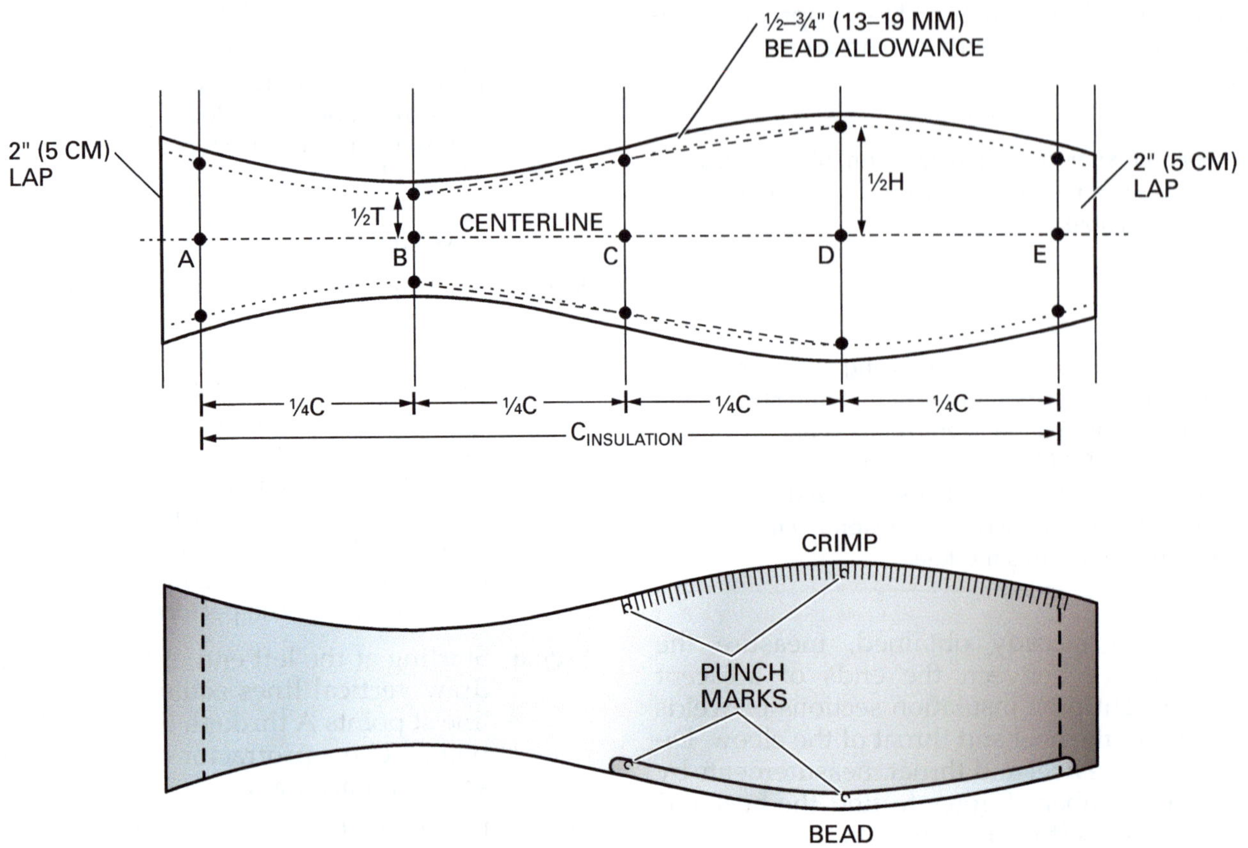

Figure 22 Developing the gore template.

Step 15 After transferring the pattern onto the roll sheet metal using a sharp scriber, carefully cut out the gore along the curved and end lines to free the gore segment.

Step 16 Carefully transfer the template pattern for the remaining gores to the roll sheet metal and cut them out. Make the number of gores determined in Step 1.

Step 17 Apply a raised bead to one edge of the heel sections and a crimp on the opposite edge. The sides to which these apply are determined by which side of the elbow you want the overlap seam to lie and at which end of the elbow the jacket will first be applied.

NOTE

Avoid applying a bead to the edge of an end gore if the adjacent straight pipe jacket must overlap the bead. Apply a crimp to that edge instead.

Before installing mitered jacket on an elbow, check the insulation surface to make sure there are no unusual ridges or bulges, as the gores must be installed over a uniform surface. The position of the lap of the gores must also be considered. For single-piece gores as described here, the single lap can be in the throat or on the left or right side of the elbow. The pattern just discussed must be adjusted if the seam will be in the throat. For large pipes, it may be convenient to have two-piece gores with laps on both sides between the heel and throat sections.

Installing the field-fabricated mitered jacket on an insulated elbow normally starts at the end that overlaps the adjacent straight pipe jacket. If the straight pipe jacket is to overlap the elbow jacket on both ends, then the elbow is jacketed first.

Refer to *Figure 23*. For a one-piece, field-fabricated gore, first position the throat end of the gore so that the overlap extends past the center line of the elbow as viewed from the side. Wrap the heel around the elbow so that it overlaps the throat end. Position the center line punch marks on the heel end over the center line running along the outside of the heel. The segment should overlap the adjacent straight pipe jacket with the allowance assumed in their fabrication.

Place the heel-end of the first gore (beaded edge up and overlapping the adjacent pipe jacket) at one end of the elbow. The longitudinal overlap of the segment should be on the side of the elbow, preferably on the side away from view. Tightly wrap temporary tie wire around the segment. It is best to wrap the wire close to the bead that overlaps the crimp of previous segment, this will pull both segments down tight.

Position remaining segments in the same manner, placing the raised bead of the new gore over the crimp of the previous one. The throat of the segment should smoothly and tightly lap the preceding segment. Continue placing segments until the complete elbow is covered. If the elbow is being covered after the adjoining piping jacket has been applied, ensure that the straight pipe section is properly overlapped. Examine the arrangement of the gores and overlap seams for a neat smooth curve around the elbow. Reposition the segments as needed to obtain the proper fit.

NOTE

For elbows with many gore segments, it may be helpful to add several marks along the side edges during fabrication to aid in centering and alignment as described previously.

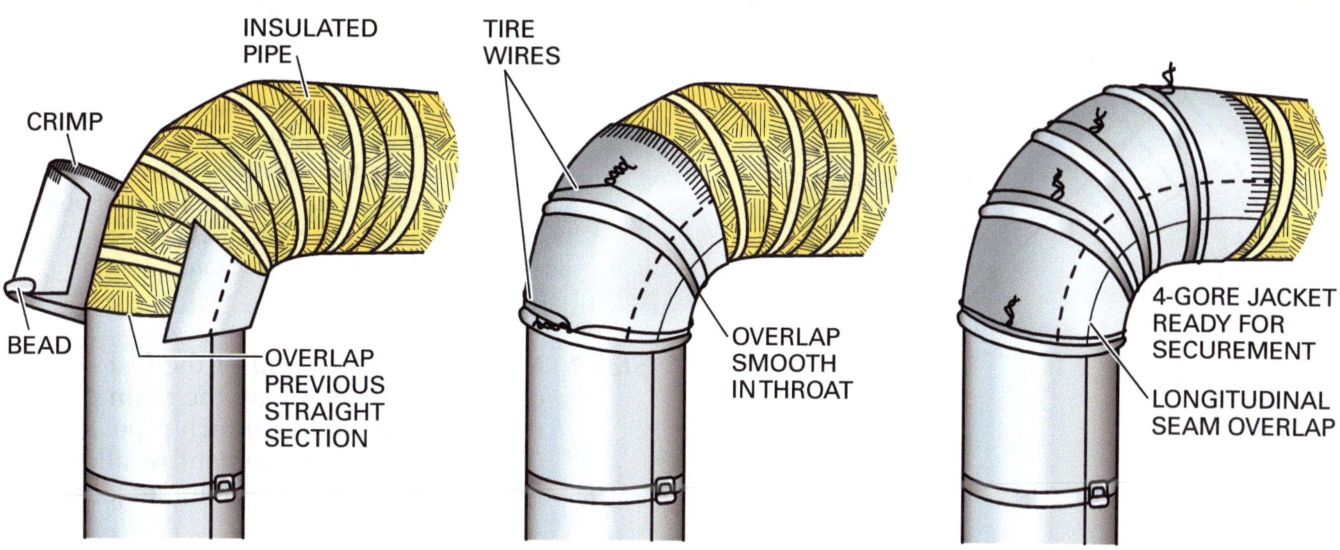

Figure 23 Installing elbow jacket gores.

In hot systems and those without vapor retarder boundaries, start at the center of the heel and install screws one per segment. Screws should be placed on each raised bead along the center line of the heel. One or two screws should secure the longitudinal overlap for each gore. Place several screws as needed at the overlaps to the adjacent straight pipe jackets to prevent gaps and provide a smooth transition between jackets. Remove the temporary wires.

For cold-service systems and those equipped with a vapor retarder, screws and rivets must not be used. Use bands to secure the jacket gores, one band at the middle for narrow gores and two along the gore seams for broader ones. Do not overtighten the bands to avoid buckling the gores and creating gaps. Band wing seals should be neatly arranged on top of the longitudinal lap.

Standard pipe elbows, especially of the welded types, are designated as either short-radius or long-radius elbows. Short-radius elbows have a curvature radius equal to the nominal pipe size of the pipe. A long-radius elbow's curvature is 1.5 times its nominal pipe size. Special-purpose sweep elbows may have radiuses 2 or more times their nominal pipe sizes, as required for the circumstances. The fabrication and installation methods for these sweep elbows are identical to standard 90° and 45° elbows. As shown in *Figure 24*, the number of gores will likely be larger.

2.5.0 Pipe Tee Jackets

Metal prefabricated pipe tee covers are available from some jacket distributors. However, tee jackets are commonly fabricated and installed at the

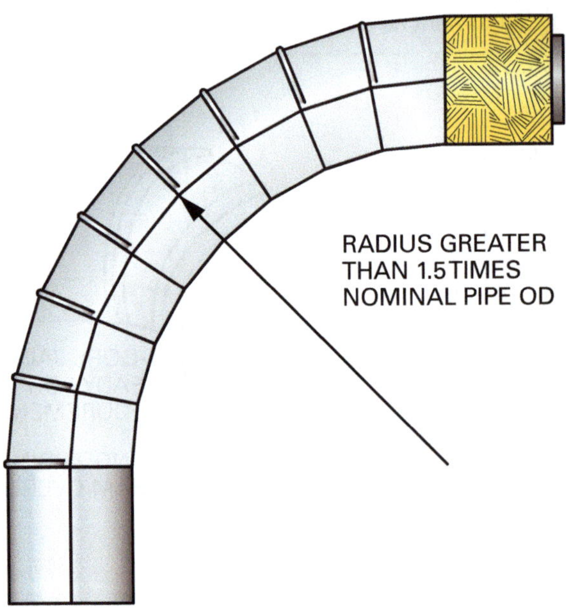

Figure 24 Jacketed sweep elbow.

work site. The following instructions are for laying out, cutting, and installing metal jacketing on pipe tees.

To help understand the discussion in this section, the term *main pipe* will refer to the top of the T, while the term *branch pipe* will refer to the stem of the T. There are two types of tees in the industry—the **equal tee** and the **unequal tee**. In equal tees, both the branch and main pipes have the same diameter. Logically, the branch and main pipe diameters are different in unequal tees. However, there are two kinds of these that require slightly different layout approaches for developing their jacketing patterns. The branch pipe diameter may be smaller than the main pipe, or the branch may be larger as illustrated in *Figure 25*. The second case is more commonly seen when applying jacket to flanged tees or to stemmed valves where the insulated bonnet has a larger diameter than the valve body. You will learn how to work with both types in this section.

2.5.1 Worksite Jacket Fabrication

There are two basic approaches to fabricating jacket for a pipe tee. One is done at the work site, and the other is a more formal layout process completed in a shop. This section presents the first method, using direct measurements off of the tee to be worked on. The approach is very similar to that previously described for applying jacket around a penetration and works well for relatively small unequal tees. The following procedure assumes a welded tee with no flanges, hangers, or other interference to deal with. If the branch pipe points downward, then the branch-pipe jacket should be installed first, followed by the main-pipe jacket to provide proper water shedding. To fabricate jacket for a pipe tee in a shop environment, use the following steps:

> **NOTE**
>
> The following steps are most applicable for an equal tee or an unequal tee with the branch pipe smaller than the main pipe. If the branch pipe is larger, then both pieces of jacket should be made from patterns using the methods described in the following section.

Step 1 Measure from the end of the last section of jacket installed along the main pipe to the center line of the branch pipe. Add 2" (5 cm) to the measurement to allow for the overlap of the last section of jacket.

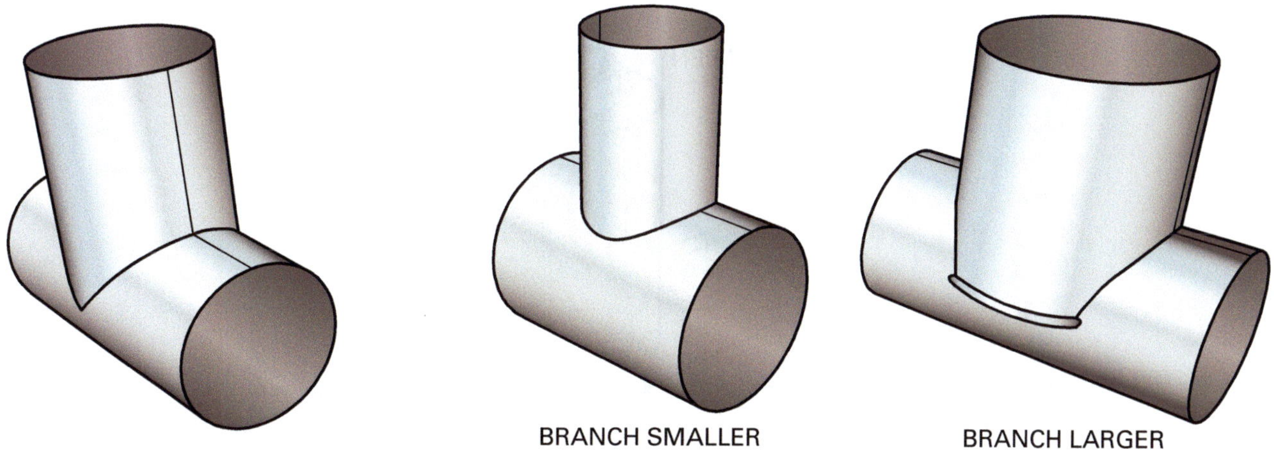

Figure 25 Types of pipe-tee jackets.

Step 2 Place a section of the jacket around the main pipe butted up to the branch pipe. Line up the longitudinal seam with the lap on the adjacent section of jacket. Then re-adjust the jacket for a snug fit before making measurements.

Step 3 Measure the shortest circumferential distance from the longitudinal lap edge to the center line of the branch pipe. If the branch pipe will lie within the lap, then also draw a line along the edge of the lap on the jacket's underlap surface.

Step 4 Remove the jacket from the pipe and place it on a cutting surface with the outside of the jacket facing up and the longitudinal lap edge facing you.

Step 5 Using the measurements from Steps 1 and 3, measure from the butt lap end of the jacket to the center line of the branch pipe cutout. Then, measure from the longitudinal overlap edge to the branch pipe center line. Mark the jacket at the intersection of these two measurements. This is the center of the branch pipe opening in the main pipe jacket. *Figure 26* shows how these measurements are made.

> **NOTE**
> If the branch pipe opening is located in the overlap area of the jacket, then the following steps will apply to both the overlap and underlap edges of the jacket. Use the edge mark made in Step 3 for reference.

Step 6 Determine the diameter of the insulated branch pipe by measuring the outside circumference of the branch pipe and insulation. Assume you measured the circumference to be $23\frac{9}{16}$" (23.56" or 59.8 cm). Then find the radius of the insulated pipe as follows:

- Calculate the diameter (*D*) by using the formula for circumference (*C*) and solving for D, as follows:

$$C = \pi D$$
$$\pi D = C$$
$$\pi D \div \pi = C \div \pi$$
$$D = C \div \pi$$
$$D = C \div 3.14$$
$$D = 23.56" \div 3.14$$
$$D = 7.5"$$

- Calculate the radius (*R*) of the insulated pipe:

$$R = D \div 2$$
$$R = 7.5" \div 2$$
$$R = 3.75" \text{ or } 3\frac{3}{4}"$$

> **NOTE**
> Technically speaking, the radius of the branch pipe hole in the main pipe jacket has the calculated value ($3\frac{3}{4}$" in this example) only in the longitudinal direction of the main pipe. The jacket's hole size is larger than this value in the direction of the main pipe's circumference. For a good estimate of the elongation of the hole, measure the circumferential distance from the main pipe's center line at the side of the branch pipe to the farthest edge of the branch pipe where it intersects the side of the main pipe (as shown in *Figure 27*). Note that as the branch pipe becomes smaller compared to the main pipe, the height of the hole becomes closer to the branch pipe's insulated diameter.

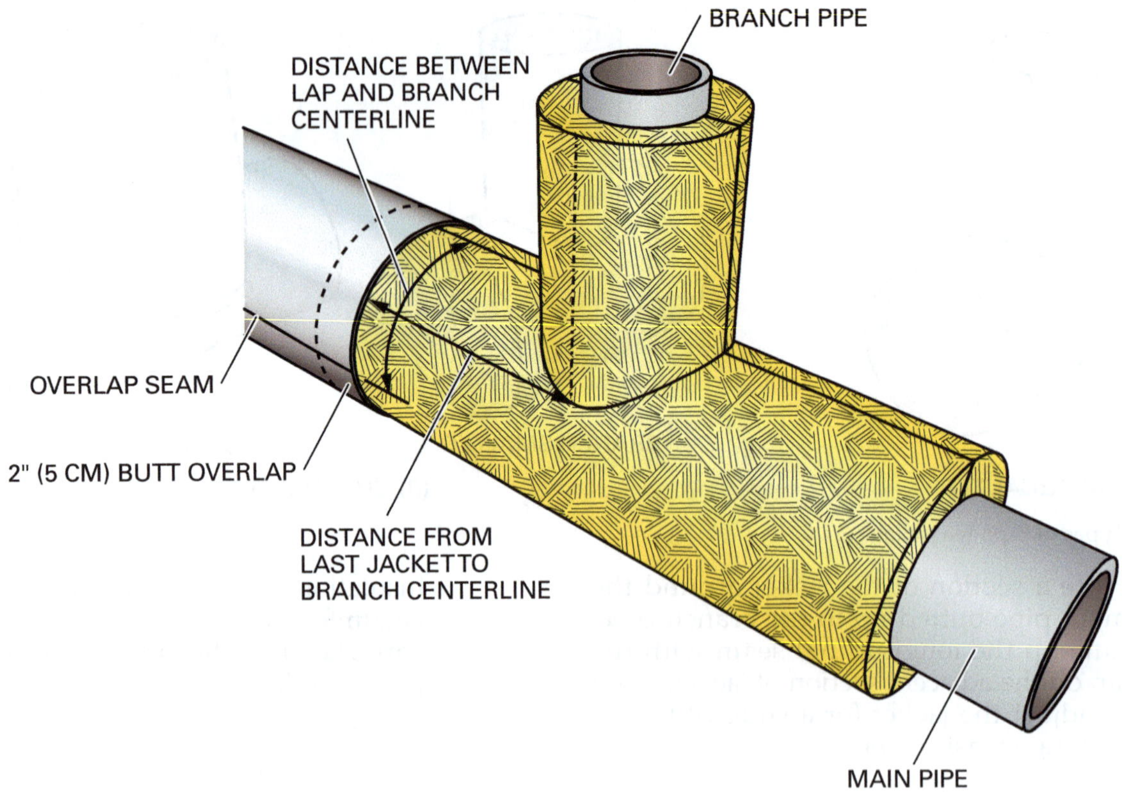

Figure 26 Locating the branch-pipe hole in the main-pipe jacket.

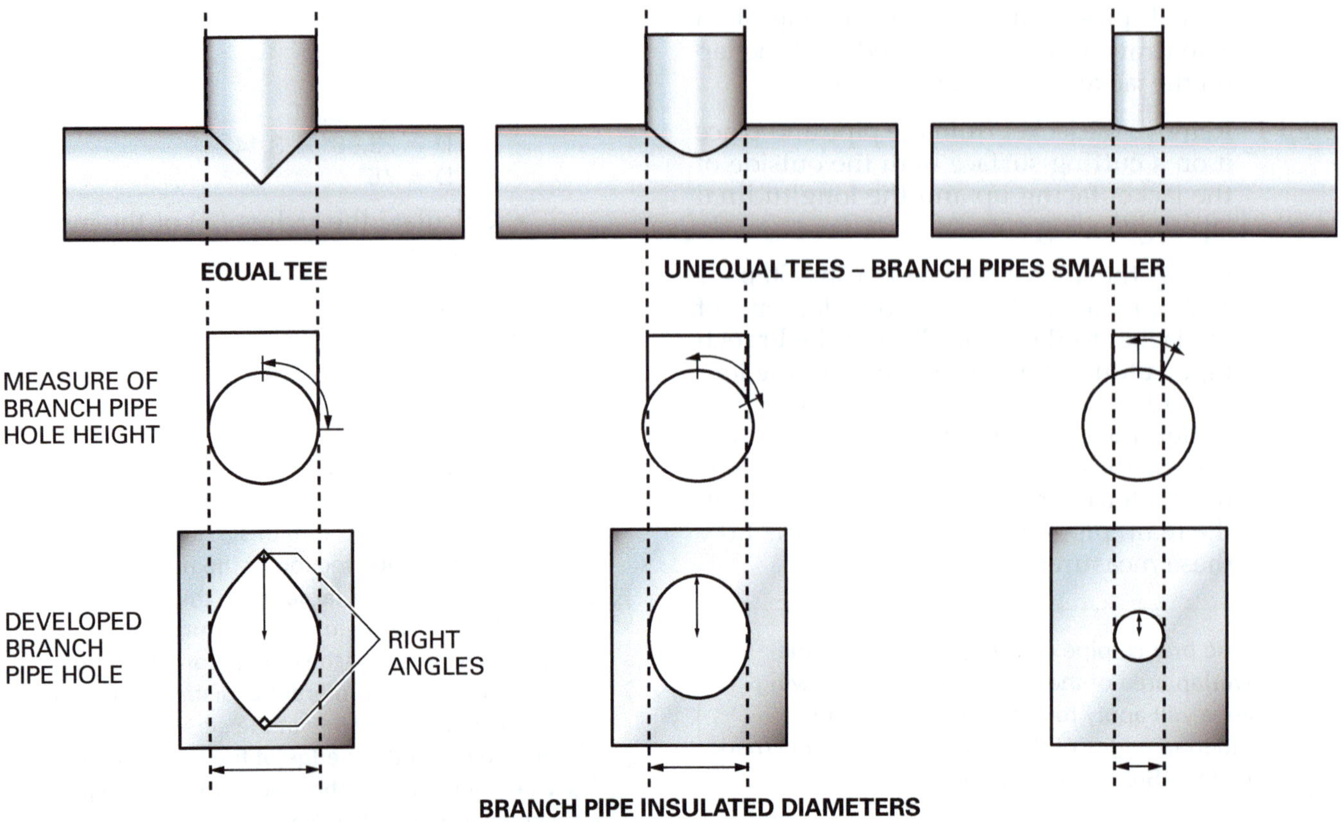

Figure 27 Estimating the circumferential height of the branch-pipe hole.

Step 7 Use the calculated radius found in Step 6 to mark the limits of the hole from its center point on a line parallel to the lap edges. Mark the estimated limits of the hole square to this line from the hole center point, then draw a smooth line to show the limits of the branch-pipe hole. If the hole spans the lap, transfer the pattern to the opposite edge of the jacket.

NOTE

An equal tee's hole will come to right-angle points at the sides of the tee. The holes for unequal tees will look like ovals. Examine *Figure 27* to see how this works.

Step 8 Mark a second line inside and parallel to the line made in Step 7 about $\frac{3}{4}$" from it.

Step 9 Use shears and cut a slit perpendicular to the longitudinal overlap edge of the jacket to the center of the circle.

Step 10 Use **offset compound shears** and carefully cut along the inside perimeter of the hole made in Step 8. When finished, press out any crimps made while cutting.

Step 11 Cut dovetails (*Figure 28*) to the outer line all around the hole. These will provide support for the end of the branch-pipe jacket after assembly. Bend the dovetails outward at the line to form a distinct crimp.

Step 12 Place the section of the jacket on the pipe and carefully spread the tabs at the slit over the branch pipe insulation. Snug up the jacket and check for a tight fit with proper butt seam lap and the longitudinal lap seams aligned. Mark and remove if necessary to adjust the fit.

Step 13 If the slit for the branch pipe opening lies on top of bare insulation, cut a scrap piece of jacket material 4" (10 cm) wider than the insulated branch pipe diameter and with a length equal to that of the slit from the lap edge. Slip the scrap under the slit and centered on the branch pipe to close the slit.

NOTE

On hot systems, screws or rivets may be used to secure the jacket around the branch pipe and along the slit. On cold systems, apply a band on either side of the branch pipe to secure the jacket laps at the slit.

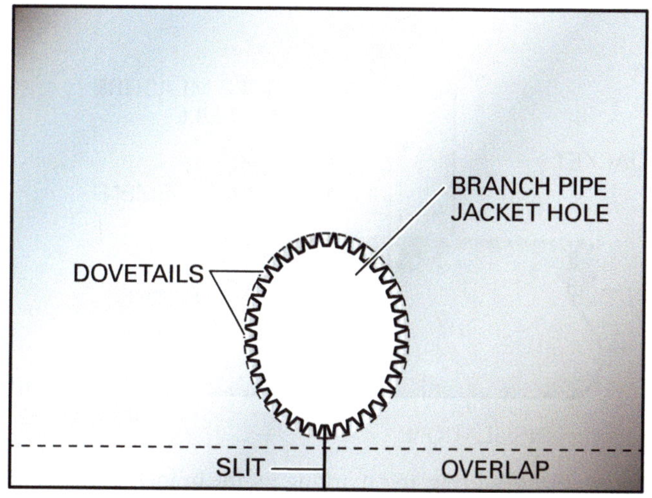

Figure 28 Jacket dovetails at the branch-pipe hole.

Step 14 Secure the jacket according to the applicable specifications. Bend the hole's dovetails back to align with the branch-pipe insulation.

Step 15 Place a section of jacket cut to the circumference of the insulated branch pipe onto the pipe at the tee and position the longitudinal lap as desired. Snugly secure the jacket to the insulation with wire while making the measurements.

Step 16 With the edge of a rule placed flat against the branch pipe jacket, measure the longest distance from the end of the branch-pipe jacket to the point where the rule just touches on the surface of the main-pipe jacket. As seen in *Figure 29*, this measurement will be on one side of the tee.

Step 17 Set your divider to this measurement. Holding the points of the divider in a line parallel to the center line of the branch pipe jacket, drag the divider around the branch pipe, following the surface of the main pipe jacket with one point while marking the branch pipe jacket with the other. After scribing completely around the branch pipe, you should have a smooth, wavy line as shown in *Figure 30*.

NOTE

With the main pipe jacket in place, the rest of the steps will determine where the branch pipe jacket should be located and the shape of jacket seam at the main pipe. Remember that all overlaps should be placed to shed the water and, when possible, should be in the least visible area of the pipe.

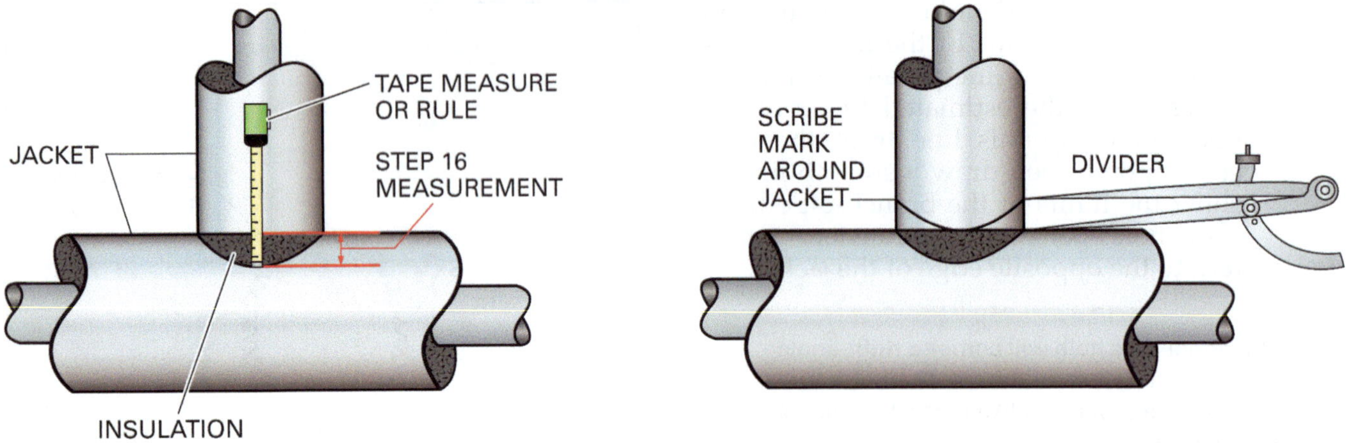

Figure 29 Measuring and marking for the branch-pipe jacket.

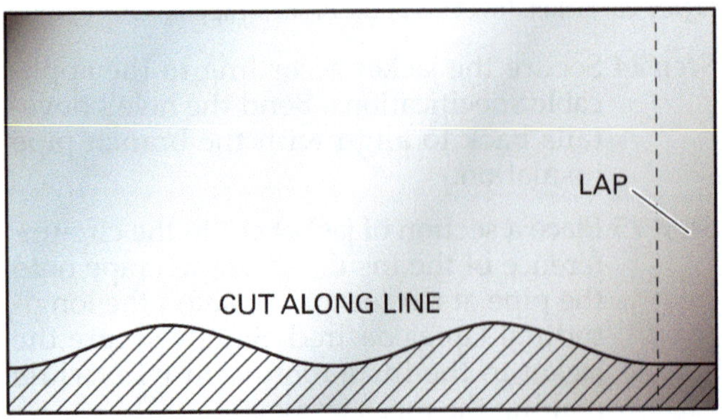

Figure 30 Cutting the branch-pipe jacket.

Step 18 Spot check the line at several points with the divider to make sure it is located correctly. When satisfied, remove the jacket from the branch pipe and lay it face up on the cutting surface.

Step 19 Use offset shears to cut along the scribed line.

Step 20 Place the jacket on the pipe and position over the dovetails in the main-pipe jacket. Check for proper fit. Mark and trim if necessary.

Step 21 When the fit is proper, apply a bead of joint sealant around the hole perimeter of the main-pipe jacket at the dovetails (if required), then permanently secure the jacket. If the jacket is part of a vapor retarder boundary, only banding is permitted. Smooth the joint sealant for a neat professional appearance.

2.5.2 Shop Layout and Fabrication of Metal Jacket for Tees

On occasions when several identical tees require jacket, or when the tee jacket is too large to be easily managed by a single worker, the jacket patterns should be created in the shop using some basic drafting methods. For most projects, it is best to draft the patterns at full size on the jacket sheet metal or on thin, stiff, template material. The following descriptions assume a full-sized pattern.

The drafting methods presented in this section only generate the shapes of the seams between the two pieces of jacket. The insulating mechanic must correctly position the patterns on the jacket material and make the cuts, taking into account the locations of the lap edges. The different kinds of tees will be treated separately to clarify the methods needed for each.

Subdividing The Branch-Pipe Circle – Before introducing the methods for developing the tee patterns in the shop, you should learn a common trade method for dividing a quarter-circle arc into six equal segments. This process is required for all three methods to be introduced here. Refer to *Figure 31* as you read through the following steps:

Step 1 Draw a horizontal baseline, then draw a vertical line square to it at its left end. The corner point is labeled O.

Step 2 Set the radius of the branch pipe (*R*) into your compass. Place the compass point at O. Draw an arc that crosses both lines. The intersection of the arc with the baseline is A; the intersection with the vertical line is B.

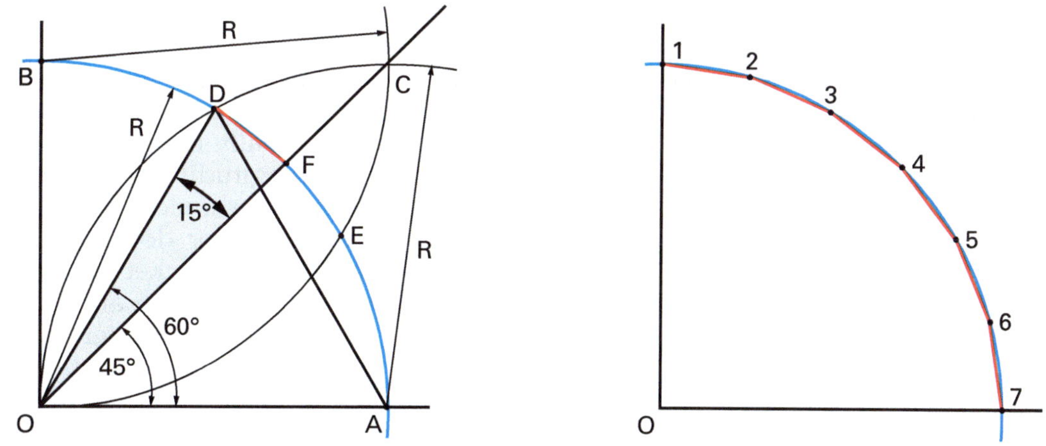

Figure 31 Subdividing a quarter-circle into six 15° segments.

Step 3 Move the point of the compass to A. Draw an arc through the quarter-circle made in Step 2. Repeat from B on the vertical line. The two new arcs should cross outside the first arc at C. They also intersect the quarter-circle at D and E.

Step 4 Using a straightedge, draw a line from O through C. This line forms a **bisector** of the 90° right angle formed by the baseline and the vertical line. The two smaller angles it makes are 45° each. The point where the bisector crosses the quarter-circle is labeled F.

Step 5 Drawing straight lines from O to D, and from D to A, creates an equilateral triangle. All interior angles in such a triangle are 60°. Therefore, the line between O and D makes a 60° angle with the baseline.

Step 6 If you subtract the 45° angle made in Step 4 at O from the 60° angle made in Step 5 at O, that leaves a 15° angle (shaded in *Figure 31*). 15° are one-sixth of 90° (90° ÷ 6 = 15°).

Step 7 Take your divider and set them to the distance between points D and F. If you have worked accurately, you should be able to step off six 15° intervals along the first arc between A and B. These segments are shown on a separate quarter-circle in *Figure 31*. These divider prick marks will be needed for the layout work to follow.

Equal Tees – For an equal tee, develop the seam patterns for the main and branch pipes, as follows (refer to *Figure 32*):

Step 1 Determine the radius of the insulated pipes. The most sure way is to measure the circumferences of the pipes with a tape measure to verify they are the same, divide this value by π, and divide by 2. (Section 2.5.1 shows the arithmetic for this step.)

Step 2 Draw a horizontal baseline and label it "A." At the left end of baseline A, draw a vertical line. Make sure it is truly square to baseline A by using a framing square or a draftsman's protractor.

Step 3 Draw a quarter-circle using the pipes' insulated radius and subdivide it into six segments as just described. Starting at the vertical line, number the points 1 through 7 ending at the baseline.

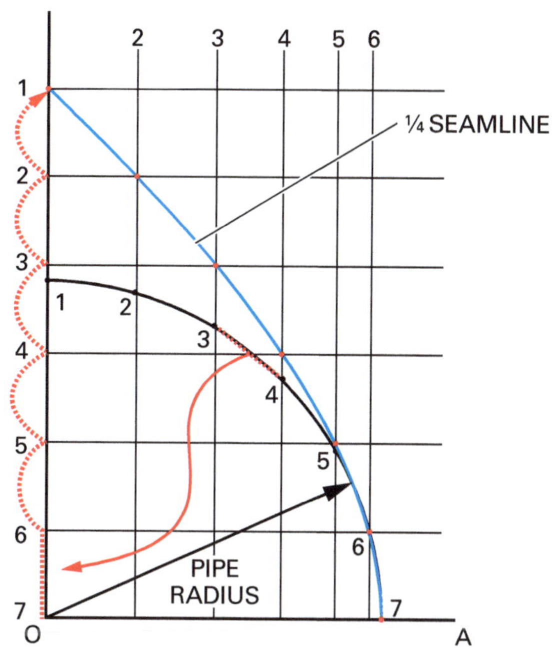

Figure 32 Developing seam patterns (equal tee).

Step 4 Draw five long vertical lines square to baseline A at points 2 through 6 on the circle made in Step 3. These need to be truly perpendicular. Use a protractor or a drafting square. Number them along the baseline to match the points they pass through on the circle.

Step 5 Set a divider to the distance between any two adjacent points on the circle. Transfer this distance to the vertical line with one point of the divider at O.

Step 6 Pick off six equal spaces on thevertical line with the divider as shown in *Figure 32*. Starting at the top mark, number them 1 through 7 to baseline A.

Step 7 Draw horizontal lines through marks 2 through 6 on the vertical line. These lines need to be truly horizontal and parallel to baseline A. Use a drafting arm or T-square.

Step 8 Insert points where each numbered vertical line drawn in Step 4 intersects the same-numbered horizontal line drawn in Step 7. For example, where the vertical and horizontal lines numbered 2 intersect, draw a point.

Step 9 Connect point 1 on the circle, the five points made in Step 8, and point 7 on the baseline with a smooth curve. It should look like the blue line in *Figure 32*.

The curve just created is one-quarter of the jacket seam for both the main and branch pipe jackets. *Figure 33* shows the general development of the jacket cutouts using this curve. Note that how you will actually make the cutouts depends on where the longitudinal jacket seams will be located. Allowance should also be made on one or the other jackets for dovetails. This will depend on the order of jacket installation to provide proper water shedding and whether the tee jacket is installed before or after the straight-pipe jacket. The approach taken for jacketing the field-fabricated tee provided in Section 2.5.1 should be followed.

Unequal Tee (Branch Pipe Smaller Than Main Pipe) – When drafting up the jacket seams for an unequal tee where the branch pipe is smaller than main pipe, the same basic approach as for an equal tee is used. The main differences are:

- Two partial circles are required to develop the seam pattern. The lower circle is made with the radius of the smaller insulated branch pipe. The upper circle in the construction uses the radius of the larger insulated main pipe.
- The curves for both pipe jackets must be constructed using auxiliary baselines.

Refer to *Figure 34* for as you read through the following process for constructing the seam line for the branch-pipe jacket:

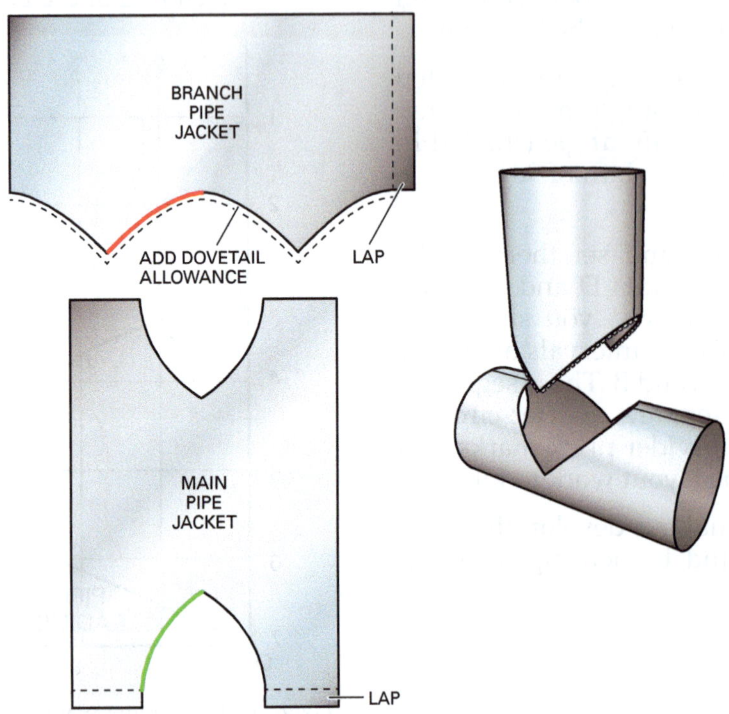

Figure 33 Equal-tee main- and branch-pipe jackets.

Step 1 Determine the radius of the insulated branch pipe and main pipe. (Section 2.5.1 shows the arithmetic for this step.)

Step 2 Draw a horizontal baseline and label it "A." At the left end of baseline A, draw a vertical line that is truly square to baseline A. Label the corner O.

Step 3 Set a compass or divider to the insulated branch-pipe radius determined in Step 1. Place the point of the compass at O and draw an arc through the vertical and base lines. Subdivide the branch-pipe circle into six equal segments as described previously. Number the points 1 through 7, starting at the vertical line and ending at baseline A.

Step 4 Draw six long vertical lines square to baseline A through points 2 through 7 made in Step 3. Number the lines to match the points they pass through on the circle.

Step 5 Draw a horizontal line perpendicular to the vertical line above the branch-pipe circle. Label this auxiliary baseline "B."

Step 6 Set the radius of the larger main pipe into your compass. Place the compass point on the vertical line above baseline B and draw another arc, as shown in *Figure 34*.

Step 7 Add points where all the vertical lines cross the main-pipe circle drawn in Step 6. Number them 1 through 7 from left to right.

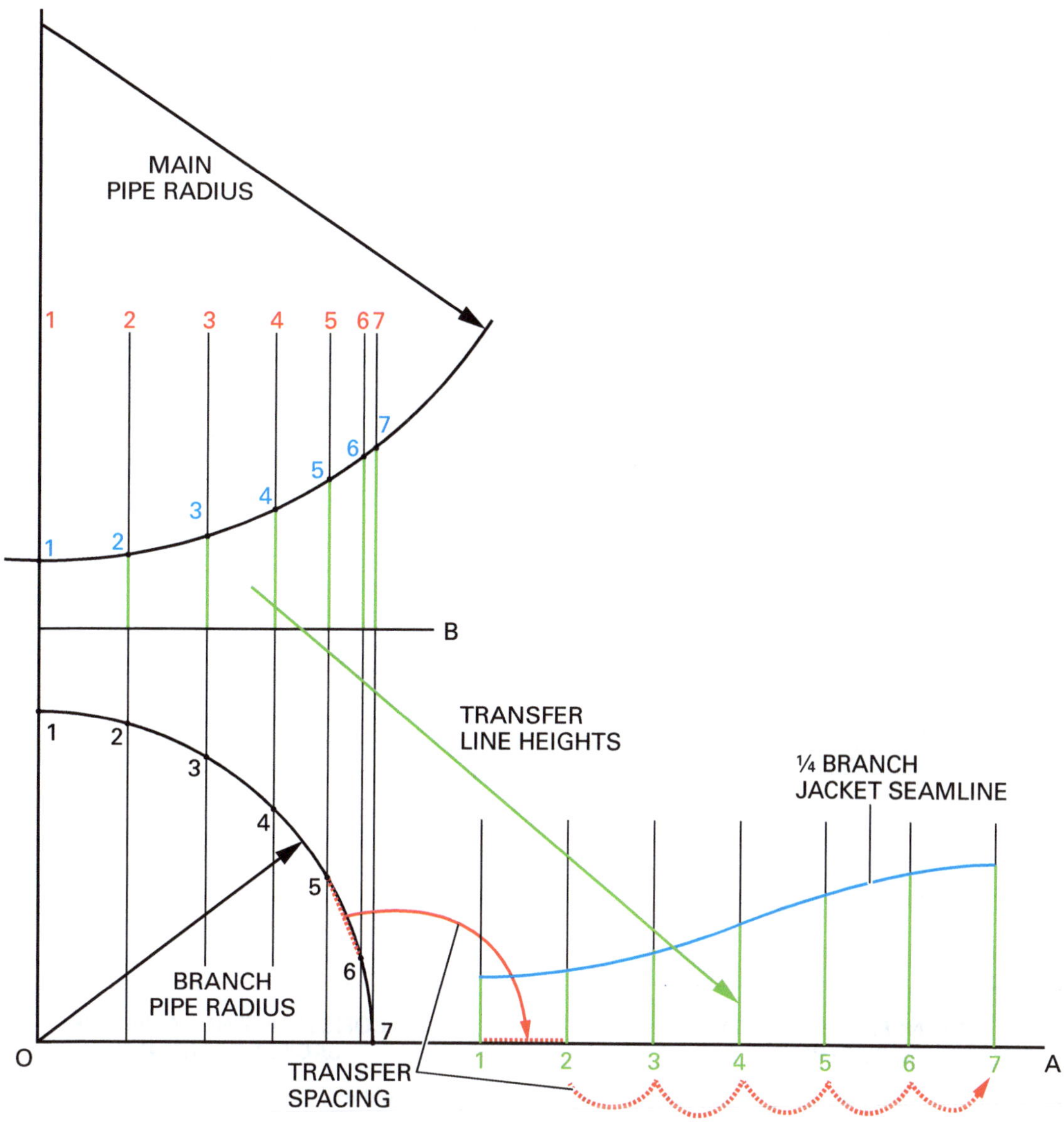

Figure 34 Developing the branch-pipe jacket seam (unequal tee—branch pipe smaller).

Step 8 Extend baseline A to the right, if required. Mark a dot on the extended baseline A and number it "1." Set your divider to the distance between any two adjacent points on the lower circle. Place a point of the divider on dot 1, then step off six spaces to the right on baseline A. Number these new marks 2 through 7.

Step 9 Draw seven vertical lines through the points made in Step 8. These lines must be truly square to the baseline.

Step 10 Using your divider, pick the distance from baseline B to point 1 on the upper circle. Transfer this distance to line 1 on baseline A drawn in Step 9. Then transfer the distance between baseline B and point 2 on the upper circle to line 2. Repeat for the heights of points 3 through 7 on the upper circle above baseline B.

Step 11 Draw a smooth, S-shaped curve through the points made in Step 10 as shown in the right-hand part of *Figure 34*.

The curve just constructed is ¼ of the pattern for the edge seam to be cut into the branch-pipe jacket. How this pattern is used will be explained shortly.

Next, the seam pattern for the hole in the main-pipe jacket must be developed. The initial setup for the construction from *Figure 34* will be needed for these steps. Refer to *Figure 35* for the following steps:

Step 1 On the upper circle, set your divider to the distance between points 1 and 2.

Step 2 Place the divider point at O on baseline A, and step off a mark. Label it "1-2."

Step 3 Repeat Step 1 for points 2 and 3 on the upper circle. Set the divider point at point 1-2 and prick a mark on the vertical line. Label this mark "2-3."

Step 4 Repeat for the remaining pairs of points on the upper circle as shown in *Figure 35*. Label the marks in order "3-4," "4-5," "5-6," and "6-7." The spans should grow smaller.

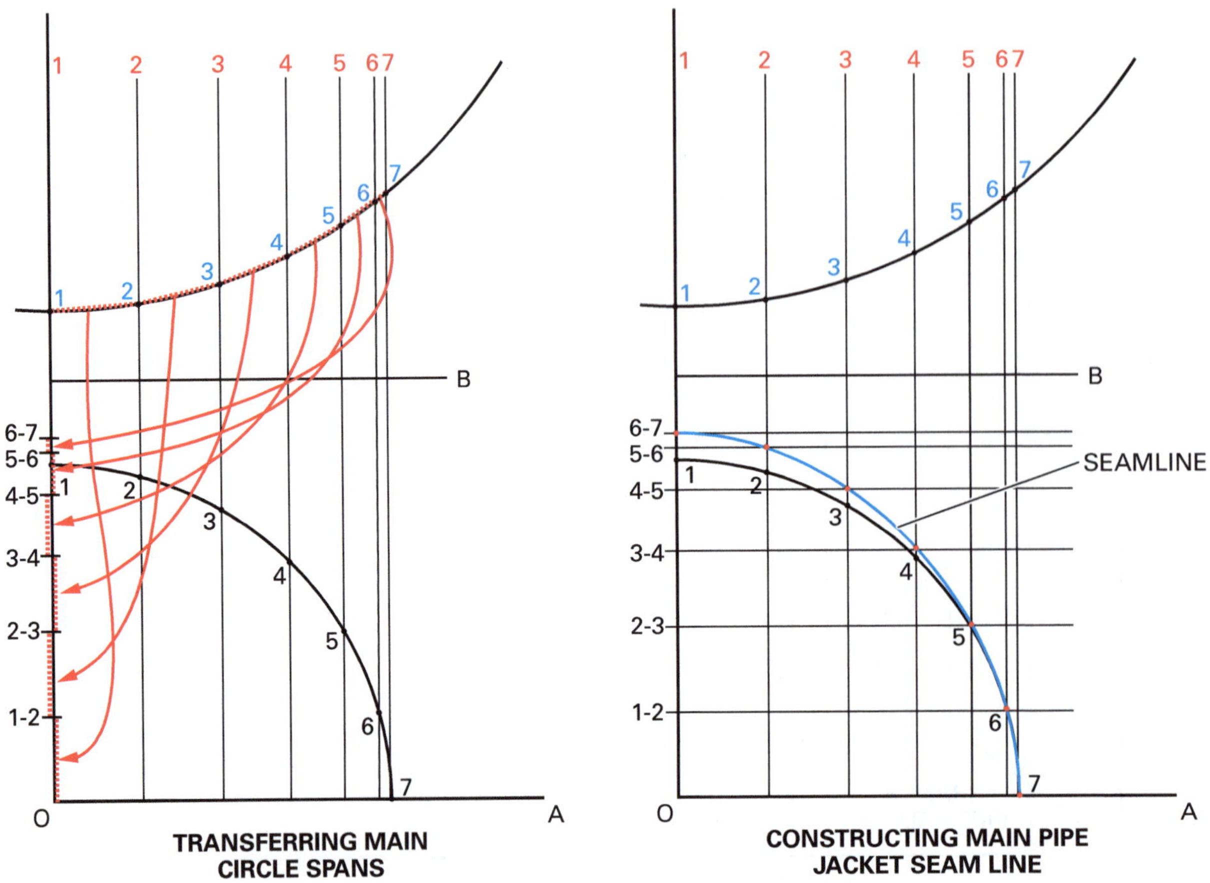

Figure 35 Developing the main-pipe jacket seam (unequal tee—branch pipe smaller).

Step 5 Draw horizontal lines through the labeled points made in Steps 3 and 4. These lines need to be truly square with the vertical lines.

Step 6 Insert dots at the intersections of the following:

- Horizontal line 1-2 and vertical line 6
- Horizontal line 2-3 and vertical line 5
- Horizontal line 3-4 and vertical line 4
- Horizontal line 4-5 and vertical line 3
- Horizontal line 5-6 and vertical line 2

Step 7 Draw a smooth curve starting with point 6-7 on the vertical axis, through the five points made in Step 6, and ending at point 7 on baseline A as shown by the blue line in *Figure 35.*

As with the branch-pipe construction, this curve is only $\frac{1}{4}$ of the seam line surrounding the hole to be cut in the main pipe jacket. Make allowances in the patterns for the seam-line dovetails before cutting. The fabricated jacket seams should appear as in *Figure 36*.

Unequal Tee (Branch Pipe Larger Than Main Pipe) – There are occasions where the insulated branch of a tee has a larger diameter than the insulated main pipe. This may result from having to insulate an adjacent flange in the branch pipe. A similar situation exists when applying jacket to an insulated valve bonnet that is larger in diameter than the valve body itself.

Setting up the layout of this kind of tee is similar to the previous unequal tee example. Part A of the construction diagram in *Figure 37* is set up with the lower branch pipe circle having a radius equal to that of the larger insulated branch pipe. The quarter-circle is subdivided into six equal arcs, and vertical lines are drawn through the points into the upper area of the diagram. Baseline B and the upper quarter-circle (with the radius of the smaller insulated main pipe) are also included. The remaining construction is outlined in the following steps:

Step 1 Draw a horizontal line to the right from the center point of the upper, main-pipe circle on the vertical line. It should extend beyond the vertical line 7 from the lower circle. Label this line "C."

Step 2 Where the vertical lines from baseline A cross the upper circle and line C, label the intersection points 1 through 7 from left to right.

Step 3 The point where line C and the main pipe circle intersect should be labeled "D" as in*Figure 37.*

Step 4 Extend baseline A to the right as in the previous unequal-tee example. This section of the figure is labeled Part B. Then mark off on the extended baseline six equal segments using the span between two points on the lower, branch-pipe circle as before.

Step 5 In Part B, draw vertical lines square to the baseline through the points made in Step 4 and number them 1 through 7 as in the previous example.

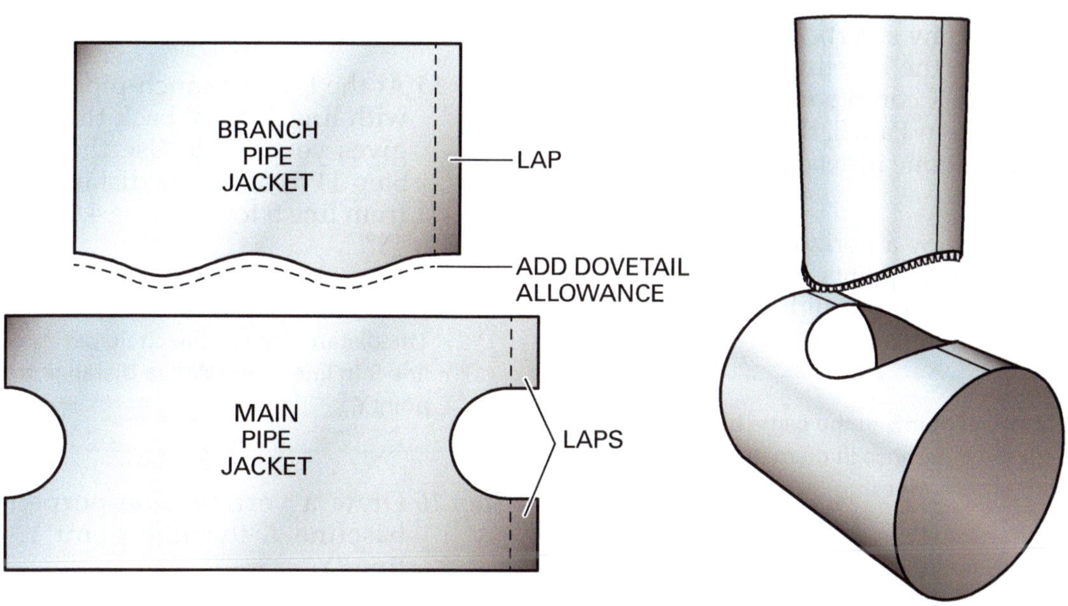

Figure 36 Sample jackets (unequal tee—branch pipe smaller).

Step 6 In Part A, transfer the heights of the vertical lines between baseline B and the upper main-pipe circle to the same-numbered lines above the baseline in Part B. Also transfer the heights of the vertical lines between baseline B and line C to their same-numbered lines in Part B. For now, mark all of these heights as points on the vertical lines in Part B.

> **NOTE**
> The remaining steps will locate the break in the seam of the larger pipe's jacket and form the saddle, which will be described later.

Step 7 At the upper, main-pipe circle, draw a vertical line downward from the point D square to baseline A. Label this line "X." (If one of previously drawn vertical lines from baseline A passes through point D, omit this step and use that line in its place for the remainder of the procedure.)

Step 8 At the lower, branch-pipe circle, note the vertical line that is nearest to where line X crosses the circle. Set your divider to the straight-line distance between where line X and this line cross the circle. In the example, since line 4 is nearest to line X, this span is indicated as "X-4".

Step 9 In Part B, mark off the distance measured in Step 8 in the same direction from the line 4. Since line X is to the left of line 4 at the branch pipe circle, make the mark in Part B to the left of line 4. Label this point "X."

Step 10 In Part B, draw a vertical line from point X square to the baseline. Label this line "X." (If line X corresponds to an existing vertical line in Part B, then omit this step and use that line instead for the following steps.)

Step 11 In Part B, draw a horizontal line through the same-height points made in Step 6 and extend this line through line X. Make a point at this crossing.

> **NOTE**
> The number of points at the same height will vary with the difference in diameters of the main and branch pipes.

Step 12 In Part B, draw a smooth curve starting at the point on line 1 through the other points on the lines to the right, ending at the point on line X (*Figure 37*).

> **NOTE**
> This curve and the horizontal line together make one-quarter of the seam and saddle for the larger pipe's jacket. The following steps will create the jacket seam for the smaller main pipe and its saddle cutout. Refer to *Figure 38* for the remaining steps.

Step 13 On the upper, main-pipe circle, starting with point 1, count the number of points to point D—but not including point D. In this example, it is three points.

> **NOTE**
> The number "3" determined in Step 13 will be used several times in the following steps for laying out the dimensions of the seam. This number will vary depending on the relative sizes of the two insulated pipes forming the tee. For emphasis, this number will be put in **bold** type in the following steps to show how it is used.

Step 14 On the lower, branch-pipe circle, set your divider to the straight-line distance between where line and line X cross the circle.

> **NOTE**
> If line X was not needed in Step 7, then replace "line X" in the following instructions with the line number that passes through point D.

Step 15 At the lower, branch-pipe circle, starting with line 7, count back **three** lines, which gives you line 5. Use the divider set in Step 14 to mark the distance on the circle from line 5 toward line 4. Label this point "Y."

> **NOTE**
> The distance along the circle perimeter from line 3 to line X equals the distance from line 5 to point Y.

Step 16 Draw a vertical line perpendicular to baseline A through point Y. Label this line "Y."

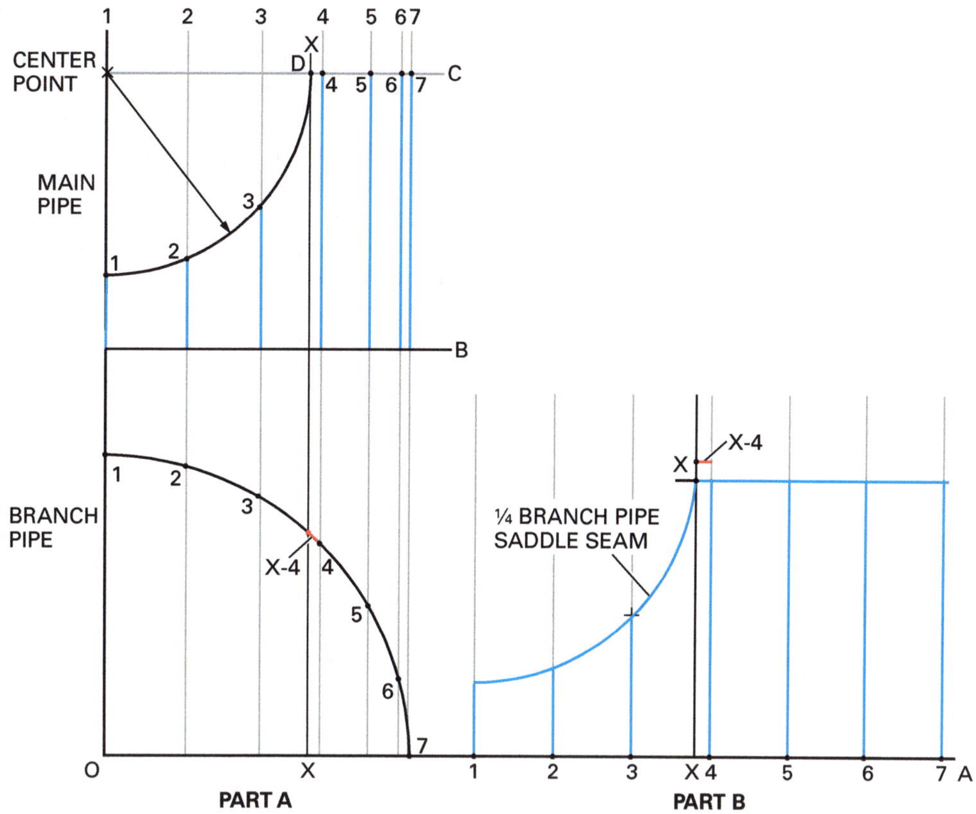

Figure 37 Layout of branch-pipe saddle seam lines (unequal tee—branch pipe larger).

Step 17 On the upper main-pipe circle, set your divider to the straight-line distance between points 1 and 2. Transfer this dimension to the vertical line at O. Mark this point and label it "1-2." Repeat for the distance between point 2 and point 3 at the upper circle, and so on. In this example, the last span will be labeled "3-D." The spans should grow larger with each measure.

Step 18 At the lower circle, draw horizontal lines through the points marked in Step 17. These must be truly square with the vertical lines in Part A. Place points at the following intersections:

- Baseline A and the circle
- Line 1-2 and line 6
- Line 2-3 and line 5
- Line 3-D and line Y

Step 19 In Part A, starting at baseline A, connect these points with a smooth line. It should form a narrow S-curve as in *Figure 38*.

Step 20 The horizontal line at point 3-D and the S-curve made in Step 19 make up one-quarter of the jacket seam and saddle cutout for the smaller main-pipe jacket.

When the jacket pieces are rolled into shape, the edges where they come together form distinct high and low spots that resemble the shape of horse saddles. To assemble the unequal tee jacket, the patterns must be combined so that the straight sections of the branch pipe jacket correspond to the straight sections in the saddle cutouts of the main pipe jacket. Dovetail allowances should be made in the patterns along the curved section portions of the jacket seams. As in the previous examples, the actual positions of these cutouts on the jacket sections will depend on where you need the longitudinal lap seams to be placed. *Figure 39* shows one way to create the jackets.

Note that there is a gap created between the sides of the branch pipe's larger jacket and the smaller main pipe jacket. These are normally filled in with small pieces of sheet metal cut to the size and shape of the gap. To help keep these pieces in place, dovetail tabs are added to the straight edges that slip inside the main pipe jacket, and a narrow bead is formed in the straight edge of the branch-pipe jacket as shown in *Figure 39*. The curved edges of the filler pieces snap into the beads when installed.

The approach to installing jacket on tees depends on the orientation of the branch pipe, especially for systems subject to weather or wetting.

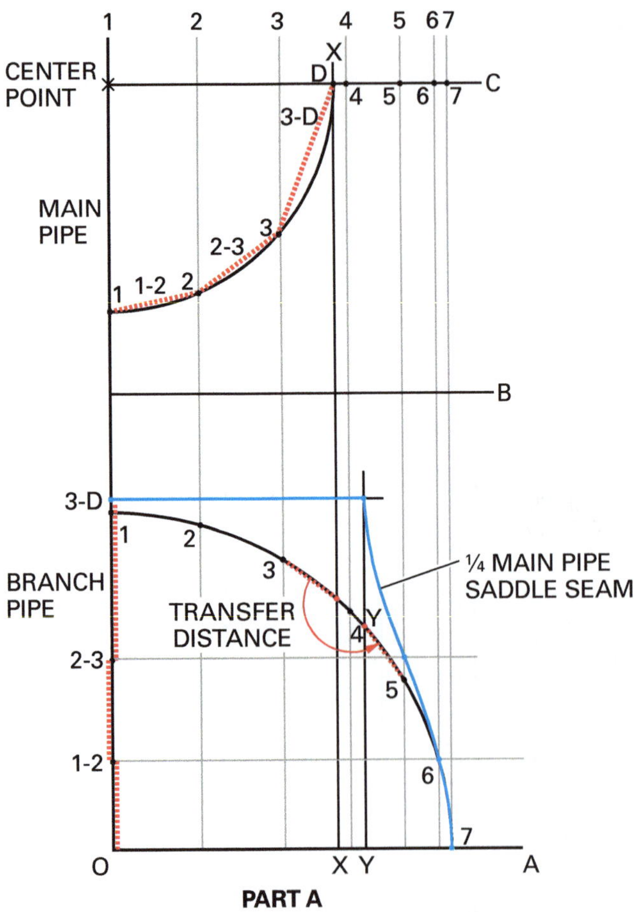

Figure 38 Developing the saddle cutout seam for the main pipe jacket.w

If the branch of the tee is located on the underside of the main pipe, the insulated branch pipe should be jacketed before the main pipe section. This jacket will also have the dovetails that insert under the main pipe jacket. This allows water to drain away from the seam. If the branch pipe is located on the upper side of the main pipe, the main pipe should be jacketed before the branch, and its hole will have the dovetails. For horizontal tees, jacket installation should progress based on the sequence of the adjoining main pipe jacketing. The first section of the tee jacket installed should be provided with dovetails that fit inside the seam of the last tee section installed.

2.6.0 Installing Jacket on Other Piping Components

Pipe reducers, end blanks of headers or mains, and other shapes may require weatherproof sheathing and jacket for insulation. Many standard configurations of covers can be ordered from jacket manufacturers and distributors. They may also be field- or shop-fabricated. With proper use of the techniques learned in this module (and other NCCER *Mechanical Insulating* modules), cover patterns for these and other components can be developed. When fabricating more than one of any type of cover, create a template to fabricate all the covers.

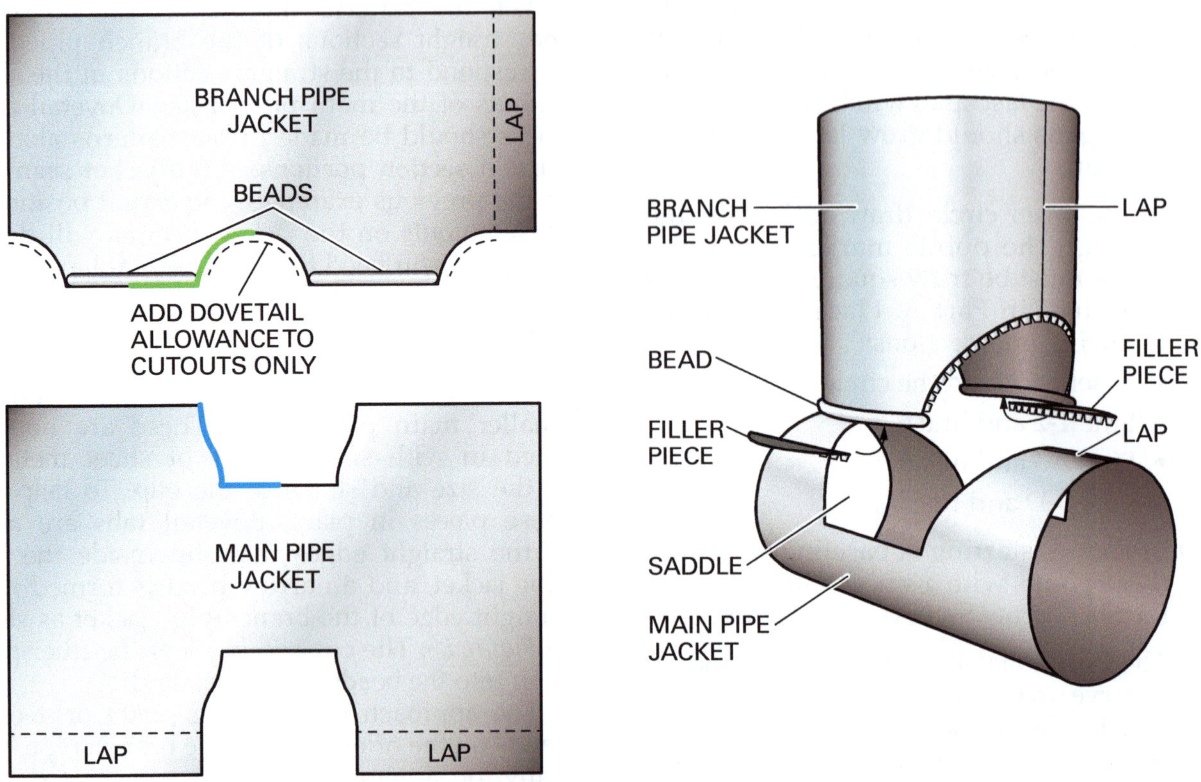

Figure 39 Fabrication of an unequal tee's jackets (branch larger than main pipe).

2.6.1 Jacket for Concentric Reducers

Figure 40 shows concentric, beveled, and eccentric reducers. In this section, you will learn how to lay out the jacket pattern for concentric beveled reducers.

There are three important measurements that must be taken before laying out the pattern for a concentric reducer cover.

- Diameter (or circumference) of the larger insulated pipe
- Diameter (or circumference) of the smaller insulated pipe
- Lateral length of the reducer

Figure 41 (*A*) illustrates that a reducer's surface is shaped like the bottom section of a cone. As shown in *Figure 41* (*B*), the lateral length of the entire cone (L_1) and the lateral length of the cone from the reducer to its apex (L_2) become the radiuses of the upper and lower edges of the reducer jacket when it is laid out flat on the pattern. Notice that the lateral length of the reducer (L) is the difference between L_1 and L_2.

The following example describes how to lay out the pattern for a concentric reducer insulated with 2" cellular foam. (For simplicity, metric equivalents will not be provided.) In this example, the larger pipe's insulated circumference is $46\frac{3}{8}$" (or 46.4"), and the smaller pipe's circumference is $33\frac{3}{8}$" (or 33.4"). The measured lateral length of the reducer is 8". Refer back to *Figure 41* often as you work through the following example:

> **NOTE**
> The most reliable method for creating the pattern for a concentric reducer's jacket involves arithmetic and some basic geometry. You may find it helpful to review the applicable sections of NCCER's Module 19212, "Trade Math and Layout".

Step 1 Find the radiuses of the larger and smaller insulated pipes. Divide their measured circumferences by π (3.14) to find their diameters, and divide the results by 2 to find the radiuses. The radius of the larger pipe is given the symbol *R*, and the radius of the smaller pipe is given the symbol *r*. (Note that it is generally acceptable to round results up to the next tenth for these types of calculations.)

- Calculate the radius of the larger pipe (*R*):

$$D_{larger} = C_{larger} \div \pi$$
$$D_{larger} = 46.4" \div 3.14$$
$$D_{larger} = 14.8"$$

$$R = D_{larger} \div 2$$
$$R = 14.8" \div 2$$
$$R = 7.4"$$

- Calculate the radius of the smaller pipe (*r*):

$$D_{smaller} = C_{smaller} \div \pi$$
$$D_{smaller} = 33.4" \div 3.14$$
$$D_{smaller} = 10.6"$$

$$r = D_{smaller} \div 2$$
$$r = 10.6" \div 2$$
$$r = 5.3"$$

Step 2 Calculate the length of L_1 using the following formula:

$$L_1 = \frac{R \times L}{R - r}$$

$$L_1 = \frac{7.4" \times 8"}{7.4" - 5.3"}$$

$$L_1 = \frac{59.2"}{2.1}$$

$$L_1 = 28.2"$$

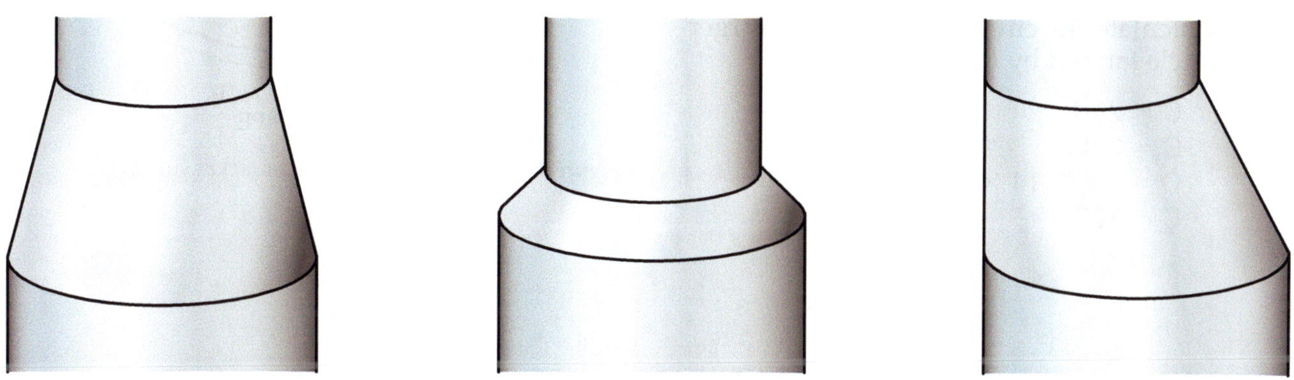

Figure 40 Types of pipe reducers.

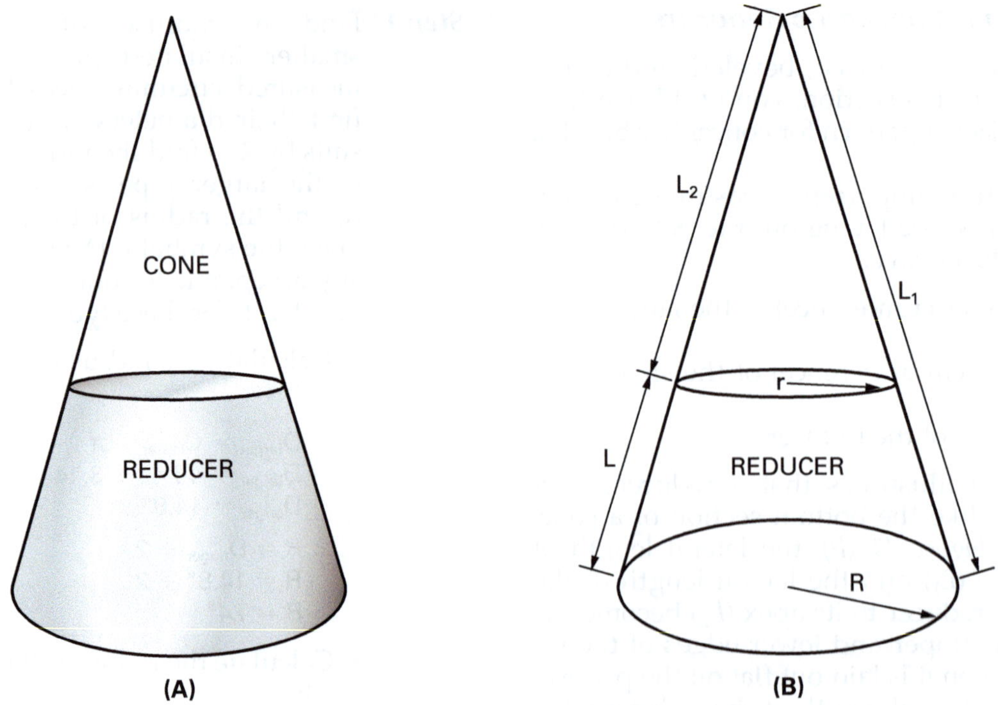

Figure 41 Pipe reducer as part of a cone.

Step 3 Calculate the length of L_2. Since L (which you measured) is the difference between L_1 and L_2, you can subtract L from L_1 to find L_2:

$$L_2 = L_1 - L$$
$$L_2 = 28.2'' - 8''$$
$$L_2 = 20.2''$$

Step 4 Refer to *Figure 42*. Draw two circles with the same centers (i.e., concentric circles). The larger circle has the radius L_1, and the smaller circle has the radius L_2. The inner circle determines the curvature of the smaller end of the reducer jacket. The outer circle determines the curvature of the larger end. You will need to find how large a sector is required to form the lateral surface of the reducer jacket.

Step 5 Calculate the circumference of the larger circle formed by L_1.

$$C = 2\pi R$$
$$C = 2\pi L_1$$
$$C = 2 \times 3.14 \times 28.2''$$
$$C = 177.1''$$

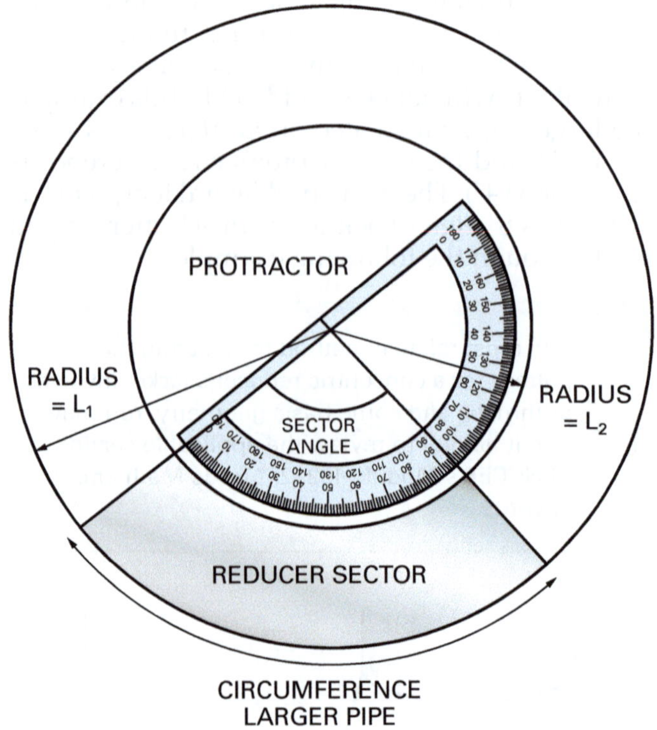

Figure 42 Developing the reducer jacket surface.

Step 6 Now you have the circumference for a circle with a radius of L_1 (177.1"). However, in the example, the insulated pipe's larger circumference is only 46⅜" (46.4" or 117.8 cm). You must next find what fraction the pipe's actual circumference is of the larger circle. Divide the larger pipe's circumference by the L_1 circumference to determine the fraction of the L_1 circle that the pipe's actual circumference takes up:

46.4 ÷ 177.1" = 0.26

Step 7 Calculate the sector angle of the reducer plan using the fraction determined in Step 5 (0.26, or $^{13}/_{50}$). Since every circle has a total of 360°:

Sector angle = 0.26 × 360°
Sector angle = 93.6° *(round up to 94°)*

NOTE

The sector angle is the angle between the radial lines that form the sides of a sector in a circle.

Step 8 Draw a straight line from the center of the circles through the outer circle. Place a protractor on this line with its index at the center and measure 94° from the line. Draw a line from the center of the circle through this angle to the outer circle. The region between the two radial lines and the two circles is the developed lateral surface of the reducer's jacket.

Step 9 Along one straight edge of the pattern, add an appropriate longitudinal overlap (2" or 5 cm).

Step 10 Draw an arc between the two radial lines that is ½" to ¾" (13–19 mm) longer than L_1 to provide a crimp edge. If necessary, do the same for the inner edge of the reducer, making the radius L_2 minus the crimp allowance.

Step 11 Cut out the pattern with sheet-metal shears along the pattern outline. Use a crimping tool to form a lip along the outer edge (and inner edge, if applicable) that will allow the straight-pipe jacket to overlap the reducer jacket.

A beveled reducer is basically a flattened concentric reducer. The same overall approach is used to develop its jacket surface. Care is needed to measure the reducer dimensions because the lateral surface is shorter, which means measuring errors will be magnified.

Because of several factors, manually developing the surface of an eccentric reducer is quite complicated. For example, eccentric reducers are typically custom-made for a given piping installation. The degree of offset (eccentricity) of the smaller to the larger pipe is different for each case, so the layout method needs to be modified depending on the shape of the reducer. Therefore, jacket patterns for such reducers are generally created today with for-the-purpose software or by skilled mechanical drafters. Industrial insulating and sheet-metal handbooks may also provide workers with instructions.

Installation of a reducer jacket is fairly simple. Install the adjoining straight pipe jacket per specifications. If screws are permitted, position the jacket tightly against the insulation and temporarily wire or tape it in place. Place screws along the longitudinal seam and several screws in each end to secure the reducer to the adjacent pipe jacket. If the jacket is installed on a cold system or one with a vapor retarder, screws must not be used. Secure with bands on the adjoining pipe jacket positioned over the underlying crimped edges. Avoid overtightening the bands to prevent puckering of the longitudinal seam.

2.6.2 Jacket for Pipe End Caps

Piping blanks are installed for various reasons. One common purpose is to stop the end of a main distribution header. Blanks and welded or threaded caps may close off pipes after removing unnecessary equipment or testing equipment, or to close off piping that was included by designers to connect planned future systems.

Blank flanges and similar components in insulated piping systems need to be protected just as with the rest of the piping. The jacket covers to accomplish this task are in general called end caps. End caps may also be used to jacket any abrupt changes in the size of a pipe. These are typically flanged connections between piping and adjacent valves other flange connected components. In these cases, the end cap is a ring of jacket joining the inner pipe jacket and outer flange jacket.

Figure 43 Beveled jacket end caps.

Jacket end caps may be flat, or beveled like the ones in *Figure 43*.

End-cap jackets can be purchased as prefabricated or field-fabricated units. Smaller, flat end-cap jacket can be made from a single piece of sheet metal. The pattern is a circle with the outside diameter of the insulated pipe or fitting. An extra $^1/_2$" to 1" (13–25 mm) is added to this diameter for a crimped edge that can be secured to the fitting jacket with a band, or screws (if permitted).

For large end caps and similar fittings, the jacket is fabricated in two halves, as shown in *Figure 44*. Draw the pattern for one half of the jacket cover as before, but add 2" (5 cm) for the lap across the cap. The second piece is just half the cap without the lap.

To install the end cap jacket, arrange the seam horizontally with the half-section overlapping the larger lower half with the lap. The crimped edges must be inserted under the adjacent pipe jacket and fastened per specifications. A beauty band may also be installed to more firmly secure the cap jacket to the pipe.

Bevel and flange end caps must be installed around points where piping or components of very different sizes come together. The main difference with these types of end caps is that a center hole with a diameter equal to the diameter of the smaller insulated pipe must be cut out. These end caps must be fabricated in two halves as described for large flat end caps. Their outer diameters must include the insulated flange perimeters. The approach for developing the conical beveled cap is very similar to that for a bevel reducer described in Section 2.6.1.

When installing these end caps, the smaller piping is usually jacketed first and the end cap jacket is installed outside the pipe jacket. The jacket end caps cover the two insulated faces of the flange and a strip of sheet metal wraps around the insulated flange perimeter. It must be the width of the flange plus the insulation thickness on both flange faces. Bands or fasteners secure the assembly in place. Flanges may be jacketed in sequence with the adjacent pipes or jacketed last, as required.

2.6.3 Fabricating and Installing Jacket on a Flanged Valve

Installing jacket on flanged valves is very similar to installing jacket on tees. The valve body corresponds to the main pipe of a tee and the valve bonnet can be viewed as the tee's branch pipe.

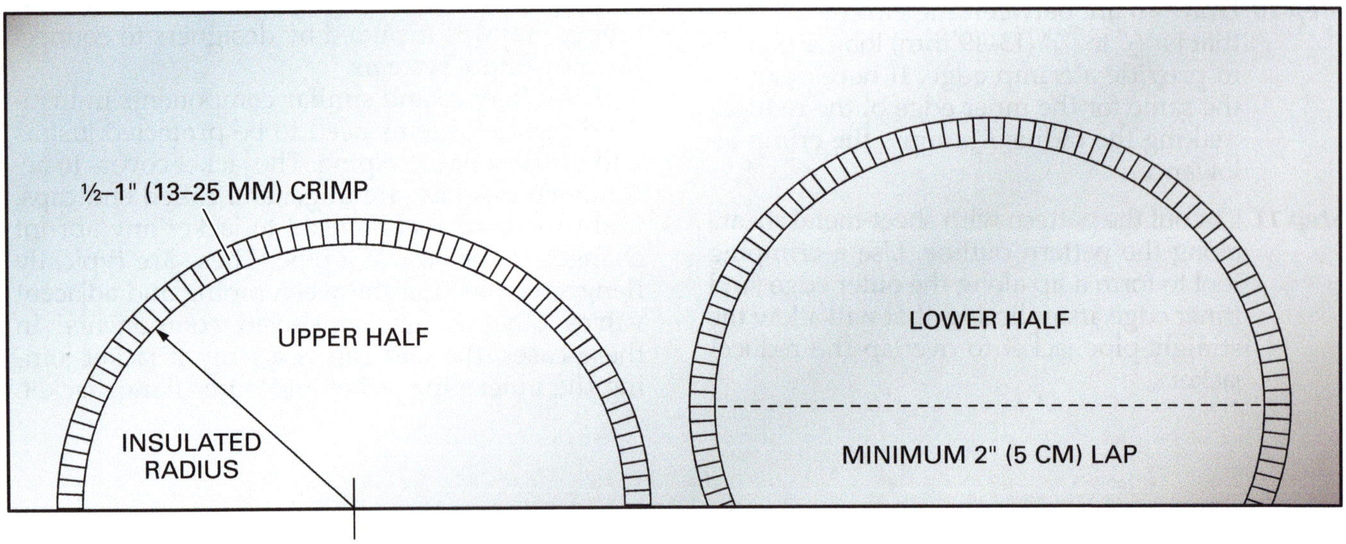

Figure 44 Jacket for a flat end cap.

Prefabricated jackets are available from some valve manufacturers. The jackets may also be field-fabricated. Refer to the instructions for jacketing pipe tees.

Typically, the valve body is jacketed as a unit between the outer faces of the connecting flanges, if applicable. The insulated valve bonnet and flanges are covered with two-piece end-cap jacket. Cutouts in the end caps for the yoke and connected pipes are required. Most valve jackets are fabricated from the same materials used on the connected pipe jacketing. *Figure 45* shows the parts and terminology of a jacketed flanged valve. The valve stem packing gland and adjustment bolts are usually left accessible. The bonnet and yoke arrangements of valves are quite variable, so only the general idea can be provided here.

An alternative to enclosing the valve entirely in sheet-metal jacket is to apply the body and bonnet sections as described, but seal the flanges and stem area of the bonnet with an approved mastic and reinforcement material.

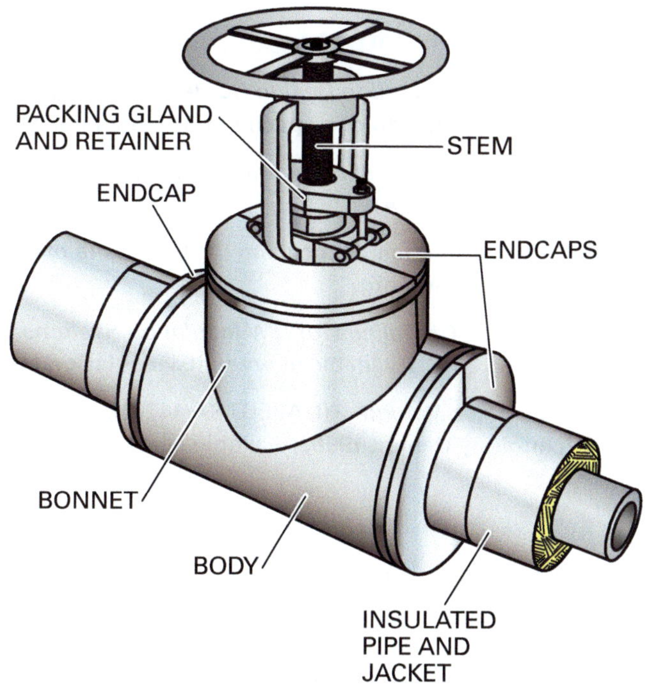

Figure 45 A jacketed flanged valve.

Additional Resources

Mechanical Insulation Design Guide, National Institute of Building Sciences. Available at **www.wbdg.org/guides-specifications/mechanical-insulation-design-guide.**

National Industrial and Commercial Insulation Standards Manual, Midwest Insulation Contractors Association (MICA).

National Insulation Association (NIA) website offers resources for products and training. **www.insulation.org.**

Installation Guidelines for ITW Insulation Systems Metal Jacketing Used on Insulated Pipe, ITW Insulation Systems. 2016. Available at **www.itwinsulation.com**

Steelworker, Volume 2 (NAVEDTRA 14251), Naval Education and Training Professional Development and Technology Center. 1996. Available at **archive.hnsa.org/doc/pdf/steelworker2.pdf**

2.0.0 Section Review

1. Which of the following actions should *not* be taken when installing jacket on horizontal insulated pipe?
   a. Ensure that longitudinal lap seams are even and straight.
   b. Allow 2" to 3" (5 to 8 cm) overlap at the butt seams.
   c. Loosen all the pipe brackets at the same time to make jacket installation more efficient.
   d. Make sure the banding seals are lined up neatly along the longitudinal lap.

2. What is a key difference between installing metal jacket on horizontal pipe and on vertical pipe?
   a. Jacket must be fitted around clamp-on pipe brackets in vertical pipe runs.
   b. Installation of jacket on vertical pipes begins at the highest point and moves to the lowest point in the riser.
   c. Bands are used for horizontal pipe jacket, but screws are required for vertical pipe jacket.
   d. Pittsburgh seams are required between the vertical pipe sections.

3. The standard approach to fitting pipe insulation jacket around a penetration involves _____.
   a. ending the section before the penetration and starting the jacket after the penetration
   b. ending the section just beyond the penetration, notching it, and starting the next section at the penetration with an overlap
   c. using a full jacket section and cutting out a hole where the penetration occurs
   d. removing fittings that can cause a penetration to keep the jacket intact

4. What is the main advantage of field-fabricated gored elbow jacket over pressed (or preformed) jacket?
   a. Field-fabricated jacket is less expensive.
   b. Pressed jacket is harder to install.
   c. Field-fabricated jacket can be made to fit any type and size of elbow, and any insulation thickness.
   d. It can be made in one piece.

5. Which tool is the best suited for cutting a round hole in jacket sheet metal?
   a. Hole punch
   b. Razor knife
   c. Scissors
   d. Offset shears

6. If the outlet of a circular pipe reducer is on the same axis as the inlet, then the reducer is _____.
   a. Eccentric
   b. Concentric
   c. Straight
   d. Linear

SUMMARY

Jacketing systems play several very important roles in insulation systems. Insulation only functions as an insulator in a dry state. For an outdoor installation, the only protection the insulation has is the jacketing system. It is imperative that the jacketing system be installed by competent insulation mechanics so the insulation will remain dry and protected over its useful life.

Some jacketing systems serve a dual purpose in that the jacket also functions as a key part of the vapor retarder boundary on cold-service installations. You have learned of the importance of an intact vapor retarder to avoiding insulation system failures.

The jacketing system is the final step of an insulation project. It is the finished product that the owner, contractor, and mechanic is able to observe. While all parts of the insulation system may technically be installed correctly so it functions as intended, if its finished appearance is shoddy, the customer won't be happy and the installer may discover finding work becomes hard. The smoothness of the sheet metal, the straight alignment of the laps, the uniform spacing of the bands, and the spacing of the screws all contribute to a finished product that is pleasing to the eye.

Creating an effective and pleasing product requires continual attention and participation by the trainee, and lots of practice. All of the knowledge learned in other modules as well as your mental and physical skills will be needed to master the procedures presented in this module.

Review Questions

1. The heaviest aluminum sheet metal typically used for insulation jacket is _____.
   a. 0.050" (1.3 mm)
   b. 0.040" (1.0 mm)
   c. 0.016" (0.4 mm)
   d. 0.012" (0.3 mm)

2. Longitudinal jacket seams on horizontal pipe should be positioned at the _____.
   a. 12 o'clock position
   b. 6 o'clock position
   c. 2 to 3 o'clock or 9 to 10 o'clock positions
   d. 3 to 4 o'clock or 8 to 9 o'clock positions

3. Which of the following statements is correct?
   a. Screws are not allowed for securing insulation jacket.
   b. Screws for elbow jacket are needed only at the ends and only when installed under straight-pipe jacket.
   c. Screws for securing straight-pipe longitudinal laps should be spaced at 6" (15 cm) apart or less.
   d. Screws are allowed for securing jacket only on below-ambient temperature systems.

4. If 8" (DN 200) pipe with an OD of $8\frac{5}{8}$" (8.6" or 21.9 cm) is insulated with 2" (5.1 cm) insulation, what is the pipe's insulated circumference?
   a. 39.6" (101 cm)
   b. 33.4" (85 cm)
   c. 31.4" (80 cm)
   d. 27.1" (69 cm)

5. What can plasma and laser cutting machines do that most other sheet-metal cutting methods cannot?
   a. They can produce complex shapes in sheet metal very precisely.
   b. They are portable.
   c. They can automatically roll the jacket.
   d. They automatically make Pittsburgh seams.

6. The proper machine for forming pipe insulation jacket to the curvature of smaller-diameter pipe is a _____.
   a. manual beader
   b. metal brake
   c. sheet-metal roller
   d. hemming machine

7. When installing jacket on horizontal pipe provided with support saddles, the correct method requires _____.
   a. installing the jacket first, then the pipe saddles
   b. cutting and fitting jacket around the saddle
   c. removing the saddle and replacing with extra-heavy jacket
   d. installing the jacket over the saddle

8. What factor determines the length of an S-clip?
   a. Strength of the metal band
   b. Width of the metal band
   c. Amount of overlap at the jacket butt seam
   d. Total weight the clip must support

9. Which of the following statements goes against good jacket installation practices on pipe risers?
   a. Jacket sections should be installed on vertical pipe starting at the bottom of a riser.
   b. Jacket sections should be temporarily wired in place as they are installed until the scaffold or lift has to be moved. Then band them in place.
   c. If an elbow is at the top of the riser, measure, cut, and install the last pipe jacket before installing the elbow jacket.
   d. Run the jacket up to a pipe bracket, then begin the next jacket section supported by the pipe bracket, leaving a gap.

10. When installing pressed sheet-metal elbow covers with screws, the first screws should be placed _____.
    a. in the middle of the heel and throat seams
    b. at the butt ends of the heel seam
    c. at the longitudinal seam where it overlaps the last-installed pipe jacket
    d. where the biggest gap in the seam occurs

11. When planning the construction of a field-fabricated mitered-elbow jacket cover, what is usually the first step?
    a. Determine the number of gores required.
    b. Lay out the gore pattern.
    c. Determine the heel dimension.
    d. Determine the throat dimension.

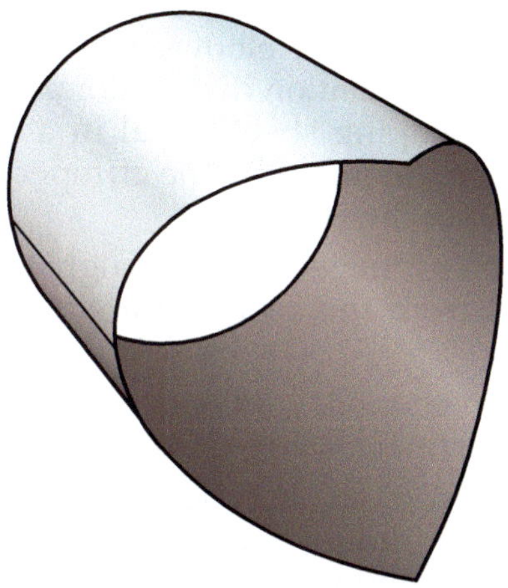

Figure RQ01

12. Based on the shape of the branch-pipe jacket shown in *Figure RQ01*, what type of tee is it made for?
    a. Equal tee
    b. Unequal tee (branch-pipe smaller)
    c. Unequal tee (branch-pipe larger)
    d. Insufficient information

13. How does the jacket form for an unequal tee with a larger branch pipe differ from other types of pipe-tee jackets?
    a. The branch-pipe jacket must be made in two parts.
    b. The branch- and main-pipe jackets have saddle cutouts.
    c. The branch-pipe jacket seam comes to a 90° point where it intersects the main-pipe jacket.
    d. Its pattern must be computer-generated.

14. The distance between the larger and smaller ends of a reducer as measured along its outer surface is called the _____.
    a. width
    b. angular height
    c. diameter
    d. lateral length

15. When installing metal jacket on flanged valves with bonnets, which of the following statements is correct?
    a. The pipe flanges are left unjacketed to allow access to the flange bolts.
    b. Flange ends may be covered with mastic and fabric reinforcement instead of sheet-metal jacket.
    c. Valve bonnet flanges are normally left unjacketed.
    d. The bonnet jacket normally covers the valve handwheel yoke or operator.

Trade Terms Introduced in This Module

Batten: In layout work, a long, thin, flexible strip of wood or other material that can evenly bend without kinks. Used for creating smooth, sweeping single- and compound-curves.

Bead: A raised ridge or groove along an edge of sheet metal.

Bisector: In geometry, a line that divides an angle in half or bisects it.

Concentric: The condition where two or more circles or circular objects have the same center point or lie on the same axis.

Crimp: A treatment of an edge of sheet metal to improve its rigidity, consisting of either many peaks and valleys along the edge, like gear teeth, or a single shallow fold parallel to the edge.

Eccentric: The condition where two circles, circular objects, or rotating objects do not have the same center point or center of rotation.

Equal tee: Identifies a piping tee where the main and branch pipes have the same diameters.

Hemmed: The folded-over edge of a sheet of material, as in a fabric hem. A hemmed edge is smooth and eliminates a safety hazard.

Laser cutter: A device that uses a form of high-energy light, usually of a single color and produced in a tight, intense beam, to cut materials. Laser cutters can be designed for extremely precise shape fabrication.

Lateral length: For a conical fitting, the top-to-bottom length of the fitting as measured along its side (or lateral) surface.

Offset compound shears: A type of hand shears with jaws designed to permit cutting small-radius curves in sheet metal without crimping the material and keeping the worker's hand clear of the cut. Most manufacturers supply color-coded tools to identify the designed direction of the cut: green is clockwise cutting; red is counterclockwise cutting; yellow is straight cutting. Also called *aviation snips*.

Pittsburgh seam: A seam between two sheet-metal edges consisting of a right-angle flanged male tab on one edge and a folded female pocket on the other edge. The flange is inserted into the pocket and the side extension of the female pocket is folded over to lock in the male flange. The various folds and closures can be manually- or machine-formed in the sheet-metal edges.

Plasma cutter: A device that uses a high-velocity, high-temperature jet of ionized air or other gas emitted from a nozzle to cut and shape materials.

Pressed cover: A type of preformed insulation jacket cover for elbows and tees available from many insulation distributors. The cover is made by shaping sheet metal in a form press and can be identified by the crimped fingers in the throat of the cover.

Radial lines: In geometry, straight lines drawn outward from the center of a circle, like a radius.

Sector: In geometry, an area contained inside a circle's perimeter between two radial lines.

Sweep elbow: A type of pipe elbow that has a longer radius than the standard short- or long-radius elbows.

Takeoff: The process of surveying, measuring, itemizing, and counting all materials and equipment needed for a construction project, as indicated by the drawings.

Unequal tee: A type of piping tee where the main and branch pipes are of unequal size. The branch pipe may have either a smaller or larger diameter than the main pipe.

Wing seal: A type of metal banding seal or buckle equipped with two tabs or wings that are bent over the folded end of the band to secure it.

X-Y controller: A computer-controlled device that precisely positions a machine or tool in the front-to-back and side-to-side directions in a plane.

Additional Resources

This module is intended as a thorough resource for task training. The following reference works are suggested for further study.

Mechanical Insulation Design Guide, National Institute of Building Sciences. Available at **www.wbdg.org/guides-specifications/mechanical-insulation-design-guide.**

National Industrial and Commercial Insulation Standards Manual, Midwest Insulation Contractors Association (MICA).

National Insulation Association (NIA) website offers resources for products and training. **www.insulation.org.**

Installation Guidelines for ITW Insulation Systems Metal Jacketing Used on Insulated Pipe, ITW Insulation Systems. 2016. Available at **www.itwinsulation.com**

Steelworker, Volume 2 (NAVEDTRA 14251), Naval Education and Training Professional Development and Technology Center. 1996. Available at **archive.hnsa.org/doc/pdf/steelworker2.pdf**

Figure Credits

Brock Services, Ltd., Module opener, Figure 19

© Clearviewstock/Dreamstime.com, Figure 1A

© Cristiano Ribeiro/Dreamstime.com, Figure 1B

© Charon/Dreamstime.com, Figure 1C

Forrest Mfg. Co., Figures 2, 3

© Dmitry Kalinovsky/Dreamstime.com, Figure 4

Bart van der Maar, Figures 5, 7

RAMS Sheet Metal Equipment, Figure 6

© Dainis Derics/Dreamstime.com, Figure 8

Gary Satterfield, Figures 13, 14, 17, 43

Section Review Answer Key

Answer	Section Reference	Objective
Section One		
1. d	1.1.1	1a
2. b	1.2.0	1b
3. d	1.3.0	1c
4. d	1.4.1	1d
Section Two		
1. c	2.1.0	2a
2. a	2.3.0	2b
3. b	2.3.0	2c
4. c	2.4.0	2d
5. d	2.5.1	2e
6. b	2.6.1	2f

NCCER CURRICULA — USER UPDATE

NCCER makes every effort to keep its textbooks up-to-date and free of technical errors. We appreciate your help in this process. If you find an error, a typographical mistake, or an inaccuracy in NCCER's curricula, please fill out this form (or a photocopy), or complete the online form at **www.nccer.org/olf**. Be sure to include the exact module ID number, page number, a detailed description, and your recommended correction. Your input will be brought to the attention of the Authoring Team. Thank you for your assistance.

Instructors – If you have an idea for improving this textbook, or have found that additional materials were necessary to teach this module effectively, please let us know so that we may present your suggestions to the Authoring Team.

NCCER Product Development and Revision
13614 Progress Blvd., Alachua, FL 32615

Email: curriculum@nccer.org
Online: www.nccer.org/olf

❑ Trainee Guide ❑ Lesson Plans ❑ Exam ❑ PowerPoints Other ____________

Craft / Level: Copyright Date:

Module ID Number / Title:

Section Number(s):

Description:

Recommended Correction:

Your Name:

Address:

Email: Phone:

Jacketing Fabrication— Vessels and Equipment

Overview

Insulating vessels and equipment can be challenging due to their sizes and locations. Those located where they are exposed to weather and other adverse environmental factors must also be protected to shield the insulation from damage. Other modules in this curriculum address insulation jacketing and its application. In this module, you will focus on fabricating and installing metal insulation jacket on vessels and equipment.

Module 19311

Trainees with successful module completions may be eligible for credentialing through the NCCER Registry. To learn more, go to **www.nccer.org** or contact us at 1.888.622.3720. Our website has information on the latest product releases and training, as well as online versions of our *Cornerstone* magazine and Pearson's product catalog.

Your feedback is welcome. You may email your comments to **curriculum@nccer.org**, send general comments and inquiries to **info@nccer.org**, or fill in the User Update form at the back of this module.

This information is general in nature and intended for training purposes only. Actual performance of activities described in this manual requires compliance with all applicable operating, service, maintenance, and safety procedures under the direction of qualified personnel. References in this manual to patented or proprietary devices do not constitute a recommendation of their use.

19311 V2

19311
Jacketing Fabrication – Vessels and Equipment

Objectives

When you have completed this module, you will be able to do the following:

1. Describe jacketing materials and general installation considerations.
   a. Describe common jacketing materials and their function.
   b. Describe general sheet-metal jacket installation considerations.
   c. Describe various methods of cutting and forming jacket materials.
2. Explain how to install jacketing on vessels and equipment.
   a. Describe general vessel sheet-metal jacketing considerations.
   b. Explain how to install roll-type jacketing.
   c. Explain how to install sheet and corrugated jacket.
   d. Explain how to fabricate and install gored and standing-seam vessel heads.

Performance Tasks

Under the supervision of your instructor, you should be able to do the following:

1. Install roll-type jacketing on a vessel using screws or bands for securement.
2. Install sheet jacketing on a vessel using screws or bands.
3. Lay out and install gores on a vessel head.

Trade Terms

Beauty band
Blind rivets
Center cap
Centerpiece
Circumference
Develop
Dished top
Expansion spring
Gauge
Knuckle
Perimeter
Pittsburgh seam
Standing seam
Wind forces

Industry Recognized Credentials

If you are training through an NCCER-accredited sponsor, you may be eligible for credentials from NCCER's Registry. The ID number for this module is 19311. Note that this module may have been used in other NCCER curricula and may apply to other level completions. Contact NCCER's Registry at 888.622.3720 or go to **www.nccer.org** for more information.

Contents

Figures

SECTION ONE

1.0.0 METAL INSULATION JACKETING AND INSTALLATION CONSIDERATIONS

Objective

Describe jacketing materials and general installation considerations.

a. Describe common jacketing materials and their function.
b. Describe general sheet-metal jacket installation considerations.
c. Describe various methods of cutting and forming jacket materials.

Trade Terms

Blind rivets: Fasteners that are deformed when installed to hold two sheets or plates of material together. A blind rivet can be installed from only one side of the surfaces. Tubular aluminum and steel blind rivets are commonly called pop rivets.

Expansion spring: A spring-loaded fitting installed in securement bands for jacketing to even out the tension in the bands as the vessel or equipment expands and contracts.

Gauge: A system of measuring the thickness of sheet materials, especially sheet metal. Often used as a general term for the thickness of such materials.

Pittsburgh seam: A seam between two sheet-metal edges consisting of a right-angle flanged male tab on one edge and a folded female pocket on the other edge. The flange is inserted into the pocket and the side extension of the female pocket is folded over to lock in the male flange. The various folds and closures can be manually- or machine-formed in the sheet-metal edges.

This module includes information discussed in Module 19313, *Jacketing Systems* and Module 19310, *Jacketing Fabrication—Piping and Fittings*. Much of the introductory information contained in Module 19310 will be repeated in this section. If you are unfamiliar with metal jacketing materials, this content will be valuable to review.

1.1.0 Common Metal Jacketing Materials

Metal jacketing material can be stainless steel, galvanized steel, or epoxy-coated aluminum or steel. Each of these materials is used for specific purposes and their use depends on the individual job specifications.

1.1.1 Jacket Sheet-metal Properties

These materials come in many different thicknesses; 0.016" (0.4 mm) aluminum is the most commonly used material for general jacketing requirements. However, individual specifications may call for heavier thicknesses in areas that are subject to natural or maintenance damage, or lighter thicknesses for areas that do not present a mechanical hazard.

Stainless steel jacket, because it is stronger than aluminum, is supplied in thicknesses from 0.010" to 0.032" (0.3–0.8 mm). Standard thicknesses are 0.010" and 0.016" (0.3 mm and 0.4 mm). Vessel and equipment aluminum jacket may be thicker, ranging up to 0.050" (1.3 mm) or thicker. Required jacket thickness is determined mainly by the diameter of the insulated vessel, and whether the insulation is rigid or non-rigid (flexible or compressible). Larger insulation diameters and non-rigid insulation require thicker jacket.

All types of metal jacketing used over insulation should contain a moisture barrier. The moisture barrier may be an acrylic paint coating, polyethylene and kraft paper sheet, or epoxy resin, polymer, or polyethylene coatings applied directly to the metal. The purpose of this barrier is to prevent galvanic corrosion of the jacketing. This kind of corrosion is especially a problem when the jacket and underlying metals are different and is made worse by direct contact with moisture.

There are many types of finishes that can be used for jacketing materials. Refer to *Figure 1*. Metal jacketing can be obtained smooth, stucco-embossed, corrugated, or epoxy finished. Metal jacketing can also be obtained painted or coated in different color schemes. Corrugated materials provide extra rigidity, but are harder to install than other types and are generally more expensive. On the other hand, corrugated jacket reduces sunlight reflection glare, shows dents less, and allows for more expansion and contraction than other forms of metal jacket. Smooth, shiny surfaces can reflect sunlight glare and dents from hail are more obvious.

Jacket sheet metals are supplied in various sizes. Sheet-metal jacket may be sold in 50' (15 m) to 300' (91 m) rolls depending on the type metal and distributor. The rolls are generally 36" or 48" (91 cm or 122 cm) wide and packed in cardboard boxes. Most suppliers will also cut the materials to the required sizes as coils or sheets of metal. Shallow-corrugated jacket materials generally supplied in either crimped or box corrugations $^{3}/_{16}$" (5 mm)

Figure 1 Common types of jacket sheet metal.

deep also come in rolls and are installed the same way as smooth sheet-metal jacket.

Heavier deep-corrugated jacket is sold in sheets. These are commonly supplied with corrugations 1¼" or 2½" (3.2 cm or 6.4 cm) wide and ½" (13 mm) or more deep. Sheets are sold in 33" (83 cm) widths, and in lengths of 8', 10', or 12' (2.4 m, 3 m, or 3.7 m). Other dimensions can be special ordered. Corrugated aluminum sheets are available in smooth or stucco finish, and may be bare, painted, or have a clear-coat finish. Generally, corrugated sheets are supplied with a polyethylene and polymer moisture barrier applied to the inner surface to prevent galvanic corrosion.

1.1.2 Jacket Accessories

Metal jacketing is installed with accessory materials. All jacketing materials can be secured with banding, which may be obtained in aluminum, galvanized steel, or stainless steel. Some applications may permit polypropylene or nylon banding. Banding width will vary from ⅜" to ¾" (10–19 mm) and their thicknesses will be between 0.015" and 0.020" (0.4–0.5 mm). In general, larger-width banding is required for larger diameter pipe and vessels. Aluminum banding generally is not used on vessel or equipment applications because of the high tensions required for securing jacketing on larger structures.

CAUTION

Some specifications may require color-coded banding. For example, older plants may signify asbestos-free insulation under metal jacket with a specific band color to distinguish it from original asbestos insulation. Any repairs involving asbestos replacement and new asbestos-free work in such a facility requires the use of these jacket band colors. Banding colors may also signify system function, or the type of jacket/banding material.

Banding materials are secured with wing seals. Properly secured bands with wing seals require a tensioning tool, which may be of the pistol grip or the ratchet type. The ratchet tensioner can apply more force to the band. Seals are sized to the banding to be secured. Seal metals should match the banding metals to avoid galvanic corrosion.

For large vessels, bands may be provided with an **expansion spring**. These fittings contain compressible springs fitted in housings with tabs that can accept standard metal banding. The spring fittings prevent overstressing securement bands and deforming jacket materials when the jacket, insulation, or the underlying vessel or equipment expands with temperature changes.

Metal jacket can be secured with many different types of fasteners. However, jacket installed on systems operating at below ambient temperatures and equipped with a vapor retarder boundary must not be secured with any kind of penetrating fastener (screws, staples, tacks, or rivets). These can puncture the vapor retarder, allowing water vapor to enter the insulation. Screws can be galvanized, coated, or stainless. They may be hex-head or pan-head, with or without neoprene washers, and can be self-tapping or self-drilling. Aluminum and steel **blind rivets** are sometimes specified for jacket securement. Rivets are not in widespread use.

Jacketing is normally sealed at its seams with some form of caulking for weather tightness. For below-ambient temperature service, jacketing is sealed with an approved vapor retarder mastic or sealant that is part of the overall vapor retarder system. Sealants and caulking will be identified in construction drawings or in other applicable specifications. Vapor retarder systems and the principles of water vapor diffusion are discussed in many other modules in the NCCER Mechanical Insulating curriculum.

1.2.0 Jacket Installation Considerations

Jacket installation is generally one of the last steps an insulator will complete in an insulating job. Before beginning jacketing work, there are a number of things to be aware of:

> **WARNING!**
> Stainless steel is very difficult to work with and represents a severe cutting hazard. Cut-resistant gloves and Kevlar® sleeves should be worn when handling this material.

- When flat roll jacket is installed on large diameter vessels, the natural curvature of rolled jacketing helps the jacket to fit around the insulation.
- The overlaps on jacketing materials are determined by the job specifications. However, it is a standard practice on vessels and equipment to allow for at least a 3" (8 cm) overlap at all edges, unless specifications dictate otherwise.
- Exposed longitudinal edges may be hemmed to eliminate a cutting hazard as well as to provide a more pleasing finish.
- Many contractors have shops where jacket sections are precut, rolled, and fabricated for jobs. Some large jobs may have an on-site shop. Manufacturers, distributors, and some large contractors have automatic machines that cut, roll, and fabricate the jackets as required. Jacketing may also be hand-cut for custom fitting and for small jobs.
- Screws and rivets may be used on hot systems or those without a vapor retarder, according to specifications. They must not be used to secure cold-system jacket where a vapor retarder is present.
- If rivets are permitted, ensure the pilot holes are drilled to match the rivets' installation instructions. Too-large holes will result in the rivets pulling out.
- When using screws, they should be spaced a maximum of 6" (15 cm) along the longitudinal overlap joints of jacket.
- If a joint sealant is required at jacket edges, it must be applied between the metal and lap before securing the jacket. Do not apply a bead of sealant along edges after the jacket is secured.
- Z- or S-clips for vertical vessel sidewall jacketing should be bent into three sections to create a seam overlap of about 3" (8 cm). These clips support the upper jacket courses.

1.3.0 Cutting and Forming Metal Jacket

Prefabricated, pre-curled jacket obtained from a distributor is ideal for most jacketing jobs, especially for piping systems. However, if the systems to be worked on vary significantly in sizes and insulation thicknesses, it is generally more practical to make on-site estimates of material requirements and submit them to the sheet-metal shop for local fabrication. Depending on the workforce organization, you may be involved in the requisitioning and even the cutting of the jacket for your job. There are several approaches introduced in this section for cutting and preparing sheet metal for insulation jacketing.

1.3.1 Cutting Sheet-Metal Jacket

A variety of equipment and tools are available for cutting jacket. Some of these include the following:

- *Automatic cut-and-roll machine* – Jacket roll is placed on the machine (*Figure 2*) and the machine is set for the length and rolled diameter of the finished pieces. The number of pieces is entered in the control unit and the jacket sections are automatically cut and rolled. These machines are the easiest and most efficient method for cutting large numbers of identical jacket sections.
- *Sheet-metal shop shears* – There are three types of shop shears used to cut jacket metal. One is a drop-shear, where an angled blade moves vertically against the cutting blade. These are heavy, floor-mounted pieces of equipment. The drop blade may be operated by a foot pedal or hydraulically. A second type is called a guillotine shear. In these cutters, one end of a heavy blade is hinged and the blade cuts by rotating around the hinge against a stationary cutter blade (*Figure 3*). Hydraulic versions are generally required for vessel and equipment steel jacket. The third type is a wheel cutting shear. For shop fabrication, a guide bar mounts a carriage on which a heavy, sharp-edged wheel bears against a cutting edge. The hand- or motor-operated carriage moves across the material and the wheel presses the metal against the blade, cutting it. The manual machines are suitable only for metals that are a lighter gauge.
- *Energy cutters* – These machines are high-end sheet-metal cutters that might be seen in some fabricator facilities. They utilize either industrial laser or plasma cutters. The X-Y controller guides the laser beam or plasma jet to specific

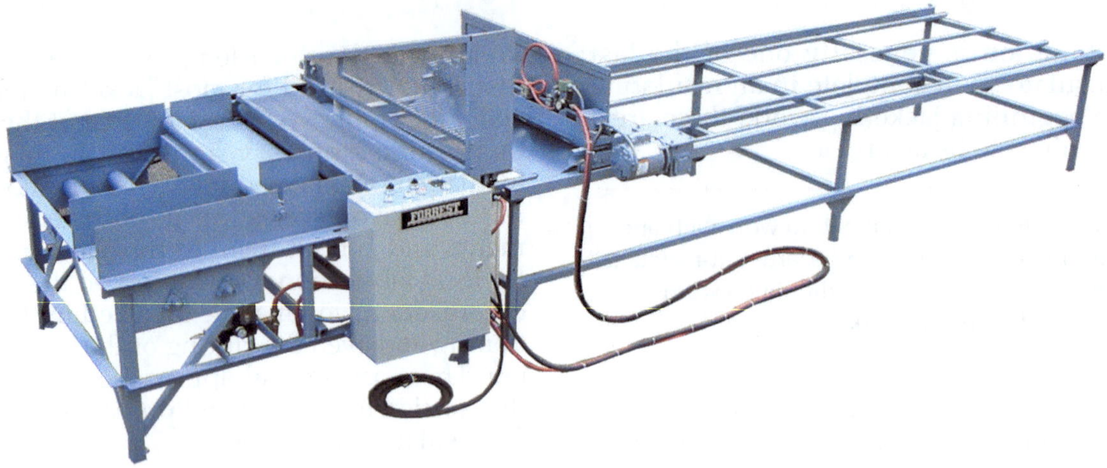

Figure 2 Automatic sheet-metal cut-and-roll machine.

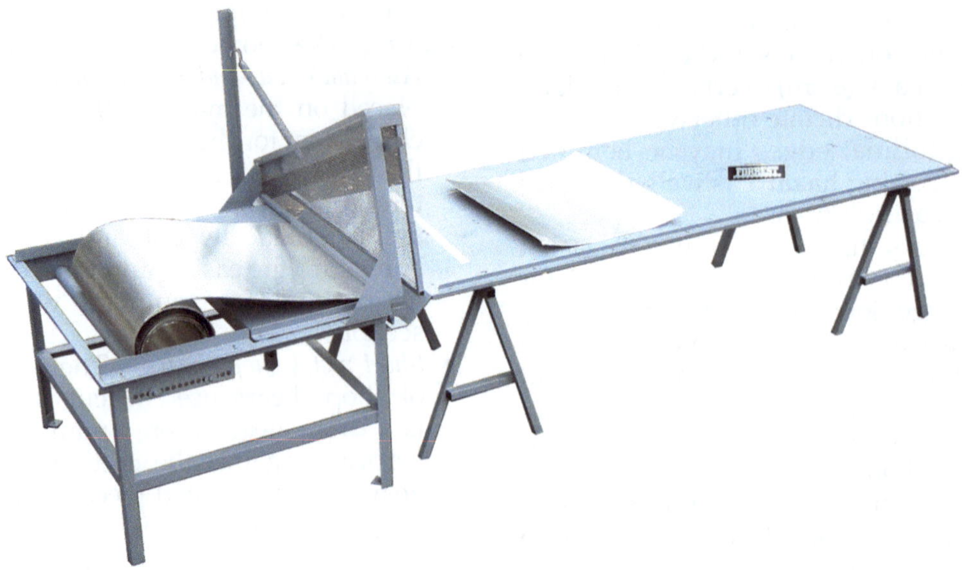

Figure 3 Manual sheet-metal shear and cutting table.

X-Y coordinates to vaporize the metal to the desired shape (*Figure 4*). The cutters are normally controlled by computer programs and are capable of cutting out complex shapes (e.g., vessel head gores and fitting covers) as well as straight cuts very precisely.

For small jobs, or when appropriate equipment is not readily available, sheet-metal jacket can be cut using basic hand tools. The following steps present one approach to hand-cutting jacket:

Step 1 Select a clear, clean area of the floor.

Step 2 Open the carton, remove the roll of metal, and place it on the floor.

> **CAUTION**
>
> To protect the floor, place a piece of cardboard or wood on the floor. It will also provide a cushioned cutting surface and minimize the dulling of the knife blade when cutting.

Step 3 Remove the tape from the roll of jacket, being careful not to allow the roll to uncurl.

Step 4 Check the factory end of the roll for a straight, square cut. Trim square if necessary with the aid of a large construction T-square.

Figure 4 Laser sheet-metal cutting machine.

Step 5 Using a tape measure, mark each end of the jacket (moisture barrier side up, if present) for the size of the piece to be cut.

Step 6 Place a straight edge between marks. Use a razor knife and cut along the straight edge. Place the knife on the jacket and pull toward the edge, pressing firmly. The point of the knife will score the jacket.

Step 7 Hold the end of the jacket and flex at the knife score. The jacket should break clean.

NOTE

Handheld electric sheet-metal shears can also be used to quickly cut sheet metal. Use a heavy straightedge as a guide and allow for the blade standoff before cutting.

1.3.2 Forming Sheet-Metal Jacket

Jacket may need to be formed or otherwise prepared for installation. The most common forming operations are beading and crimping jacket to make attractive and secure seams between jacket sections on curved surfaces. Some specifications may require hemming or other forms of edge preparation. Some jacket preparation and fabrication methods include the following:

- *Beading and crimping* – Some jacket fabrication methods require a means of reinforcing the connections between adjacent sections. To ensure a close, friction fit, each section receives crimped and beaded edges, depending on the kind of seam. Special machines, called *beaders*, *crimpers*, or *beader-crimpers* if they can perform both functions (*Figure 5*), are required to form neat, effective seams. The sizes and depths of the beads and crimps are chosen based on the thickness of the sheet metal and the type of installation. Beading and crimping rollers come in pairs to press the required forms. Refer to the jacket specifications for this information.

(A) SHEET METAL BEADER

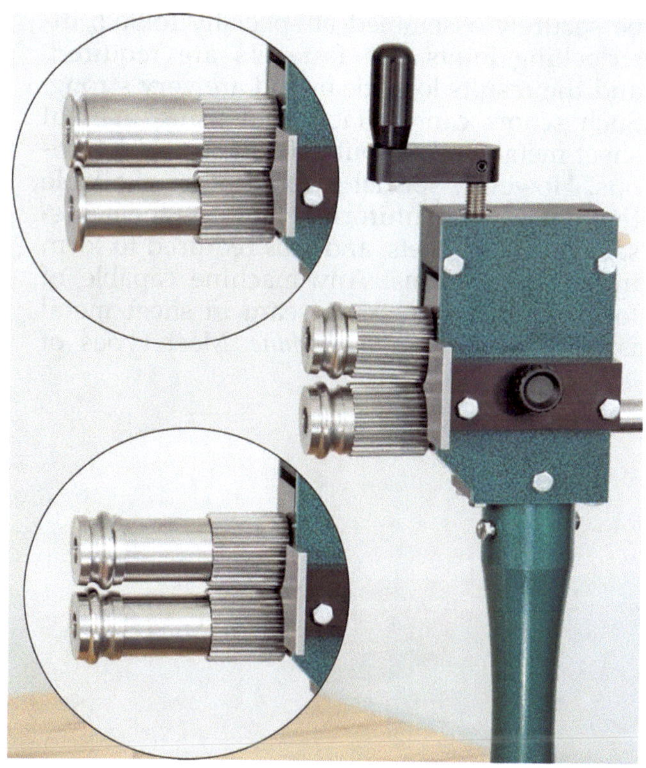

(B) BEADER-CRIMPER

Figure 5 Sheet-metal beader and beader-crimper tools.

- *Bending* – Jacket may need to be fabricated with folded corners of various angles at the edges. Neat crisp corners are formed with a standard shop machine called a sheet-metal brake. Large motor- or hydraulic-operated brakes may be used for bending long sections of ductwork and rectangular jacket. Some sheet-metal fabricators use brakes for hemming jacket edges. Smaller, hand-operated brakes (*Figure 6*) may be used for fittings and their covers.
- *Hemming* – Some specifications may require finishing one or more exposed edges of jacketing in a way that reduces or eliminates the cutting hazard of raw-edged sheet-metal jacketing. Bending over the edge presents a smooth seam that minimizes the hazard to people rubbing up against the jacketing while walking by or when performing maintenance. Hemming sheet metal is similar to creating a hem in a piece of cloth fabric. The edge is folded under the exposed surface. Sheet metal is stiff enough that no other securement for the hem is needed. Hemming can be formed by manual tools or large floor-mounted shop machines (*Figure 7*), as well as sheet-metal brakes. Hemming requires at least two steps to create the folded edge.
- **Pittsburgh seam** – While pieces of sheet metal can be assembled by using screws or rivets in overlaps at seams and edges, they can also be securely assembled at specially folded, interlocking joints. No fasteners are required, and the results look clean and are very strong. Such seams can be fabricated using manual sheet-metal brakes, but the process is laborious. However, special machines are available that quickly and uniformly fabricate the necessary bends, pockets, and tabs required to form interlocking seams. Any machine capable of forming an interlocking seam in sheet metal is called a *Pittsburgh machine*. Most types of interlocking sheet-metal seams are often called *Pittsburgh seams*, even though there are different types of seams besides true Pittsburgh seams. Interlocking seams may be used in various jacket applications where sheet metal has to be assembled in complex shapes.

Figure 6 Hand-operated sheet-metal brake for creating folds and corners in sheet metal.

Figure 7 Shop sheet-metal hemming machine making the first crimp.

Additional Resources

National Industrial and Commercial Insulation Standards Manual, Midwest Insulation Contractors Association (MICA).

National Insulation Association (NIA) website offers resources for products and training. **www.insulation.org**.

Mechanical Insulation Design Guide, National Institute of Building Sciences. Available at **www.wbdg.org/guides-specifications/mechanical-insulation-design-guide**.

Installation Guidelines for ITW Insulation Systems Metal Jacketing Used on Insulated Pipe, ITW Insulation Systems. 2016. Available at **www.itwinsulation.com**

Aluminum Deep Corrugated Sheets, ITW Insulation Systems. 2015. Available at **www.itwinsulation.com**

1.0.0 Section Review

1. All other factors being the same, which of the following would require heavier jacket sheet metal?
   a. Smaller vessels
   b. Flexible or compressible insulation
   c. Smaller insulated diameter
   d. More rigid insulation

2. Which of the following would *not* be a technical reason for using colored banding to secure jacket?
   a. Indicate an asbestos-free system
   b. Contribute to a pleasant working environment
   c. Indicate the purpose of a system
   d. Identify the banding material

3. Which of the following PPE is mandatory for working with stainless steel sheet metal?
   a. Cut-resistant gloves
   b. Respirators
   c. Hearing protection
   d. Steel-toed shoes

4. Which type of shop machine would most likely be used for cutting fifty pieces of sheet jacket for vessels and equipment?
   a. Sheet-metal brake
   b. Sheet-metal shop shears
   c. Pittsburgh seamer
   d. Hand sheet-metal shears

Section Two

2.0.0 Installing Vessel Sheet-Metal Jacket

Objective

Explain how to install jacketing on vessels and equipment.

a. Describe general vessel sheet-metal jacketing considerations.
b. Explain how to install roll-type jacketing.
c. Explain how to install sheet and corrugated jacket.
d. Explain how to fabricate and install gored and standing-seam vessel heads.

Performance Tasks

1. Install roll-type jacketing on a vessel using screws or bands for securement.
2. Install sheet jacketing on a vessel using screws or bands.
3. Lay out and install gores on a vessel head.

Trade Terms

Beauty band: A narrow strip of sheet-metal jacket that is installed over the rough edges of adjacent jacket to provide a finished appearance. Often supplied with beading, crimped edges, or both.

Center cap: A circular or cone-shaped piece of jacket sheet metal installed at the center of a vessel head to finish off the jacket installation there.

Centerpiece: A disk of jacket material fastened to the insulation at the center of a vessel's head to anchor the inner ends of the gore segments during installation.

Circumference: The distance around the perimeter of a circle or a circular object. Measured in units of length, such as inches, feet, centimeters, or meters.

Develop: As used in geometric layouts, to draw the true shape of a three-dimensional surface in two dimensions. It is an important part of fabricating jacket for three-dimensional vessels and equipment using two-dimensional sheet metal.

Dished top: When referring to a vessel head, it is a shape similar to an upside-down bowl; a low, domed appearance compared to a flat-topped vessel.

Knuckle: In general, an area of sharper curvature in a surface compared to the surrounding surfaces. For vessel heads, the surface near the head perimeter that is more sharply curved than the central dish-part of the head or its perimeter.

Perimeter: The outside boundary of any object or region, especially one that encloses a space inside. The line forming a circle and the sidewall of a vessel are examples of perimeters.

Standing seam: A method for connecting two pieces of sheet metal. Squared flanges are formed in the adjacent edges of the sheet-metal sections and brought together. At the top of the flanges, the pieces are folded together, locking them into a tight joint.

Wind forces: Forces generated by air flowing over a surface, like the lifting forces on an airplane wing or wind on a sail. Also called *aerodynamic forces*.

The methods for fabricating and installing vessel and equipment insulation jacket are very similar to those for piping systems. Generally the sizes of the materials and the scale of work are much larger. This section provides an overview of the considerations and methods for creating and installing jacket on these types of mechanical system components.

2.1.0 General Considerations for Installing Vessel Jacket

In addition to the general sheet-metal jacket considerations listed in Section 1.2.0 above, there are things the mechanical insulator should keep in mind when planning jobs involving vessel and equipment jacketing:

- Deep-corrugated jacket for vessels and equipment is generally recommended for vertical vessels and towers with insulated diameters over 8' (2.4 m), and for flat-surfaces. Sheets should be installed with the corrugations running vertical.
- For very large flat surfaces, especially associated with heavy industry and power plant boiler facilities, heavy box-ribbed lagging should be specified. See NCCER Module 19312, *Sheet Metal Lagging*."
- Deep- and shallow-corrugated jacket should generally not be installed on horizontal vessels because water can pool within the corrugations, possibly leading to corrosion.

- Deep-corrugated jacket is usually supplied with corrugations 1¼" or 2½" (3.2 cm or 6.4 cm) wide. Such sheets should overlap adjacent sections at longitudinal seams by at least two corrugations. $^{3}/_{16}$" (5 mm) corrugated jacket should be installed like flat jacket.
- In all insulation jacket installations, jacket must lap so as to shed water. Lower sections and courses must be installed before higher sections. Longitudinal laps at the very top of horizontal vessels should be avoided if possible.
- Securement bands should be applied directly over all circumferential joints. Specifications will determine band spacing, but a maximum of 12" to 18" (30–45 cm) between bands is standard.
- Banding metal gauge should always be equal to or heavier than that of the underlying jacket material.
- On hot systems and those without a vapor retarder, screws should secure the longitudinal seams between bands. No screws should be used in the circumferential seams of flat jacket.
- S- or Z-clips are used to support the bottoms of jacket courses on vertical vessels at the circumferential seams. The securement band located at the circumferential seam should lie inside the S-clip.
- Check specifications for longest allowed band length. Depending on the size of the vessel, expansion springs may need to be used.
- Band expansion spring fittings are generally installed no farther apart than 25' (7.6 m) around the **perimeter** of cylindrical vessels. Check the specifications for each unit worked on.
- Some types of insulated equipment or vessels may include bolted or deep support flanges. Jacketing must encase the insulation adjacent to bolted flanges to maintain the required setback to allow bolt removal. If a flange is insulated, then jacket must be fabricated to cover the flange insulation as well.

2.2.0 Installing Roll-Type Jacket

The installation of vessel and equipment jacket involves first estimating the kind and amount of materials and accessories, then the installation of the jacket itself. The approaches, to jacketing vertical and horizontal vessels are similar, but different temporary support methods are required.

2.2.1 Estimating Vessel Sidewall Jacket and Accessories

The estimation process is the same for vertical or horizontal cylindrical vessels. Two key dimensions must be obtained by direct measurement or from engineering plans: the insulated diameter of the vessel and the distance between the tangent lines of the insulated heads (the tan-to-tan distance). The vessel insulated diameter can be determined by taking the diameter of the vessel listed on the drawings and adding twice the insulation thickness. A smaller vessel's insulated **circumference** may be measured directly, which is generally more accurate.

The worker also has to know the planned overlaps at both the longitudinal and circumferential jacket seams. For most vessels, this is usually 3" (8 cm) unless specifications state otherwise. Finally, the worker must know the width of the roll material to be installed.

To go through an estimation example, the following assumptions will be made (refer to *Figure 8*):

- The vessel to be jacketed is vertical with elliptical top and bottom heads.
- The insulated tan-to-tan sidewall height is 40'–8" (488" or 1,240 cm).

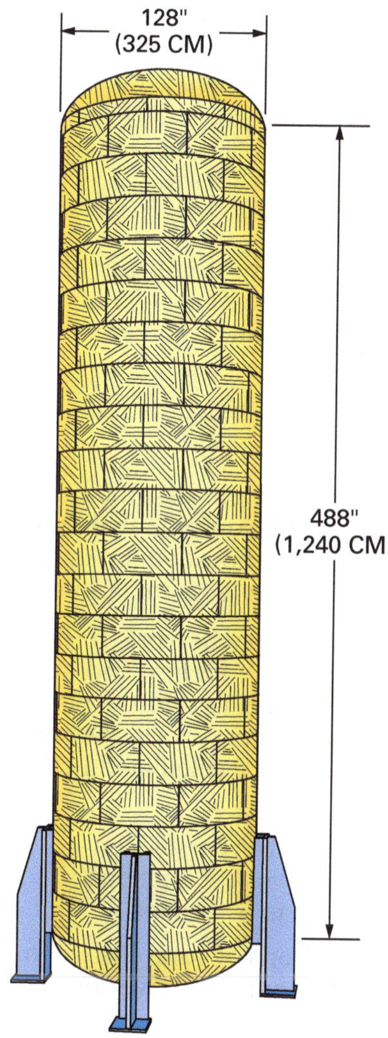

Figure 8 Example vessel and dimensions.

- The insulated vessel diameter is 10'–8" (128" or 325 cm).
- Jacket seam overlaps are 3" (8 cm).
- Roll jacket width is 48" (122 cm).

One objective to jacket installation is to minimize the number of vertical seams. This goal has to be balanced with the difficulty of handling larger jacket sections, especially on scaffolding or from a lift. An approach to determining the number of jacket sections per course is to calculate the insulated circumference and divide this dimension into enough equal sections so that an individual section has a length workers feel safe handling.

Estimating Amount of Roll Jacketing

The following steps are an example for estimating the number of sheets of jacket (for clarity, only Imperial units of measurement will be used in this example):

> **NOTE**
> It is a good practice when estimating or making other calculations in mechanical insulating work to round up your answers to avoid coming up short with materials or fabricated dimensions. It's always easier to return excess materials or trim something that's too large. With experience, you will learn what kind of rounding is appropriate in each case.

Step 1 If not already done, calculate the insulated circumference (C_I) of the vessel using the insulated diameter (D_I).

$C_I = D_I \times \pi$
$C_I = 128" \times 3.14$
$C_I = 402"$

For this example, assume the cut roll jacket will be ordered from the shop in 7' (2.1 m) sections, which will minimize waste and are easy to handle.

Step 2 To verify this will work, find the length of a 7' section in inches, and subtract the 3" (8 cm) overlap to find the visible length of a section. Then divide the result into the circumference calculated in Step 1.

7' = 84"
84" − 3" = 81" per section
402" ÷ 81" = 4.96 sections
Round up to 5 sections

Five sections of cut roll sheet metal 84" (214 cm) long will cover one course of jacket.

> **NOTE**
> In some cases it may be better to round down to keep the section length manageable. In any case, divide the insulated circumference by the whole number of sections determined, and then add the end lap to obtain the actual cut length of the jacket to be used.

Step 3 Determine the number of jacket courses. Divide the tan-to-tan height of the vessel by the visible width of the roll jacket (jacket width minus the overlap).

48" − 3" = 45" course height
488" ÷ 45" = 10.8 courses
Round up to 11 courses

Step 4 Calculate the total number of cut jacket sections and the total length from roll material required to jacket the vessel sidewall.

5 sections/course × 11 courses = 55 sections
55 sections × 84" per section = 4,620" *or* 385'

Therefore, to jacket the vessel sidewall in the example, you should order 55 sections of 48" (122 cm) cut roll jacket 84" (213.4 cm) long. Of course you need to also specify the type, gauge, and finish of the material if that hasn't been determined beforehand.

Estimating the amounts of securement accessories is also part of the job. You will need to determine the length of banding and number of screws to be used. Banding is generally ordered by the coiled length, but some distributors sell by weight. Screws are generally sold in bulk by number of screws per box.

Estimating Length and Weight of Banding

The following steps are an example for estimating the length and weight of banding:

Step 1 Assume that specifications for the example vessel require band spacing at 12" (30.5 cm). The tan-to-tan height is 488" (1,240 cm). Divide this height by the band spacing.

488" ÷ 12" = 40.7 band spaces
Round up to 41 band spaces

Note that the number of band spaces does not equal the number of bands required to secure the sidewall jacket. This calculation leaves you one band short. An additional band is required.

Add 1 band for top band:
41 bands + 1 = 42 bands

Step 2 An individual band length is equal to the jacketed circumference of the vessel (402 inches or 1,021 cm). Calculate the total banding required.

42 bands × 402" = 16,884" *or* 1,407'

> **NOTE**
> It is always a good practice to add 5–10 percent of the calculated amount to the banding order to account for cuttings, overlaps, and S-clip fabrication.

Step 3 Assuming you are using ¾", 0.020" thick stainless steel banding, which is rated at 19.6' per lb (13.1 m/kg). Calculate the weight of the banding required.

1,407' ÷ 19.6' per lb = 71.8 lb

> **NOTE**
> You can obtain length-per-weight specifications for banding of various widths and gauges from banding suppliers.

Estimating Number of Banding Seals

The following steps are an example for estimating the number of banding seals:

The number of band seals required will equal the number of bands, 41 seals.

> **NOTE**
> If the specifications require expansion springs for the example vessel, then estimate two banding clips per expansion spring.

Estimating Number of Fasteners

The following steps are an example for estimating the number of fasteners:

Longitudinal (that is, vertical) seams are secured with screws at the overlaps, when allowed by specifications. Horizontal seams in flat jacket materials are normally secured by banding and not with screws. Determining the number of screws requires calculating the total length of vertical seams and dividing by the specified screw spacing (assumed to be 6" or 15.2 cm).

Step 1 Determine the number of screws per vertical seam:

Tan-to-tan length divided by screw spacing:
488" ÷ 6" = 81.3 (round to 82 spaces)
Add 1 for number of screws:
82 +1 = 83 screws per seam

Step 2 Determine the number of longitudinal seams:

Number of jacket sections per course = 5
therefore,
Number of vertical seams = 5

Step 3 Calculate the total number of screws:

Total number of screws =
Number of seams × screws per seam =
5 seams × 83 screws per seam =
415 screws (minimum)

A box of 500 screws should be adequate for this example.

2.2.2 Installing Roll Jacketing: Vertical Vessel

Installing roll jacketing on cylindrical vessels is very similar to the methods used for jacketing insulated piping. Flat and shallow-corrugated roll jacket is appropriate for vessels up to about 8' (2.4 m) in diameter. For larger vessels, even low bumps and hollow spots in the insulation can make it difficult for these materials to lie flat without causing gaps at the overlap seams. The use of these materials should be evaluated on a case basis for larger vessels, depending on the smoothness and rigidity of the underlying insulation.

As penetrations and fittings are encountered, cutouts must be made to ensure a smooth application of jacket. Make sure that at least a 2" (5 cm) overlap is included in flashing or jacket on connected systems, flanges, and access hole covers.

The following describes a general approach for installing jacket on vertical vessels.

Step 1 Stage the pre-cut jacket sections and securement accessories convenient to the work site.

Step 2 Stretch a bungee cord around the perimeter of the vessel about half the width of a jacket section above the lower tangent line of the insulation. Make it taught enough to hold jacket sections against the insulation without damaging the insulation or creasing the sheet metal.

Step 3 Position the first section at the proper height. Have assistants pull out the bungee cord while slipping the sheet under it into position.

> **CAUTION**
> The level and plumb condition of the first course of jacket sections is critical to the successful completion of the entire job. Take your time, and check and recheck the level of the first-course seams before securing it in its final position.

Step 4 Continue positioning additional sections along the lower tangent line of the vessel. Stretch the bungee cord to allow for room to slip the next section in, making sure that the correct end overlap is made.

Step 5 The final section will likely have to be cut to fit. Measure between the ends of the last-placed jacket section and the first section. Add two times the overlap to this measure, then cut the final section of jacket to total length. Make sure that the cut is square to the length of the jacket. Then position it under the bungee cord.

Step 6 When all sections of the first course are in position, go back and recheck the level of each piece and that all overlaps are correct. Adjust the sections as needed.

Step 7 Measure out the banding to secure the first course. This can be done by walking around the base of the vessel while unreeling the banding from the spool. With the help of other workers, position the band along the bottom edge of the first course of jacket and cut the band with about 6" (15 cm) of overlap. Install a band seal and apply tension to the bottom band, but do not finish because this band will need to be loosened to complete the bottom tank head.

NOTE

For a neat installation, band seals should be positioned at the vertical edge lap between sections and all seals should align in a straight row up the seam. The bands themselves should be leveled and made parallel to the adjacent jacket edge before tightening to produce a pleasing appearance.

Step 8 Install additional bands above the bottom band to secure the first course of jacket, observing the maximum-allowed spacing between bands. Tension the bands without creasing the jacket. Do not band the top edge of the jacket.

Step 9 Fabricate S-clips from banding to provide the specified overlap at the circumferential edges. Take extra care to ensure the S-clip lengths are the same, because they will determine how level each course is. Enough S-clips should be made to provide adequate support to the jacketing system.

Step 10 After hanging the S-clips along the top edge of the first course of jacket at the recommended spacing, install the second course of jacket supported by the clips following Steps 2 through 8. Again, check for level and adjust where needed. Make sure the vertical seams are correctly aligned and vertical by level. Place the bottom band for the second course inside the S-clips along the lower edge before tightening and securing.

Step 11 Install the remaining full courses of jacket following Steps 2 through 10. For vertical vessels, scaffolding or lifts may be required to reach the upper courses. If scaffolding is used, preposition the required jacket sections on each level before installing a course.

CAUTION

When working at elevated heights, always wear proper fall-protection equipment as required by job site requirements.

Step 12 The width of the top course of sidewall vessel jacket will likely have to be trimmed to fit the remaining space. Measure from the top of the last-installed jacket to the top edge of the sidewall insulation. Add the circumferential edge overlap, then cut all the top-course sections to this width.

Step 13 Install the top course of jacket as in the preceding steps. Leave the top band off. The top edge will be banded when the vessel head jacket or a **beauty band** is installed. See Section 2.4.2.

Step 14 On hot vessels and those without a vapor retarder boundary, secure the vertical seams with screws at the specified spacing between bands. Screw installation may be done as a separate step after all sidewall jacket is in place or completed with each course of jacket.

S-Clip Jig

Large jacketing projects will require lots of S-clips. To easily make many S-clips the same size, create a jig from a 2 × 4 and two nails spaced apart the required lap between courses. The number of S-clips required can be quickly made and stored in a bag until use.

During the process of installing jacket on a vertical vessel, it is likely that you will have to deal with brackets, pipes, flanges, manholes, and other jacket penetrations. The methods for locating and cutting the necessary holes in jacket for these obstructions are discussed in detail in NCCER Module 19310, *Jacketing Fabrication—Piping and Fittings*. Review those methods, if needed. Vessel jacket is normally installed before jacket for attached piping and other fittings to provide for water shedding. Also, make sure the tank jacketing is not in contact with the surface of piping or fittings because of heat transfer. In all cases, jacket penetrations must be flashed and caulked to provide a continuous protective seal for the jacket.

2.2.3 Installing Roll Jacketing: Horizontal Vessel

The general arrangement of jacketing on horizontal vessels is similar to that on vertical vessels. The measurement and securement of the jacket sections are the same. However, some changes to the order of installation are needed because the direction of gravity in relation to a course of jacket is different.

Jacketing may begin at either end of the vessel. The first section in a course is positioned at the underside of the vessel and is held in place by a one or more bungee cords. Additional sections are positioned on both sides of the vessel, building from the bottom to the top. This ensures that the horizontal overlaps will shed water as in *Figure 9*.

NOTE

For a horizontal vessel, the worker should consider planning the jacket installation so that there is an even number of equal-length sections per course. This approach will permit arranging the sections to avoid horizontal seams at the very top and bottom of the cylindrical vessel. Seams at these locations can trap water.

For a horizontal vessel, the end laps between sections must be checked for perfect level, since any misalignment or inclination of the completed seams will be readily visible to the eye. As each course is completed and secured, the next course is installed with the proper overlap in the same way. No S-clips are required between courses as is the case with vertical vessels, since the banding

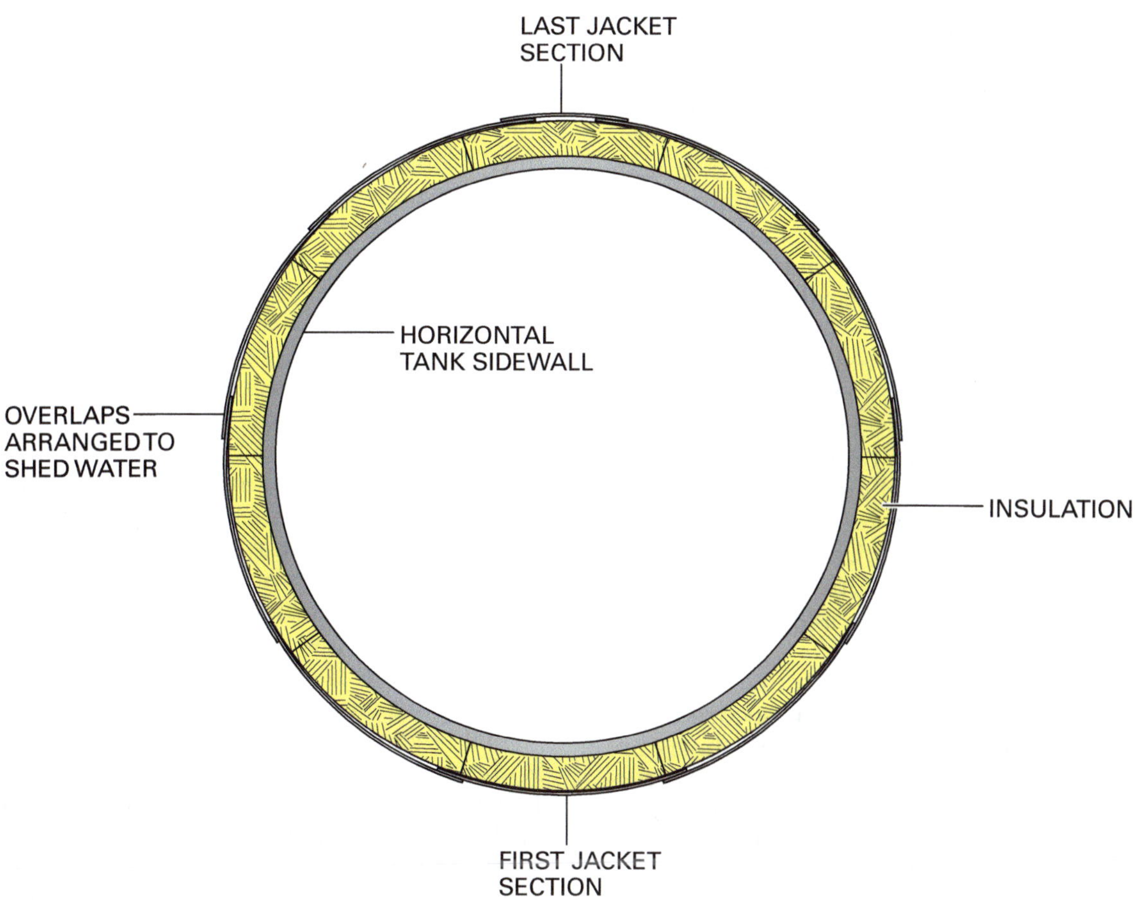

Figure 9 Jacket installation sequence on a horizontal vessel.

and screws prevent the movement of jacket sections around the vessel perimeter. Complete the sidewall jacketing as previously described.

Horizontal vessel jacket circumferential seams can't rely entirely on lap seams to prevent water from penetrating through the lap. Some form of mastic or sealant is required to make the circumferential seams water- and weather-tight. The sealant must be applied to the lap surface before the next course of jacket is installed. Do not apply a bead of sealant to the joint after installation. The bead will eventually fail, allowing water entry.

2.3.0 Installing Sheet and Corrugated Jacket

Heavier gauge flat-sheet and corrugated jacket materials are usually specified for large-diameter vessels. These have greater rigidity and can better resist the highly-tensioned securement bands required for larger vessels. Another factor for outside installation is resistance to **wind forces**. Stiffer flat metal sheets and corrugated sheet metal are better able to avoid deforming under high winds than thinner roll materials.

As noted earlier, corrugated jacket is usually not specified for horizontal vessel installations because it tends to trap water on the upper surfaces of the vessel within the corrugations. Even so, this may be acceptable if the corrugated material is otherwise suitable for the application.

Corrugated sheet and heavy flat jacket is typically applied to the sidewalls of large-diameter vertical vessels. The methods for estimating and installing these materials are similar to those for roll jacket. Flat jacket sheets may be installed with the long dimension vertical or horizontal. Corrugated jacket can only be installed with the corrugations parallel to the axis of the vessel so that it can flex to follow the surface.

Since the sizes of sheet jacket sections are typically larger than for roll materials, it is even more critical to make sure the sections are plumb and level before final securement. Any departure from level or plumb will quickly show as additional courses are installed. Out-of-plump corrugated jacket is especially noticeable because of the numerous parallel ridges.

In vertical vessel installations, courses are usually supported by S-clips as well as secured with banding and screws. Deep-corrugated sheets are usually given a longitudinal overlap of two corrugations and 3" (8 cm) overlaps at the ends. Sheet-metal screws sized for the thickness of the material may be installed in the valleys of corrugations around the circumferential seams under the securement bands, since the bands ride on top of the corrugations. When using screws at the horizontal laps, they must be installed before the banding.

> **NOTE**
>
> It is generally not a good practice to use corrugated jacket from different manufacturers. Each company has its own corrugation equipment and there are often large-enough differences that corrugations from different sources will not overlap properly.

2.4.0 Installing Jacket on Vessel Heads

Some of the possible types of vessel heads are described in NCCER Module 19207, *Board and Block Insulation*. *Figure 10* shows that heads of vessels may be flat, domed, spheroidal, have a cone roof, or other shapes.

The arrangement of insulation jacket on the vessel head depends on the form of the head. It will also depend to some extent on how the insulation is arranged where the vessel head and sidewall meet. For flat-topped vessels and those without head flanges or insulation support rings near the tangent line, the sidewall insulation should extend above the vessel's tangent line to overlap the edges of the board insulation on top. For a conical, domed, or other form of **dished top**, especially in larger vessels, the sidewall insulation may end at a support ring at or below the tangent line. Some domed vessel heads may contain a **knuckle** near their perimeters, which has a constant-radius curve that is a small fraction of the radius of the dished part of the head. Typically, the knuckle is welded to a short cylindrical skirt having the same diameter as the vessel as part of the head fabrication. The whole assembly is then welded to the sidewall or bolted to a flange. Elliptical vessel heads generally have a central, gently-curved, dished region toward the middle of the head that becomes more sharply-curved near the tangent line, as shown in *Figure 11*.

Vertical vessels may also have a bottom head if the vessel is supported on legs. For horizontal vessels and equipment, the heads at both ends may be similar. Vessel equipment heads may be heavily flanged and have complex shapes that will require custom jacket layout and fabrication.

Flat heads on insulated vessels may be jacketed with smooth or stucco-embossed sheet jacket. Fabrication and installation can usually be done at the worksite, using measurements taken as the work proceeds. Flat sheet material may be fastened together with a **standing seam** to improve

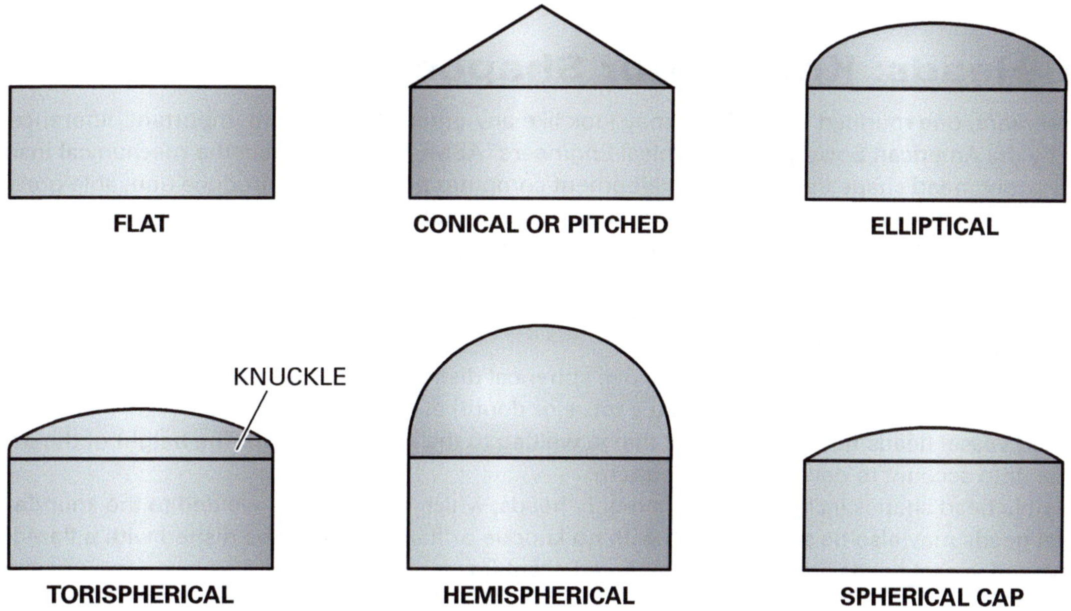

Figure 10 Examples of vessel head forms.

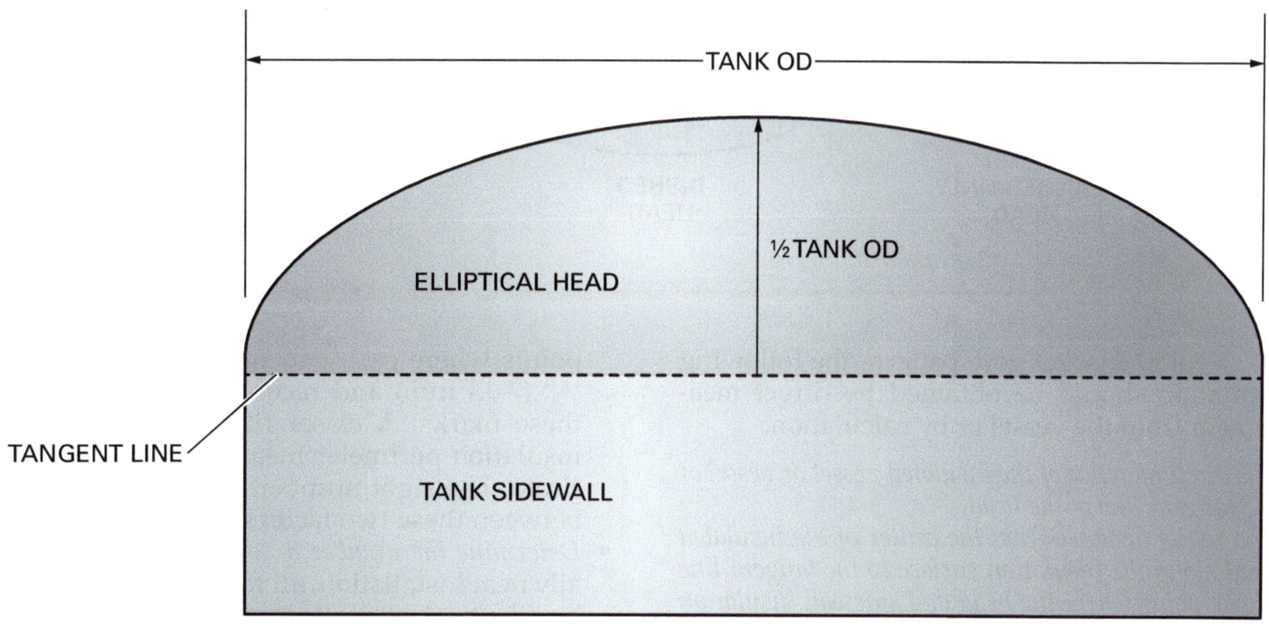

Figure 11 Profile of a typical elliptical vessel head.

rigidity and weather proofing of the seams. Generally, large, flat-roofed vessels should have sub-girts or stud rods welded to the vessel head, similar to lagging. The insulation jacket can be secured to these accessories to resist wind forces.

Mechanical insulators will need to develop gore patterns for vessels with some form of dished or conical head before the jacket can be cut and installed. The layout and fabrication of gores usually must be done in a shop having special sheet-metal forming equipment. The development and installation of jacket gores for an elliptical vessel head will be the subject of the next sections.

2.4.1 Developing the Jacket Plan for an Elliptical Vessel Head

The general approach to developing insulation gores for an elliptical head is very similar to developing the insulation patterns found in NCCER Module 19207, *Board and Block Insulation*. The preferred method for generating a pattern for a vessel head gore is by the use of special engineering software. These programs are typically used by custom insulation and sheet-metal fabricators. If a gore-pattern program is not available, the mechanical insulator needs to know how to draft up a pattern from field measurements.

Vessel Heads: Know Your Shapes

To the unobservant, one rounded vessel head may look like any other, but there are important differences established by the American Society of Mechanical Engineers (ASME) that can affect the mechanical insulator. Entering the wrong head shape into a gore-development computer program can produce unusable gore segments at the job site. As with anything having to do with computers, "garbage in, garbage out."

The *hemispherical head* is the simplest domed vessel head shape to understand. It's half a sphere. The radius of the head curvature is constant at every point.

The *elliptical head* is geometrically a flattened hemisphere. The vertical height of the dish at its center is about half that of the vessel's horizontal radius.

A *torispherical head* is a combination of a shallow, spherical dish region in the middle merging with a sharper constant curve or knuckle (part of a shape called a torus or donut) at the head perimeter.

Many domed vessel heads have a cylindrical flange welded to the knuckle or dish. The height of the flange must often be taken into account to develop a gore pattern.

Other possible head shapes include the flanged-only heads, which are flat disks welded to the knuckle and its flange. Vessel heads may also be simply a dish with no knuckle or flange, or may be dished with a flared flange. These last type of vessel heads are illustrated in the adjacent figure.

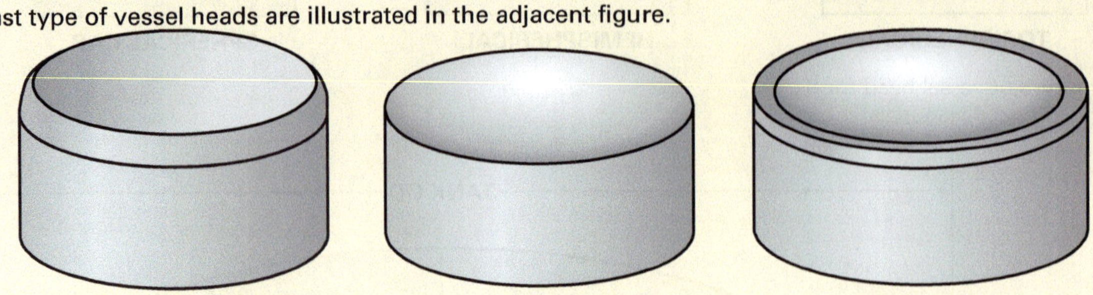

To develop a jacket gore pattern, the following information should be obtained by direct measurement from the vessel or by calculation:

- *The circumference of the insulated vessel or vessel at the tangent line of the head.*
- *The radial distance from the center of the insulated head along the insulation surface to the tangent line or the point where the head and sidewall insulation meet.* If there is a pipe or other obstruction at the head's center, then measure the radial distance from its surface. Measure its diameter or circumference to find its radius and add this value to the head's radial measurement.
- *The width of the gore at the vessel head perimeter.* Place a straight edge against the insulation tangent line and parallel to it. Note the sizes of the gaps between the straight edge and the insulation surface on both sides of where it touches. To minimize the size of the voids under the jacket, these gaps should be less than about $\frac{1}{2}$" (13 mm). Mark the insulation at the points where each gap under the ruler is $\frac{1}{4}$"–$\frac{1}{2}$" (7–13 mm) and record the width between these marks. A closer fit of the jacket to the insulation perimeter means more gores to fabricate. The right number of gores is a tradeoff between these two factors.
- *Determine the number of whole gores.* For a visually neat installation, all the jacket gores should be identical. Therefore, you need to have a whole number of gores. Divide the circumference of the insulated vessel head by the width

NOTE

While normally not a problem when developing an insulation jacket pattern, the maximum width of the gore can't be any wider than the width of the jacket material. More to the point, the maximum gore width, including the allowances for crimp and beading, must be narrower than the jacket material.

measurement obtained in Item 3 above. Round down or up to the nearest whole number.

- *The width of the gore at certain radial distances from the head center.* Divide the circumference of the vessel head by the whole number of gores to be used to find the width of a gore at the head perimeter. Mark off this distance along the head perimeter on the insulation. Snap chalk lines from these marks to the center of the vessel head. Measure the distances between the gore side lines at regular distances from the center of the vessel head. Depending on the size and shape of the vessel head, fewer measurements are needed where the vessel curvature is gentle; more measurements are needed where the curvature is sharp. Record the radial distance for each measurement and the distance between the chalk lines. Refer to *Figure 12*.

> **NOTE**
> A valid gore pattern can be generated with at least four measurements not including the knuckle, if present. At the vessel head knuckle, take a measurement just above the knuckle. For a sharp knuckle, you can assume that there is little or no spread between the chalk lines from the knuckle to the insulated sidewall. For more gradual curves, such as for elliptical heads, you can make additional width measurements. Again, record the radial surface distance from the head's center for each measurement. The more measurements taken, the more accurate will be the shape of the gore and the better it will fit.

The following method assumes you are creating a full-sized pattern, which will allow direct measurements and estimates. It is also assumed that there are no obstructions or penetrations to work around. Refer to *Figure 13*. The pattern should be created on a stiff durable material, such as sheet metal or thin plywood. If you do not have the space or materials to lay out a full-sized pattern, then you will need to make an accurate scale drawing to provide cutting dimensions marked directly on the jacket material.

> **NOTE**
> See NCCER Module 19212, *Trade Math and Layout*, for a review of scale-drawing methods.

Step 1 Draw a straight line representing the center line of the gore. It should be longer than any radial distance recorded in your field notes.

Step 2 Select one end of the center line to represent the center of the vessel head. This end will be considered the head of the gore pattern. The bottom of the pattern will be called the heel. Stretch a tape measure from the top of the center line to the

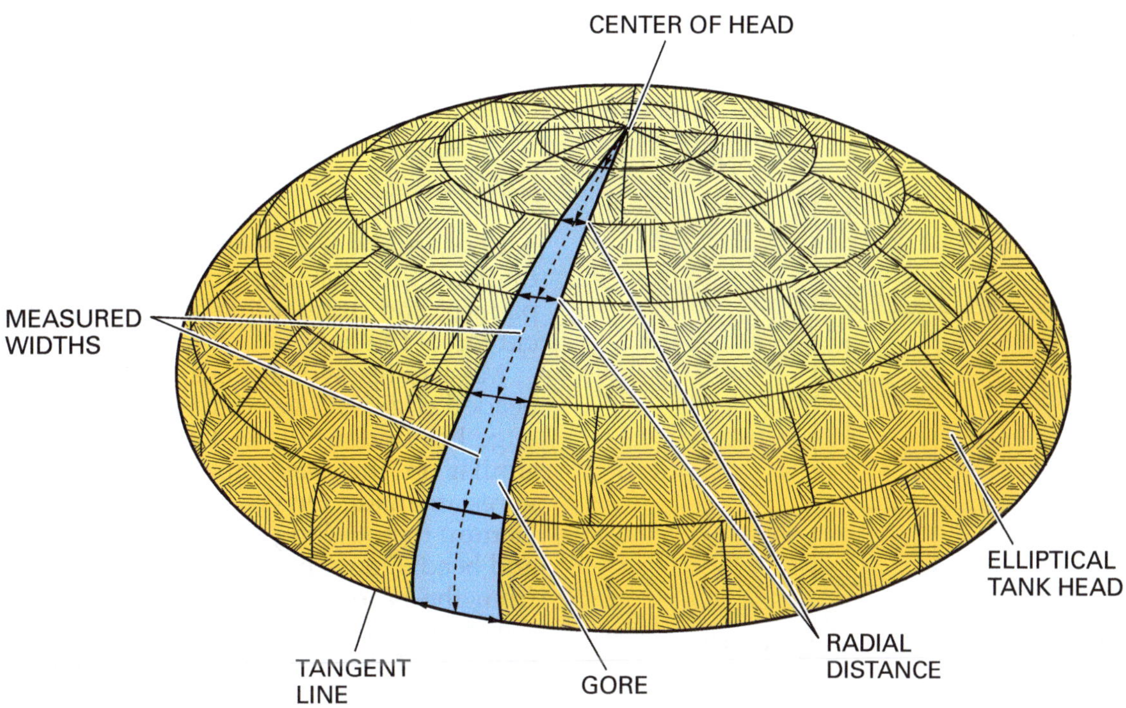

Figure 12 Taking measurements for a gore pattern.

first radial distance and mark the center line. Repeat for all the other radial distances you recorded.

Step 3 For each radial distance, draw a line squared to the center line. Measure half the corresponding gore width on both sides of the center line. Mark these points on the perpendicular line. Repeat for all the radial distance marks.

Step 4 Draw a smooth curve from the top of the center line to the bottom mark on one side through the marks made in Step 3. A flexible wooden batten, metal yardstick, or tinner's rule will be useful for this. Then repeat to create mirror-image curve on the other side of the center line. Some professional mechanics make it a practice to create the curve for one side of the gore as described in Steps 1 through 4. Then they cut out the pattern, and flip it over to make the curve on the other side. This ensures that the two curves are identical.

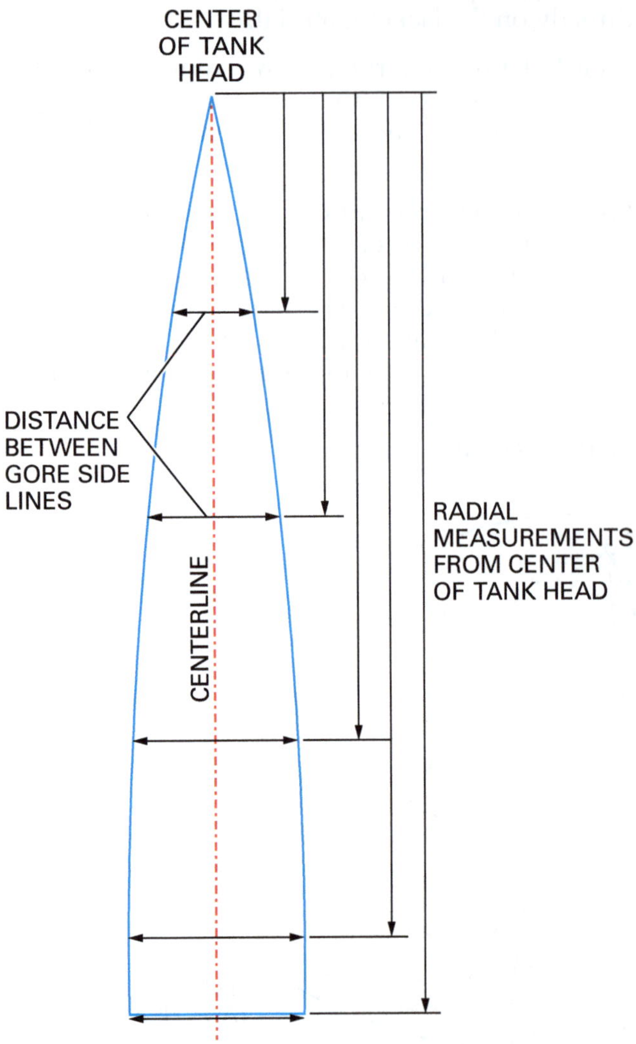

Figure 13 Layout of a gore pattern for a vessel's elliptical head.

The pattern just created is for the finished dimensions of the jacket gore. However, standard installation methods will require some adjustments to the pattern. Normally, vessel gores are anchored at the head's center using a **centerpiece**—a disk of jacket material. There are no set rules for the size of the centerpiece. But there are several things to consider:

- The head ends of the gores should be several inches wide. This will help ease their installation and provide room for making any adjustments necessary.
- You can use one or two screws to secure the top end of the gore. Insert the screws next to the overlapping beads and crimps.
- The centerpiece should extend beyond the location of the screws far enough to provide material to hold the screws. *Figure 14* shows these considerations.

Taking into account these factors, the centerpiece radius should be figured on the dimensions shown in *Figure 14*. The size of the centerpiece will be different for each vessel head installation.

The head end of the gore needs to overlap the centerpiece only 2" (5 cm). Back to the gore pattern, subtract 2" from the centerpiece radius, measure this distance from the point of the gore pattern, and draw a line across it to indicate the cut top of the gore.

> **NOTE**
>
> If the vessel has a central pipe, manhole, or other obstruction, then the original pattern will have to be shortened by the radius of the interference. Normally, a ring of metal jacket is cut to fit around the perimeter of the obstruction to which the heads of the gores are secured.

Most gores will have a crimped edge and a beaded edge as a standard practice. The crimped edge lies under the bead of the adjacent section, similar to the way pipe mitered elbow jacket gores are made. On the gore pattern add ¼"–1" (6–25 mm) to both long edges to allow for the crimp and bead. Draw the final cut edge at this distance from the base gore pattern. Designate one edge the crimp and the other edge the bead on the pattern for reference. *Figure 15* shows an example of a complete gore template pattern. Carefully cut

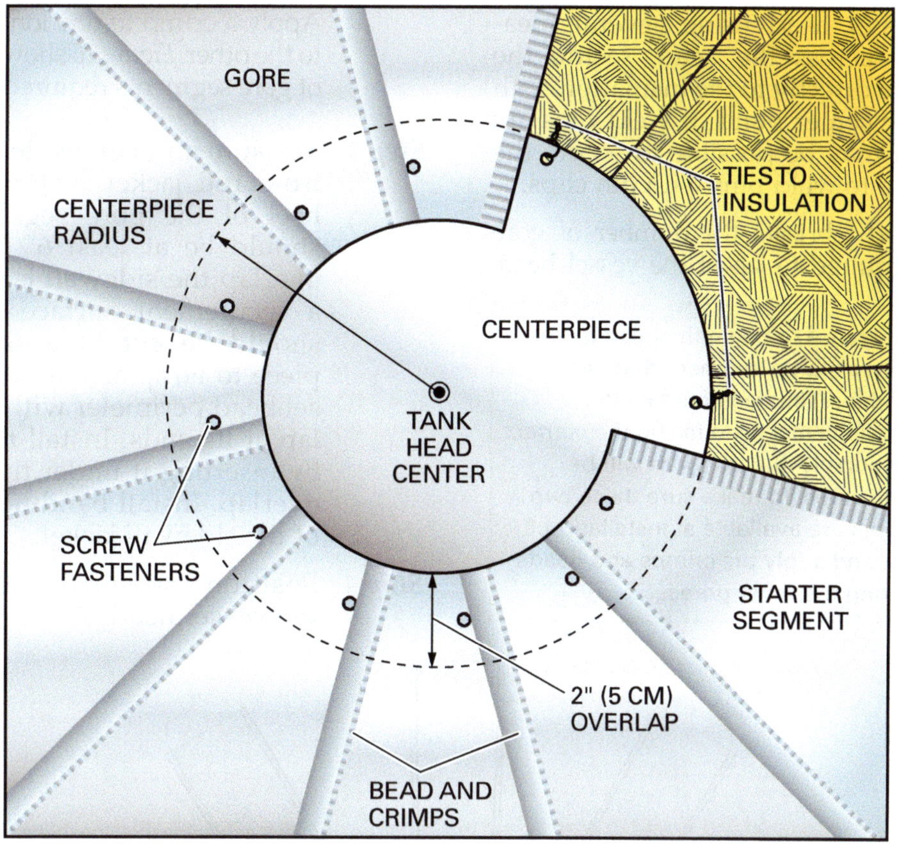

Figure 14 Factors to consider when sizing the jacket centerpiece.

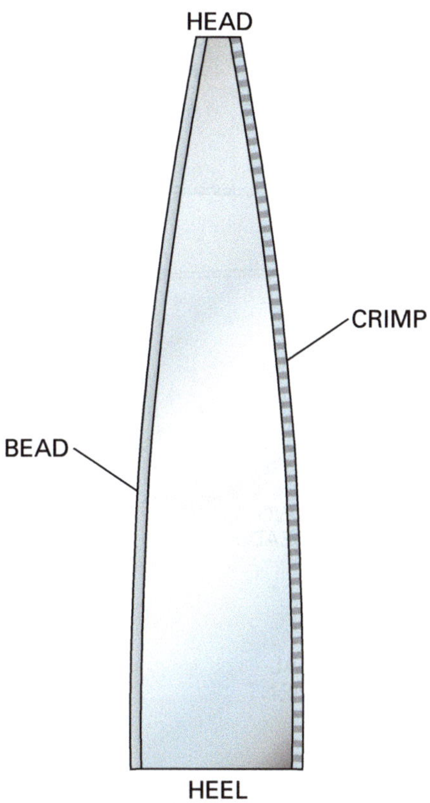

Figure 15 Completed vessel head jacket gore template.

NOTE

For accurately determining the amount of edge required for a full bead, you may want to do a trial bead on a scrap piece of sheet metal. Scribe parallel lines along a straight edge of metal at $^{1}/_{16}$" intervals out to an inch or more from the edge. Then run the scrap through the beader with an edge guide to find exactly how much metal is required to form a bead or crimp with that machine. Use this value for extending the gore pattern on the beaded edge.

Crimpers usually have an adjustable width guide, so setting the width of the crimp is straight-forward.

out the template and smooth the edges to prepare it for use.

2.4.2 Fabrication and Installation of Jacket for a Vertical Vessel Head

The following process is suggested as a method for installing jacket on the elliptical head of a vertical vessel.

Step 1 Measure and cut a circular centerpiece from jacket stock material as described above. After verifying the center point of the vessel head using several measurements across the head, secure the centerpiece directly to the insulation with wire at the center point. If weld pins are present, it may be possible to secure the centerpiece to weld pins with pin clips.

Step 2 Cut out and fabricate the number of gore sections required to jacket the vessel head.

> **NOTE**
> It is standard practice to begin the head jacket installation with a starter segment that has two crimped edges. This requires that the second piece will have a bead facing the starter segment. Additionally, the last gore will be beaded on both sides. To make sure these two special segments are available at installation, set them aside and apply the crimps and beads before fabricating the other gores.

Apply a crimp to one long edge and a bead to the other. *Figure 16* shows the arrangement of gore segments required for installation.

Step 3 Measure and cut a sidewall anchor strip from roll jacket material to secure the heel of the gore segments. The band should be at least 6" (15 cm) wide to overlap the sidewall jacket and provide a securement surface for the gore. It should be cut in a single continuous piece to wrap around the insulated vessel head perimeter with a 3" (8 cm) overlap at its ends. Install the sidewall strip by inserting it under the sidewall jacket overlap. Install banding at the top edge of the sidewall jacket.

Step 4 Position the starter segment gore on the vessel head. After checking the de-

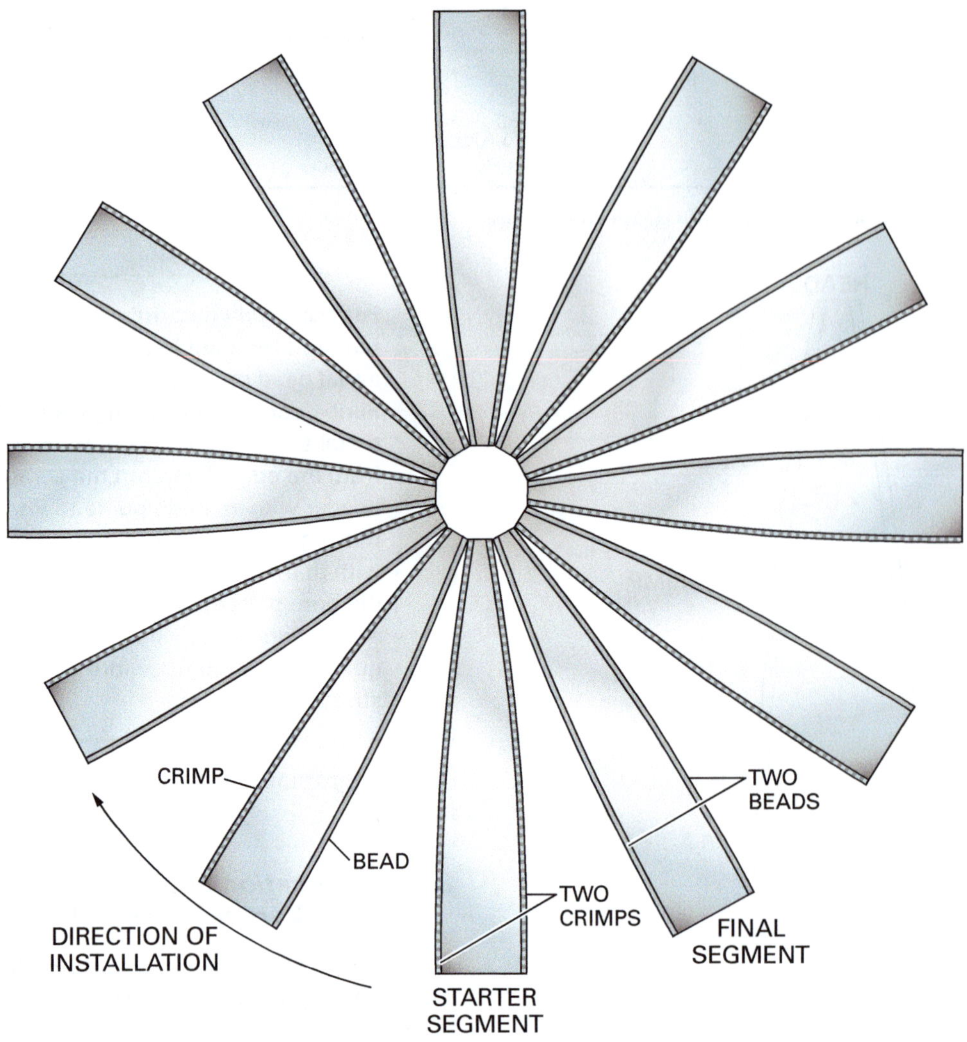

Figure 16 Arrangement of gore segments for jacketing the head of a vertical vessel. Note the positions of edge crimps and beads.

CAUTION

Be careful when working over applied insulation when installing head jacketing, especially on larger vessels. To avoid damaging compressible insulation (and impaling yourself on insulation pins, if present), you may need to uses sheets of plywood to help distribute your weight when working near the center of the vessel head.

signed overlap exists, secure the head of the segment to the centerpiece with screws.

Step 5 Smooth the starter segment down to its heel edge and align it radially on the vessel head. Secure the heel of the gore to the sidewall anchor strip with screws.

Step 6 Position the second gore segment next to the first. The beaded edge should overlap the crimp of the starter segment. Secure the top of the gore at the centerpiece by inserting a screw next to the bead and crimp of the starter piece. Carefully wrap the segment over the insulation surface to the heel of the segment. Secure the heel to the sidewall band with screws. For a neat and uniform appearance, care should be taken that all gores are evenly spaced around the head perimeter.

Step 7 Secure the beaded edge of the gore segment to the underlying crimped edge by driving self-drilling screws next to the bead. Set the screws at equal spacing no more than 6" (15 cm) apart.

Step 8 Proceed with installing the remaining gore segments as in Steps 6 and 7.

Step 9 The final gore segment has two beaded edges. Center the segment over the remaining space and the crimped edges, then secure it as the other segments.

Step 10 Neatly cut a circular center cap that covers the narrow ends of the gores and the centerpiece. The center cap may be a simply flat disk or it can be slightly conical to help shed water [*Figure 17(A)*]. Some fabricators apply fancy beading, crimping, and other features for a unique look [*Figure 17(B)*]. The center cap is secured with screws to the jacket underneath. For weatherproofing, a layer of sealant should be applied before installing.

Step 11 Install a beauty band over the heels of the gores to provide a finished appearance for the head jacket installation. Like the sidewall anchor strip, it should be cut from a single continuous piece of roll jacket. One or both of its edges may be beaded and/or crimped as in *Figure 18*. The beauty band may be secured with bands, screws (on hot systems), or both.

2.4.3 Fabrication and Installation of Jacket for a Horizontal Vessel Head

The jacket applied to the heads of vertical vessels naturally sheds water because of the arrangement of beads and crimps at the overlapping seams. However, installing jacket on a horizontal vessel head using the same method will create seams where water entry can easily occur. The fabrication of the gores and their order of installation must take this problem into account so that proper water shedding is maintained.

After the sidewall of the vessel is jacketed, the head gore segments are developed and cut out

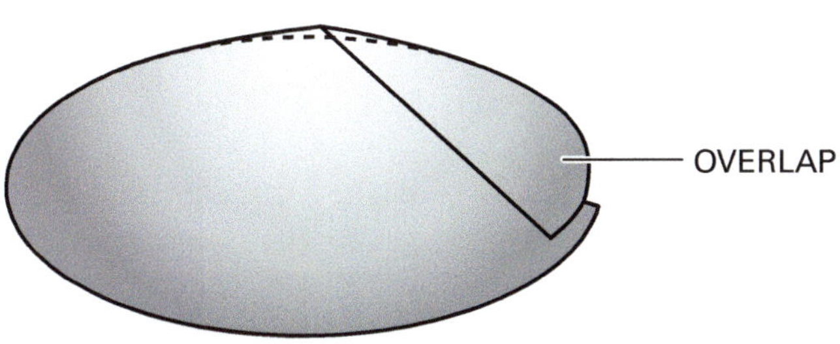

(A) CONICAL CENTER CAP

(B) FANCY CENTER CAP

Figure 17 Examples of vessel head jacket center caps.

Figure 18 Beauty band on a vessel head.

as described in Section 2.4.2 and the first gore to be installed will have crimps applied to both edges. However, this starter segment will be installed at the lowest point on the underside of the horizontal head, secured to the centerpiece and sidewall jacket. The next two gores will be installed on opposite sides of the starter segment. The crimp and beads will be on opposite sides for these segments. When installed, the beaded edges overlap the crimps of the starter segment. The remaining gore segments are installed in a similar way, with the beads at the lower edges of the segments overlapping the crimps on the upper edges of the gores below, as in *Figure 19*. To finish, the final gore segment has beads on both edges, which overlap the crimped edges of the last-installed gore segments. This approach will ensure that no falling water can penetrate under the beaded edge and enter the insulation.

As always, a proper sealant should be applied under the gore laps before securing them in place. Finishing off the installation with the center cap and beauty band must be done carefully to avoid leaving paths for water entry.

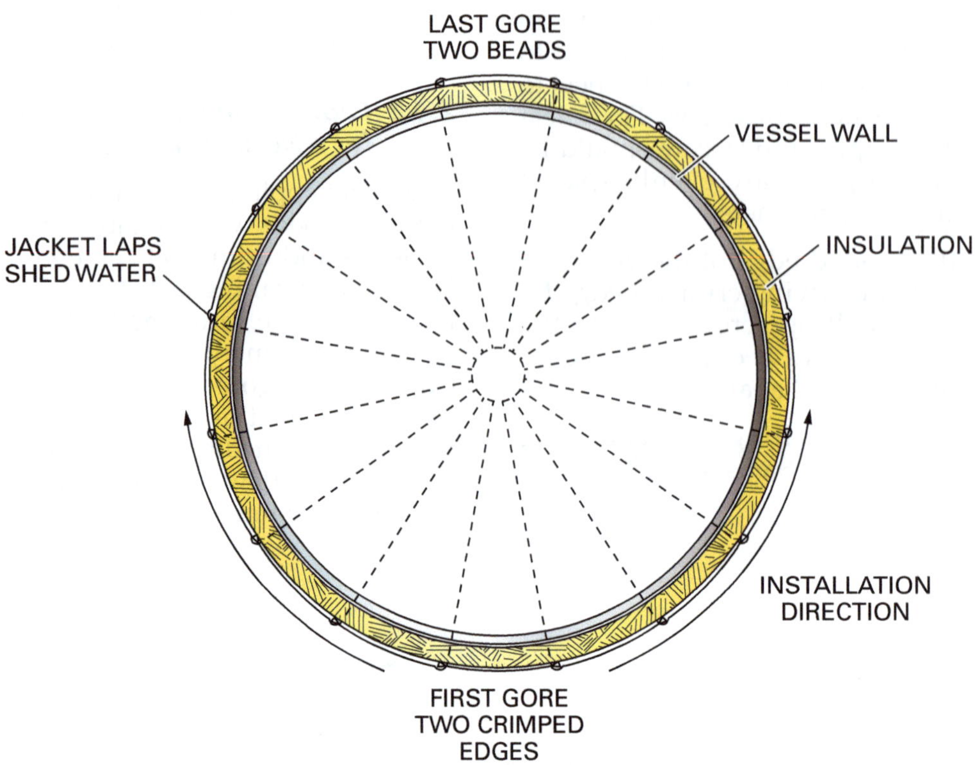

Figure 19 Jacket sequence on the head of a horizontal vessel.

2.4.4 Jacket Penetrations

Ideally, a vessel head will have no penetrations and the jacketing installation will proceed as described in the steps above. However, vessel heads often have a variety of pipes, support stanchions, lifting pad eyes, and so on. *Figure 20* shows an extreme example with many such penetrations. Jacket gores must be fabricated to provide openings for these obstructions while creating a smooth protective sheathing for the insulation.

NCCER Module 19310, *Jacketing Fabrication—Piping and Fittings,* provides detailed instructions for measuring and cutting jacket for penetrations

Figure 20 A jacketed vessel head with many penetrations.

in pipe jacket, and the same principles apply to vessel heads. Work carefully to create close-fitting covers, then seal all seams and joints between jacket and the penetrating fittings with an approved sealant or mastic.

Additional Resources

National Industrial and Commercial Insulation Standards Manual, Midwest Insulation Contractors Association (MICA).

National Insulation Association (NIA) website offers resources for products and training. **www.insulation.org**

Mechanical Insulation Design Guide, National Institute of Building Sciences. Available at **www.wbdg.org/guides-specifications/mechanical-insulation-design-guide**

2.0.0 Section Review

1. For large-diameter cylindrical vessels, it is common for jacket installations to have which of the following features?
   a. Elastic securement bands spaced no more than 18" (46 cm) apart to take up vessel expansion
   b. Narrow-corrugated sheet jacket installed with corrugations horizontal
   c. Securement band expansion springs at a maximum of 25' (7.6 m) apart
   d. Large cryogenic vessel jacket must be secured with screws no more than 12" (30 cm) apart

2. Which of the listed quantities is *not* necessary for estimating a vessel's sidewall jacket quantities?
   a. The thickness of the jacket
   b. The required section overlaps
   c. The width of the jacket sections
   d. The insulated circumference of the vessel

3. If the tan-to-tan height of an upright vessel is 215" (546 cm), how many securement bands will be required for the sidewall jacket of the vessel, assuming the standard spacing of 12" (30.5 cm) for bands?
   a. 13
   b. 19
   c. 25
   d. 31

4. Which of the listed items would *not* be considered an advantage of corrugated insulation jacket?
   a. It is more rigid and has better resistance to wind forces.
   b. Produces less glare from reflected sunlight.
   c. It is the preferred jacket for horizontal vessel installations.
   d. Better-suited for large-diameter vessel insulation jacket.

5. Which of the listed factors is the limiting one when deciding how many gores to fabricate for a vessel head jacket installation?
   a. The diameter of the vessel head
   b. The width of the sheet or roll jacket material
   c. The length of the gore
   d. The size of the head knuckle

SUMMARY

The primary reason to install jacketing material is to protect the integrity of the insulation system. This concept has been reviewed in many of the modules throughout the NCCER Mechanical Insulating curriculum. If the insulation gets wet, the insulating system is ruined and corrosion under insulation can occur. In cold systems, the jacket is the first boundary to protect the overall vapor retarder system.

Installing jacketing in a professional manner is also important because the finished jacketing is what everyone sees. Unprofessional workmanship can reflect negatively on the entire installation. This module covers in detail the different forms of metal jacketing that can be used on vessels and equipment, the different thicknesses of materials, the types of securements, useful layout procedures, and applicable installation methods. Jacket fabrication and installation is partly mechanical but it is also part artistry. It takes knowledge, a sense of shape and form, as well as much practice to do this kind of work well.

Review Questions

1. One important advantage of corrugated jacket over flat jacket materials in outside applications is that it _____.
   a. is more corrosion-proof
   b. is less rigid and more easily installed
   c. reflects less sunlight glare into a person's eyes
   d. is has less expansion and contraction

2. Which of the following statements is correct regarding sheet-metal jacket?
   a. Jacket for large vessels normally requires rolling before installation.
   b. Joints between jacket sections should be caulked with a bead of sealant after installation.
   c. It is standard for allowing at least a 3" (8 cm) overlap at all edges on vessels.
   d. Screw fasteners should be spaced at least 12" (31 cm) apart.

3. For which of the following vessel jacket applications would S-clips be used?
   a. Supporting vertical vessel sidewall jacketing
   b. Positioning vessel head gores
   c. Supporting horizontal vessel sidewall jacketing
   d. Sealing jacket bands after they are tensioned

4. Which of the following is *not* considered a type of sheet-metal forming?
   a. Beading
   b. Crimping
   c. Cutting
   d. Hemming

5. To secure a hemmed edge in sheet metal, which of the following is required?
   a. Screws
   b. Rivets
   c. Crimps
   d. Nothing else

6. Longitudinal overlaps for deep-corrugated sheet jacket should be at least _____.
   a. one corrugation
   b. two corrugations
   c. three corrugations
   d. 3" (8 cm)

7. $^3/_{16}$" (4.8 mm) corrugated jacket should overlap longitudinal seams by _____.
   a. the same as flat jacket
   b. at least one corrugation
   c. at least two corrugations
   d. at least three corrugations

8. As a minimum, the insulation mechanic must know which of the listed dimensions to estimate a vessel's sidewall jacket requirements?
   a. Overall length of the vessel and its outer diameter
   b. The vessel's insulated diameter and insulated tan-to-tan length
   c. The vessel's tan-to-tan length and the type of vessel head
   d. The thickness of insulation and the vessel's outer diameter

9. Estimate the minimum number of screws required to secure the longitudinal seams of sidewall flat roll jacket on a horizontal vessel. The vessel has a tan-to-tan length of 184" (467 cm). There are 6 sections of jacket per course. Assume the standard maximum screw spacing of 6" (15.2 cm).
   a. 102
   b. 122
   c. 162
   d. 192

10. Which of the choices is an important limitation for corrugated sheet jacket?
    a. It is more rigid than standard flat jacket.
    b. It can only be installed on cylindrical vessels with its corrugations parallel to the vessel's axis.
    c. It shows hail damage more visibly than flat jacket.
    d. S-clips do not work with corrugated sheets.

11. When installing deep-corrugated sheet jacket on vertical vessels, _____.
    a. only banding may be used to secure the jacket
    b. corrugations may run horizontal or vertical, as desired
    c. screws must be installed at horizontal seams before banding
    d. all seams have 3" (8 cm) overlaps

12. Which of the choices best describes a dished vessel head?
    a. Upright-bowl shaped
    b. Shallow plate shaped
    c. Flat disk shaped
    d. Upside-down bowl shaped

13. When installing vessel head jacket, to what piece of jacket are the heads of the gore segments secured?
    a. Starter segment
    b. Sidewall anchor strip
    c. Center cap
    d. Centerpiece

14. A neat finished appearance is given to the center of a dished vessel head jacket installation with a _____.
    a. beauty band
    b. centerpiece
    c. center cap
    d. coating of mastic

15. Which of the following is the biggest difference between the approaches to jacketing horizontal and vertical vessels?
    a. The order of jacket section installation
    b. The types of jacket used
    c. The types of fasteners required
    d. The spacing of the securement bands

Trade Terms Introduced in This Module

Beauty band: A narrow strip of sheet-metal jacket that is installed over the rough edges of adjacent jacket to provide a finished appearance. Often supplied with beading, crimped edges, or both.

Blind rivets: Fasteners that are deformed when installed to hold two sheets or plates of material together. A blind rivet can be installed from only one side of the surfaces. Tubular aluminum and steel blind rivets are commonly called pop rivets.

Center cap: A circular or cone-shaped piece of jacket sheet metal installed at the center of a vessel head to finish off the jacket installation there.

Centerpiece: A disk of jacket material fastened to the insulation at the center of a vessel's head to anchor the inner ends of the gore segments during installation.

Circumference: The distance around the perimeter of a circle or a circular object. Measured in units of length, such as inches, feet, centimeters, or meters.

Develop: As used in geometric layouts, to draw the true shape of a three-dimensional surface in two dimensions. It is an important part of fabricating jacket for three-dimensional vessels and equipment using two-dimensional sheet metal.

Dished top: When referring to a vessel head, it is a shape similar to an upside-down bowl; a low, domed appearance compared to a flat-topped vessel.

Expansion spring: A spring-loaded fitting installed in securement bands for jacketing to even out the tension in the bands as the vessel or equipment expands and contracts.

Gauge: A system of measuring the thickness of sheet materials, especially sheet metal. Often used as a general term for the thickness of such materials.

Knuckle: In general, an area of sharper curvature in a surface compared to the surrounding surfaces. For vessel heads, the surface near the head perimeter that is more sharply curved than the central dish-part of the head or its perimeter.

Perimeter: The outside boundary of any object or region, especially one that encloses a space inside. The line forming a circle and the sidewall of a vessel are examples of perimeters.

Pittsburgh seam: A seam between two sheet-metal edges consisting of a right-angle flanged male tab on one edge and a folded female pocket on the other edge. The flange is inserted into the pocket and the side extension of the female pocket is folded over to lock in the male flange. The various folds and closures can be manually- or machine-formed in the sheet-metal edges.

Standing seam: A method for connecting two pieces of sheet metal. Squared flanges are formed in the adjacent edges of the sheet-metal sections and brought together. At the top of the flanges, the pieces are folded together, locking them into a tight joint.

Wind forces: Forces generated by air flowing over a surface, like the lifting forces on an airplane wing or wind on a sail. Also called *aerodynamic forces*.

Additional Resources

This module presents thorough resources for task training. The following reference material is recommended for further study.

National Industrial and Commercial Insulation Standards Manual, Midwest Insulation Contractors Association (MICA).

National Insulation Association (NIA) website offers resources for products and training. **www.insulation.org**.

Mechanical Insulation Design Guide, National Institute of Building Sciences. Available at **www.wbdg.org/guides-specifications/mechanical-insulation-design-guide**.

Installation Guidelines for ITW Insulation Systems Metal Jacketing Used on Insulated Pipe, ITW Insulation Systems. 2016. Available at **www.itwinsulation.com**

Aluminum Deep Corrugated Sheets, ITW Insulation Systems. 2015. Available at **www.itwinsulation.com**

Figure Credits

Brock Services, Ltd., Module opener, Figures 17B, 18, 20

© Clearviewstock/Dreamstime.com, Figure 1A

© Cristiano Ribeiro/Dreamstime.com, Figure 1B

© Charon/Dreamstime.com, Figure 1C

© Adolfolazo/Dreamstime.com, Figure 1D

Forrest Mfg. Co., Figures 2, 3

© Dmitry Kalinovsky/Dreamstime.com, Figure 4

Bart van der Maar, Figures 5A, 6

RAMS Sheet Metal Equipment, Figure 5B

© Dainis Derics/Dreamstime.com, Figure 7

Section Review Answer Key

Answer	Section Reference	Objective
Section One		
1. b	1.1.1	1a
2. b	1.1.2	1a
3. a	1.2.0	1b
4. b	1.3.1	1c
Section Two		
1. c	2.1.0	2a
2. a	2.2.1	2b
3. b	2.2.1	2b
4. c	2.3.0	2c
5. b	2.4.1	2d

2.0.0 Section Review

Question 3

Use the height of the upright vessel and the standard spacing to find the number of securement bands needed:

Standard maximum spacing for bands is 12"
Divide tan-to-tan height by spacing:
215" ÷ 12" = 17.8 (round up to 18 band spaces)
Add one for the top band:
18 bands + 1 band = **19 bands**

The number of bands needed is **19**.

NCCER CURRICULA — USER UPDATE

NCCER makes every effort to keep its textbooks up-to-date and free of technical errors. We appreciate your help in this process. If you find an error, a typographical mistake, or an inaccuracy in NCCER's curricula, please fill out this form (or a photocopy), or complete the online form at **www.nccer.org/olf**. Be sure to include the exact module ID number, page number, a detailed description, and your recommended correction. Your input will be brought to the attention of the Authoring Team. Thank you for your assistance.

Instructors – If you have an idea for improving this textbook, or have found that additional materials were necessary to teach this module effectively, please let us know so that we may present your suggestions to the Authoring Team.

NCCER Product Development and Revision

13614 Progress Blvd., Alachua, FL 32615

Email: curriculum@nccer.org

Online: www.nccer.org/olf

❏ Trainee Guide ❏ Lesson Plans ❏ Exam ❏ PowerPoints Other ______________

Craft / Level: Copyright Date:

Module ID Number / Title:

Section Number(s):

Description:

Recommended Correction:

Your Name:

Address:

Email: Phone:

Removable and Reusable Flexible Insulation Covers

Overview

Insulated valves, piping, ductwork and equipment may need to be routinely accessed for inspection or maintenance. When such a need is anticipated, it may be preferable to use insulation materials that can be easily removed and reused, rather than destroying and then replacing the original insulation. In this module, you will learn about removable and reusable insulation, and its fabrication and installation.

Module 19314

Trainees with successful module completions may be eligible for credentialing through the NCCER Registry. To learn more, go to **www.nccer.org** or contact us at 1.888.622.3720. Our website has information on the latest product releases and training, as well as online versions of our *Cornerstone* magazine and Pearson's product catalog.

Your feedback is welcome. You may email your comments to **curriculum@nccer.org**, send general comments and inquiries to **info@nccer.org**, or fill in the User Update form at the back of this module.

This information is general in nature and intended for training purposes only. Actual performance of activities described in this manual requires compliance with all applicable operating, service, maintenance, and safety procedures under the direction of qualified personnel. References in this manual to patented or proprietary devices do not constitute a recommendation of their use.

19314 V2

19314

Removable and Reusable Flexible Insulation Covers

Objectives

When you have completed this module, you will be able to do the following:

1. Describe the uses and fabrication of removable and reusable covers.
   a. Describe the materials and methods used in fabricating removable and reusable covers.
   b. Describe how to fabricate and install a removable and reusable cover using an insulation kit.

Performance Tasks

Under the supervision of the instructor, you should be able to do the following:

1. Fabricate a removable and reusable flexible insulation cover for a flange using an insulating materials kit.
2. Install a field-fabricated removable and reusable flexible insulation cover on a flange.

Trade Terms

Amorphous silica
E-glass
Encasement
Hog rings
Hook-and-loop
Needled
Noise abatement

Industry Recognized Credentials

If you are training through an NCCER-accredited sponsor, you may be eligible for credentials from NCCER's Registry. The ID number for this module is 19314. Note that this module may be used in other NCCER curricula and may apply to other level completions. Contact NCCER's Registry at 888.622.3720 or go to **www.nccer.org** for more information.

Contents

Figures and Tables

SECTION ONE

1.0.0 REMOVABLE AND REUSABLE COVERS

Objective

Describe the uses and fabrication of removable and reusable covers.

a. Describe the materials and methods used in fabricating removable and reusable covers.
b. Describe how to fabricate and install a removable and reusable cover using an insulation kit.

Performance Tasks

1. Fabricate a removable and reusable flexible insulation cover for a flange using an insulating materials kit.
2. Install a field-fabricated removable and reusable flexible insulation cover on a flange.

Trade Terms

Encasement: The fabric that encloses the insulation in a reusable insulation cover.

Hog rings: Open, c-shaped rings with both ends beveled so that they interlock when clamped into place. Special pliers or pneumatic tools are used to install and remove them. Also called *C-rings*.

Noise abatement: Noise reduction; in this case, by use of insulating material.

Amorphous silica: A term identifying especially pure, noncrystalline (that is, amorphous) artificial mineral wool fibers made from molten silica.

E-glass: One of several chemical categories of glass. E-glass is most commonly used in fiberglass and glass-reinforced products.

Hook-and-loop: A material used to temporarily and repeatedly attach fabrics and other kinds of objects together. One surface is covered with a material containing tiny hooks and the other surface has a material containing tiny loops. When brought together, the hooks and loops engage. Originally manufactured under the Velcro® brand name.

Needled: A term describing a manufacturing method for nonwoven fabrics with randomly oriented fibers. These fabrics are created by rapidly and repeatedly inserting numerous barbed needles into a mass of fibers, during which they are pulled into a tightly tangled, cloth-like mass.

Removable/reusable flexible insulation covers are used on systems or parts that must be frequently accessed for maintenance or inspection. They are often referred to as *insulation pads, insulation blankets, reusable covers*, or *flexible covers*. For simplicity, this module will use the term *reusable insulation covers*.

1.1.0 Materials and Fabrication Methods

Reusable insulation covers are manufactured or fabricated insulation products typically used in industrial, commercial, and marine applications where insulation must be periodically removed for inspection and maintenance. They are also used for the insulation of irregular or complex surfaces such as dryers, heat exchangers, pipe fittings, valves, flanges, flexible connectors, instruments, manway covers, manifolds, mufflers, pumps, strainers, and turbines. *Figure 1* shows examples of reusable insulation covers.

Reusable insulation covers are used in applications having above- and below-ambient thermal conditions, and can be used for condensation control, **noise abatement**, and other specialty purposes. The cover's manufacturer or fabricator should provide specific recommendations regarding use of their product. Such information should be available in their catalogs or product data sheets, which are generally available online. Reusable insulation covers are often used as primary insulation for bare pipes, valves, and fittings (*Figure 2*) that may never have been insulated, and in cases where the original insulation was damaged or removed for maintenance or inspection.

> **NOTE**
> Many insulators believe that removable/reusable insulation covers are not suitable for below-ambient service. The covers' porous construction and intended use are not compatible with the need to prevent water vapor entry into the insulation on cold-service systems. It is possible to construct removable insulation covers that protect the insulation from water vapor intrusion, but these covers can be quite expensive.

Reusable insulation covers differ from many other insulation systems in that they are made from insulation and **encasement** materials for the intended service and can be custom fit to accommodate penetrations and irregular shapes. The covers include an attachment system suitable for the conditions. The attachment system is designed to make the covers easy to install, remove, and reuse, or replace.

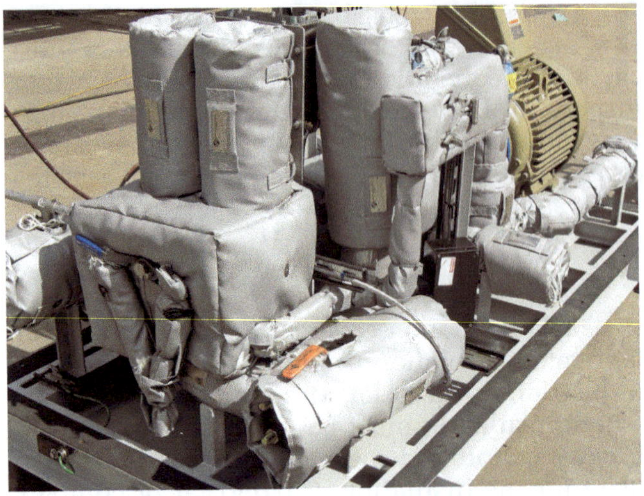

Figure 1 Examples of reusable insulation covers.

The encasement and insulation core materials are manufactured by many companies. The actual fabrication of the reusable insulation covers is usually performed by fabricators specializing in that area and by contractors and other companies who include that capability in their services. However, newly-available insulating kits make it possible to fabricate effective reusable insulation covers on site. Therefore, two types of reusable insulation covers are available: commercially fabricated covers and reusable insulation kits. The fabricated covers are designed to fit a specific component and application. Reusable insulation covers manufactured from general-purpose kits were originally intended for temporary or minor insulation fabrication by workers in the field. However, newer products permit fabrication of long-term, reusable insulation covers for piping system fittings.

1.1.1 Commercially Fabricated Reusable Insulation Covers

Commercially-fabricated reusable insulation covers are custom-made to fit a specific equipment, pipe component, or fitting. They are constructed from flexible insulation materials encased in fabric. The seams are usually sewn with an industrial sewing machine. Their assembly may also include other fastenings, such as metal C-rings known as hog rings (*Figure 3*), that make the cover more durable.

Reusable insulation covers are specifically designed to fit tightly around penetrations or adjoining insulation covers. For that reason, covers require the fabricator to have the specific measurements of the object to be covered. Fabricators may use several methods to obtain the required measurements. These methods can vary from providing drawing templates so measurements can be added (*Figure 4*), taking measurements from the field site or from technical drawings, and reviewing photos.

From the measurements obtained, fabricators may use any of the following to develop the patterns from which the covers are constructed:

- Computer-designed patterns to run automated cutting machines
- Software to nest or tile blanket patterns
- CAD programs to convert 3D shapes to flat patterns
- Other engineering tools to design and fabricate the reusable insulation covers

Simple patterns, especially for kit-based covers, can be developed using manual layout methods.

The following terms identify key parts of commercially fabricated reusable insulation covers:

- *Hot face* – The inside surface area of the cover that is in direct contact with the item being covered.
- *Cold face* – The outside surface of the cover, which is exposed to atmospheric and ambient temperature conditions.
- *Identification tag* – A tag (*Figure 5*) used to identify the specific item being insulated, part number, fabricator, etc.
- *Core insulation* – The insulating material inside a flexible cover.

Various methods are used to stabilize and secure the removable insulation covers. Some of these features include the following (*Figure 6*):

- *Insulation securements and stabilizers* – These can be lacing hooks, insulation anchors, or quilt pins to prevent insulation settlement or slippage within the cover.

Hog Rings

The original use of the term *hog ring* dates back to the mid-seventeenth century. According to the Oxford Dictionary, a hog ring is: "A ring or piece of bent wire put into the snout of a pig to prevent it from grubbing in the ground." The ring is open and the ends are beveled to aid piercing and so that they will align when inserted and crimped shut.

Figure 2 Removable insulation covers on piping.

Figure 3 Hog ring assembly.

- *Hook-and-loop securement* – The **hook-and-loop** securement method, shown in *Figure 6 (A)*, uses a material commonly called Velcro®.
- *D-rings* – These are rings for use when securing with a fabric strap, shown in *Figure 6 (A)*.
- *Quilting pins* – Quilting pins, also shown in *Figure 6 (A)*, are disk-headed pins that are pushed through the cover layers and secured with a clinch washer on the underside. The pins stabilize the insulation material like quilting in a blanket.
- *Lacing hooks* – Lacing hooks, shown in *Figure 6 (B)*, are installed to serve as lacing anchor points when securing the cover by wire
- *Tie-down anchor straps* – These are straps used in conjunction with buckles to secure the covering in place when it is installed, shown in *Figure 6* (C).
- *Draw strings* – Draw strings, shown in *Figure 6* (C), are used with an overlap area to seal the cover tightly against adjoining surfaces; primarily on outdoor applications
- *Overlap* – There is usually a minimum of 2" (5 cm) to overlap adjacent insulation or surface, as shown in *Figure 6* (C).

There are a wide variety of suitable encasement fabrics for the cold or hot face of reusable insulation covers. The fabric selection will depend upon the service temperature and other considerations. Quite often, one fabric will be used on the interior or hot side and another on the exterior or cold side.

The following are a few examples of cold-face fabrics:

- Silicone-coated, impregnated fiberglass or **amorphous silica**
- PTFE-coated (Teflon®-coated), impregnated fiberglass or amorphous silica
- Aluminum-foil-laminated/aluminum-film-laminated fiberglass or amorphous silica
- Plain and texturized fiberglass

The following are a few examples of hot-face fabrics:

- Silicone-coated, impregnated fiberglass or amorphous silica
- PTFE -coated, impregnated fiberglass or amorphous silica
- Aluminum-foil-laminated/aluminum-film-laminated fiberglass or amorphous silica
- Plain and texturized fiberglass
- Vermiculite-coated, impregnated fiberglass or amorphous silica
- Stainless steel or Inconel® foil
- Stainless steel or Inconel® blanket mesh

A variety of insulation materials may be used in the fabrication of reusable insulation covers. The selection will depend on the application, ambient conditions, and other considerations. Typical examples of insulation materials are aerogel, ceramic fiber, fiberglass, flexible foam, **E-glass** in **needled** bats, and mineral fiber.

GOING GREEN

Reduce, Reuse, Recycle

When common insulation is removed for any reason (such as maintenance or inspection), it is usually damaged in the process and must be discarded. For this reason, reusable insulation covers have a positive effect on the environment because they do not result in wasted material.

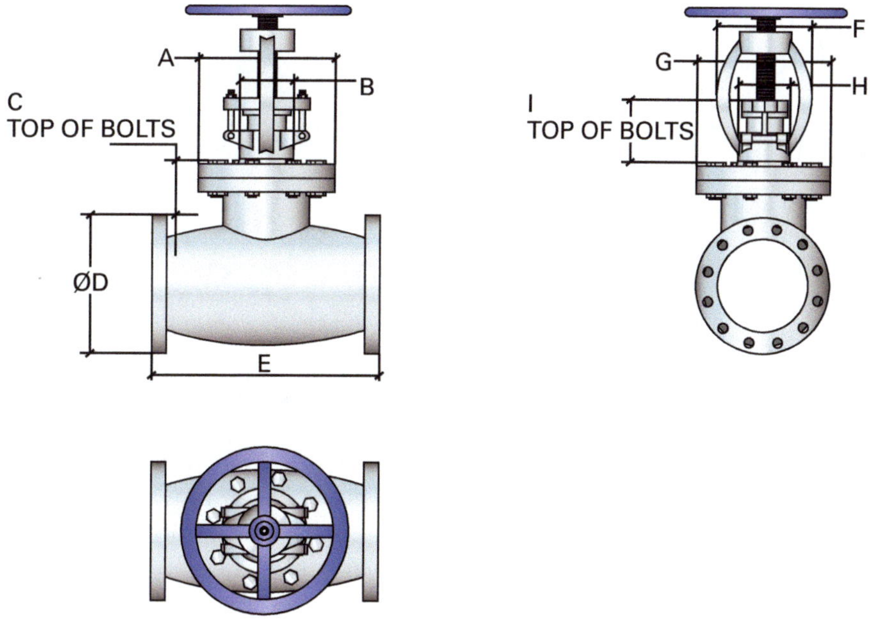

FIELD TAKEOFF DATA						
DIMENSION				VALVE SPECS		ADDITIONAL INFORMATION/DETAILS
A		G		NPS:		
B		H		TYPE:		
C		I		CLASS:		
D		J		MODEL:		
E		K		BONNET SHAPE:	ROUND ☐ RECTANGLE ☐ OVAL ☐	
F		L				

Figure 4 Valve dimensions.

Figure 5 Identification tag.

Reusable insulation covers can be fabricated in any thickness, depending on the size and weight of the insulation blanket. However, as a general rule, thicknesses over 2" (5 cm) are usually made in multiple layers.

1.2.0 Fabrication and Installation

Insulation covers that are commercially fabricated must be installed, used, and stored appropriately. Insulation covers can also be fabricated using kits consisting of the necessary materials. This section describes the installation of commercially fabricated insulation covers, as well as how to fabricate and install kit insulation covers.

1.2.1 Installation of Commercially Fabricated Reusable Insulation Covers

The fabrication of reusable insulation covers is normally done by commercial companies specializing in this kind of work. They have access

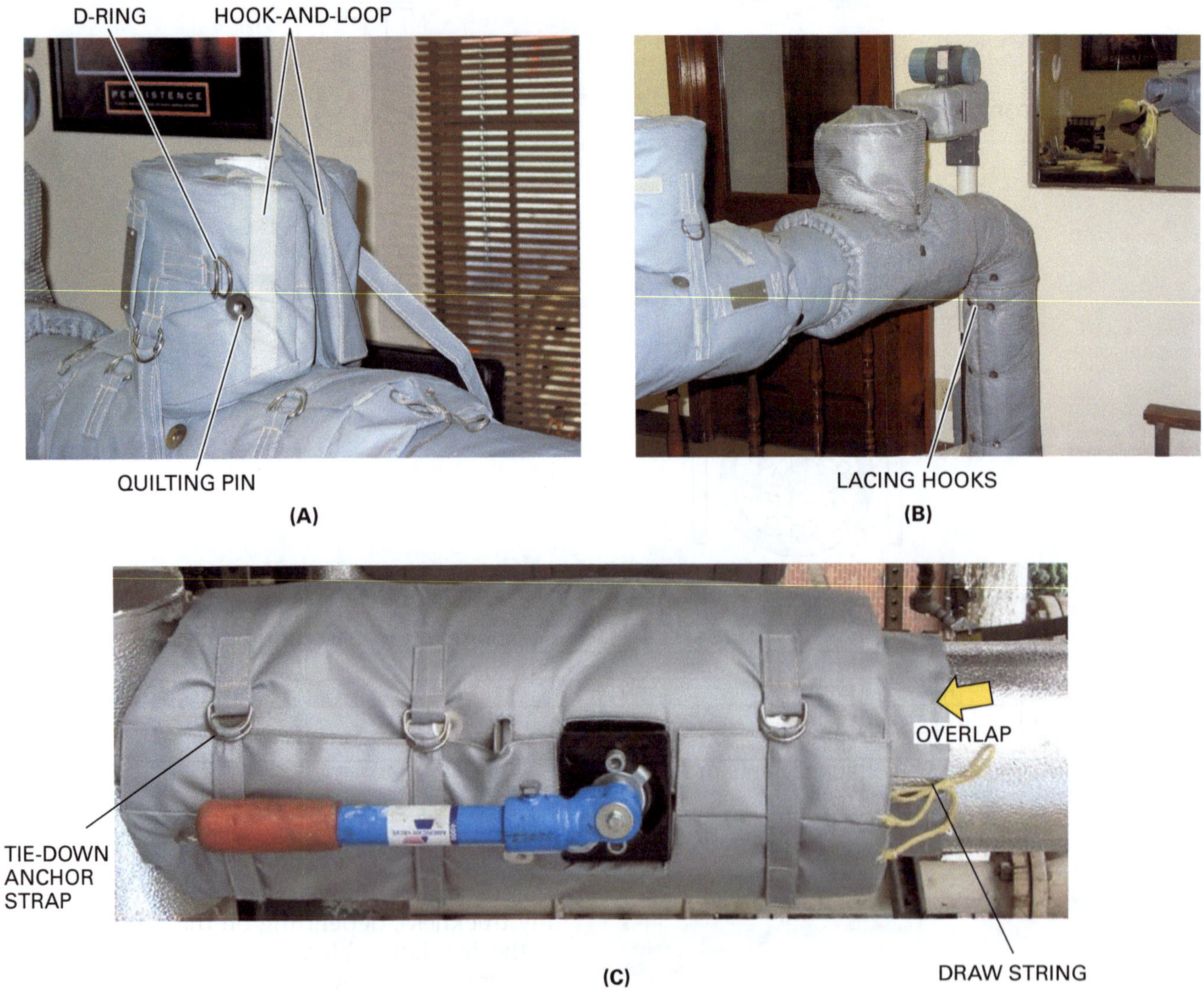

Figure 6 Reusable cover securement methods.

to specialized design software, fabric cutting machines, supplies of stabilizing and fastening hardware, and the expertise to quickly complete the job. Typically, the mechanical insulation contractor will provide the fabricating company with the required information and dimensions of the equipment to be insulated, and the covers arrive at the worksite ready to install.

For large or complex reusable insulation cover contracts, the fabricator may also perform the installation. This section provides examples of several installation methods for several prefabricated reusable insulation.

The first example is a one-piece sewn cover for a flanged valve (*Figure 7*). This cover is designed to fit a specific valve while overlapping the adjoining pipe insulation. It is simply installed around the valve and secured with hook-and-loop strips sewn into the overlapping sections on the longitudinal and vertical joints on top of the cover. It takes minutes to install and remove, and it can be reused and reinstalled in the same manner as the original installation.

One-piece sewn flanged valve covers can also be designed for conditions that require more extensive securement systems. A hook-and-loop securement system closes the longitudinal valve body and bonnet seams, as shown in *Figure 7*. To reduce the tension across the longitudinal seam, tie-down/anchor straps and D-rings built into the cover sides are used to tightly secure the cover. In addition, there are draw strings in the overlaps on the adjacent pipe insulation that are used to tightly seal the lap. Even with the additional securement features, installation and removal are simple and fast.

The same general principles apply to all types of fittings. *Figure 8* is a cover has been designed to fit a 90-degree elbow with simple straps run through D-rings that have hook-and-loop securement points. Again, removal and replacement is an easy task.

Figure 7 One-piece sewn valve flange cover.

Some covers are designed to be stored, used, and removed as needed for universal applications such as freeze protection. The spiral-wrap cover in *Figure 8* wraps around the pipe and can be secured in place with wire, hook-and-loop tape, or other types of tape, and simply removed again until needed. One or multiple sections can be used. This form may also serve as temporary insulation until permanent insulation can be installed.

Reusable covers can be installed on larger piping, pieces of equipment, turbines, exhaust hoods, etc. In *Figure 9*, lacing hooks have been incorporated into the design, thus making securement by wire possible. Potential examples for reusable insulation covers to fit specific contour configurations and design variables are virtually limitless.

It is important to remember that, after a cover has been removed, it should be handled and stored carefully to avoid soiling and damage. It is not uncommon for covers to be removed, misplaced, handled roughly, and even discarded. This typically occurs when workers are not familiar with the value and reusable feature of these covers. For a variety of reasons, many are not reinstalled.

1.2.2 Fabrication and Installation of a Reusable Pipe Fitting Cover

If specifications permit, it may be more economical to field-fabricate a reusable insulation cover from insulation material kits for small, uncomplicated insulation jobs. These may even serve as primary insulation in some applications, though they aren't suitable for below-ambient service. The process for constructing an effective cover from readily-available kit materials is straight-forward and easy to master.

Comparing Cost

When comparing the cost of reusable insulation covers to the cost of field-fabricated insulation for the same applications, you need to compare the cost of labor as well as the materials. While the materials for conventional insulation may be less expensive than for a reusable cover, it takes much longer to field-fabricate the insulation for a valve or fitting. Also, where occasional removal is required, the life-cycle cost of the reusable cover may be significantly less than that of field-fabricated insulation. Once purchased, the reusable cover can be used many times, while the field-fabricated insulation must be remade each time it is removed.

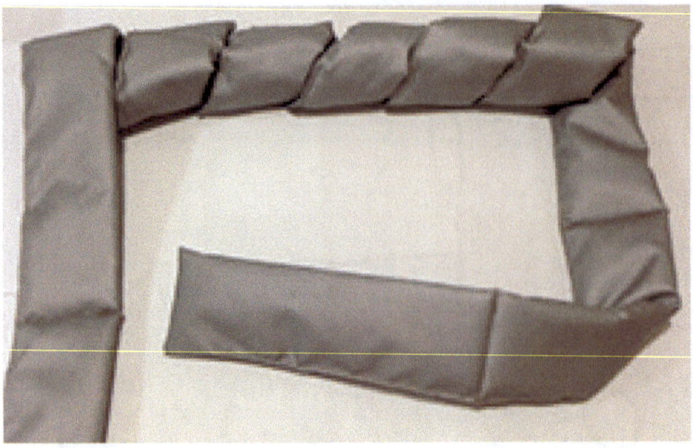

Figure 8 Spiral-wrapped cover.

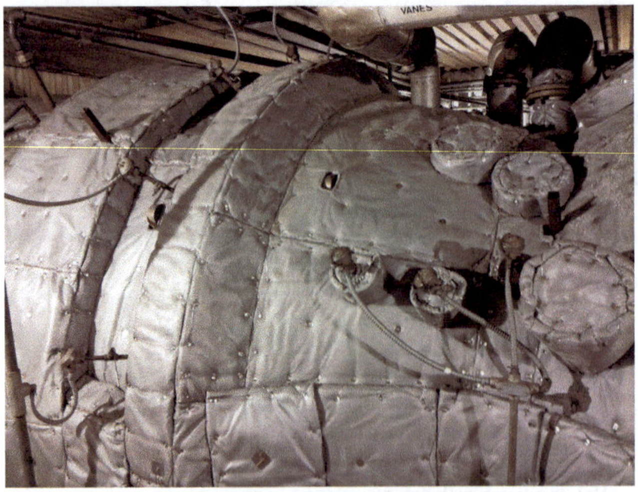

Figure 9 Fabricated covers for larger assemblies.

There are a variety of removable/reusable insulation cover materials kits available on the market. Some are clearly assumed to be temporary removable insulation solutions. A modular blanket insulation kit, such as CUT 'N WRAP™ by EVER GREEN®, is a good example of a kit for field-fabrication and installation of truly reusable insulation covers. The kit consists of blankets of fiberglass insulation encased in a stitch-quilted outer covering made of polymer-coated glass fiber fabric (*Figure 10*). Other kit-form reusable insulation products may contain loose or batt fibrous insulation that is encased by a prefabricated fabric cover when installed on the component.

The modular quilted covers are cut to shape and then assembled with locally purchased $^3/_8$" (9.5 mm) galvanized staples. They are secured with double-sided hook-and-loop tape. Tools required for fabrication include a tape measure, a heavy-duty stapler, sheet-metal shears, and a utility knife or industrial scissors that can be resharpened, if required.

Fabrication begins by examining the pipe component to be insulated to determine the most

Figure 10 A modular reusable insulation cover kit featuring quilted material.

efficient way to cover it by wrapping. For instance, a stemmed valve would need a blanket section to wrap the valve body and the flanged connections to the adjoining pipes, another blanket to wrap the bonnet, and possibly a third to wrap the yoke, if required. A pipe flange would require only a single piece that spans the flange and overlaps the insulation on the adjoining pipes. When installed, the wraps should overlap at least 4" (10 cm).

Measure the component and lay out the dimensions for each wrap. While the wrap will work using whole squares of the blanket fabric, a better and more visually pleasing fit will result if the pattern is trimmed so that closely follows the shape of the component. The blanket material can be cut out using a utility knife, scissors, or sheet-metal shears (*Figure 11*).

If the cut section has raw edges not stabilized by the manufacturer's stitching, carefully staple the edge closed from the outside surface (*Figure 12*). Place staples about 1" (2.5 cm) apart. Test fit the blanket sections on the component. Measure around the wrapped blanket to determine the length of hook-and-loop tape to use for securing the wrap. Enough tape securements should be provided to make sure the wrap is a snug fit. Peel off from the roll a length of tape required plus 6" (15 cm). Cut the number of tapes needed. Fasten one end of the tape to the cover section with at least two staples (*Figure 13*). Make sure the tape is square to the

Figure 11 Cutting the blanket with sheet-metal shears.

edge of the cover. At the other end of the tape, double it over and staple it to provide an easily-grasped tab or handle for removal.

Figure 12 Stapling the raw edge of the blanket wrap.

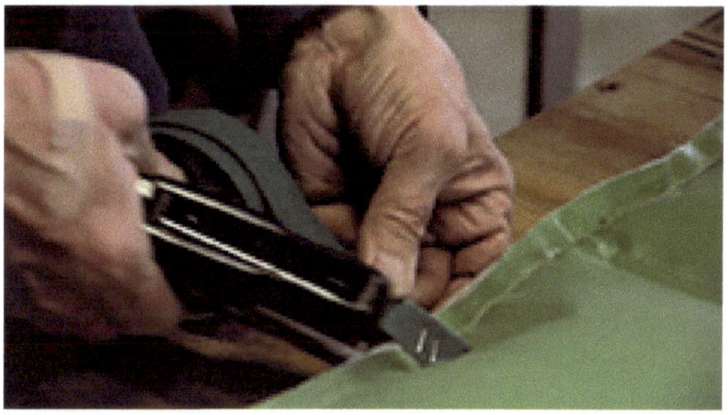

Figure 13 Stapling the securement tape to the cover.

Installing the wrap cover is simple. For multiple-piece covers, start with the innermost cover. Position it around the component so that the laps with the hook-and-loop tapes are on the outside. Wrap the tapes around the component and adhere them to the backside of the tape (*Figure 14*). Check the fit of the wrap for a close-fitting cover. Separate the tape, adjust, and retighten if necessary. When satisfied with the fit, it is a good practice to remove the cover, neatly write the component's identifying information on the inside of the cover, and then reinstall it permanently. This will make it more likely that the reusable cover will be returned to its component when it is removed at a later date.

Some installations may require more than one layer of reusable covers to attain the required insulating specification. If more layers are needed, repeat the process above to fabricate the additional layers.

The modular wrap can be removed as easily as it is installed. Simply pull the doubled tape ends to separate the hook-and-loop fabric, and then slip the cover off. As with commercially fabricated covers, be sure to properly store the covers when removed so they will be available to be reinstalled when needed.

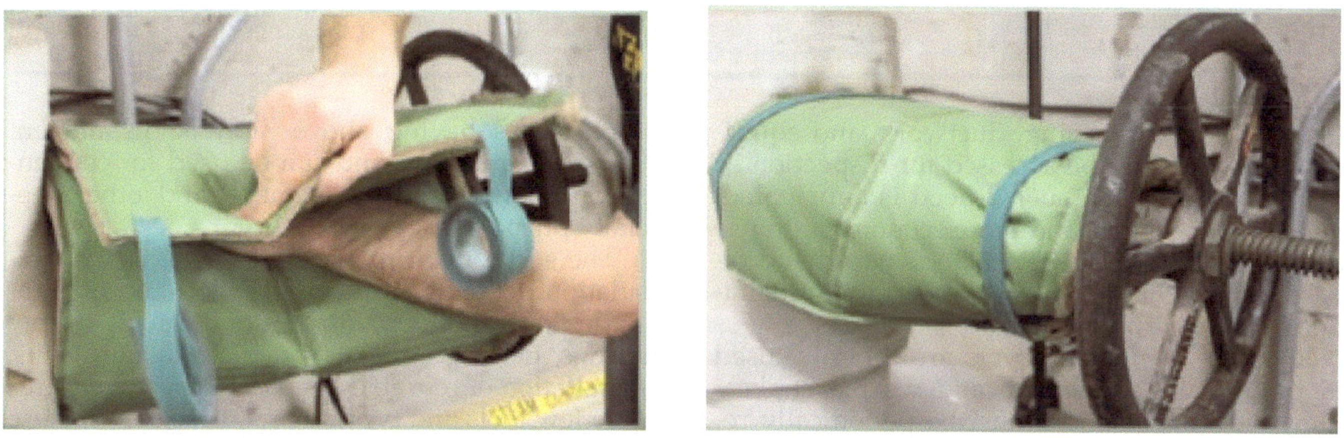

Figure 14 Installing the modular blanket reusable cover.

Additional Resources

Mechanical Insulation Design Guide, National Institute of Building Sciences. Available at **www.wbdg.org/guides-specifications/mechanical-insulation-design-guide**

National Industrial and Commercial Insulation Standards Manual, Midwest Insulation Contractors Association (MICA).

The following websites offer resources for products and training:

National Insulation Association (NIA). **www.insulation.org**

Firwin Corporation *(a manufacturer of custom reusable insulation products).* **www.firwinthermowrap.com**

Auburn Manufacturing, Inc. *(a manufacturer of modular insulation blanket wrap kits).* **www.auburnmfg.com**

1.0.0 Section Review

1. Which of the following is a correct statement regarding reusable insulation covers?
   a. They are always fabricated in the field.
   b. They cannot be used on valves.
   c. They must be discarded after a specified number of uses.
   d. They can be used for noise abatement.

2. Tools typically needed for installing reusable insulation covers are wire cutters, scissors, and _____.
   a. a screwdriver
   b. a hammer
   c. needle-nose pliers
   d. a hacksaw

SUMMARY

Insulated valves, piping, ductwork, and equipment may need to be accessed for inspection or maintenance. When such a need is anticipated, it is often preferable and more economical to use insulation materials that can be removed and reused, rather than replacing the original insulation.

Reusable insulation covers are typically used in industrial, commercial, and marine applications where insulation must be periodically removed for inspection and maintenance. They are also used for the insulation of irregular or complex surfaces of industrial and commercial equipment. Reusable insulation covers are available for applications having above- and below-ambient thermal conditions, and can be used for condensation control, noise abatement, and other specialty purposes.

The securement method used for reusable insulation depends on the application and may include hook-and-loop, D-rings and buckles, draw strings, tie-down anchor straps, or lacing hooks. Installation of reusable insulation covers requires very few tools and typically requires only the use of the securement method built into the cover. New insulation products are available that permit the insulation mechanic to easily fabricate effective removable covers on-site.

Review Questions

1. Reusable insulation covers are made from _____.
   a. insulation and encasement materials
   b. fiberglass and sheet metal
   c. insulation and lagging
   d. cotton and mineral wool

2. Which of the following is a correct statement regarding reusable insulation covers?
   a. They can only be used in above-ambient conditions.
   b. They can only be used in below-ambient conditions.
   c. They are easier to install than permanent insulation.
   d. They are one-size-fits-all devices.

3. The flexible material that surrounds the insulation in a reusable insulation cover is the _____.
   a. hot face
   b. cold face
   c. encasement
   d. overlap

4. The metal rings used to secure the seams of some types of reusable insulation covers are _____.
   a. quilting pins
   b. hog rings
   c. encasement rings
   d. hooks-and-loops

5. The inside surface area of a reusable insulation cover is called the _____.
   a. hot face
   b. cold face
   c. encasement
   d. vapor retarder

6. The securement method used with buckles to secure the cover is _____.
   a. lacing hooks
   b. hook-and-loop
   c. tie-down anchor straps
   d. draw strings

7. Which of the following is *not* a cold-face fabric?
   a. Silicone coated, impregnated fiberglass and amorphous silica
   b. PTFE-coated, impregnated fiberglass and amorphous silica
   c. Aluminum-foil-laminated fiberglass and amorphous silica
   d. Stainless steel knitted blanket mesh

8. Impregnated fiberglass cloth is used in removable/reusable insulation _____.
   a. only as the cold face
   b. only as the hot face
   c. as both hot face and cold face
   d. only for noise abatement

Figure RQ01

9. The securement method used in *Figure RQ01* is _____.
   a. hook-and-loop
   b. D-rings
   c. lacing hooks
   d. anchor straps

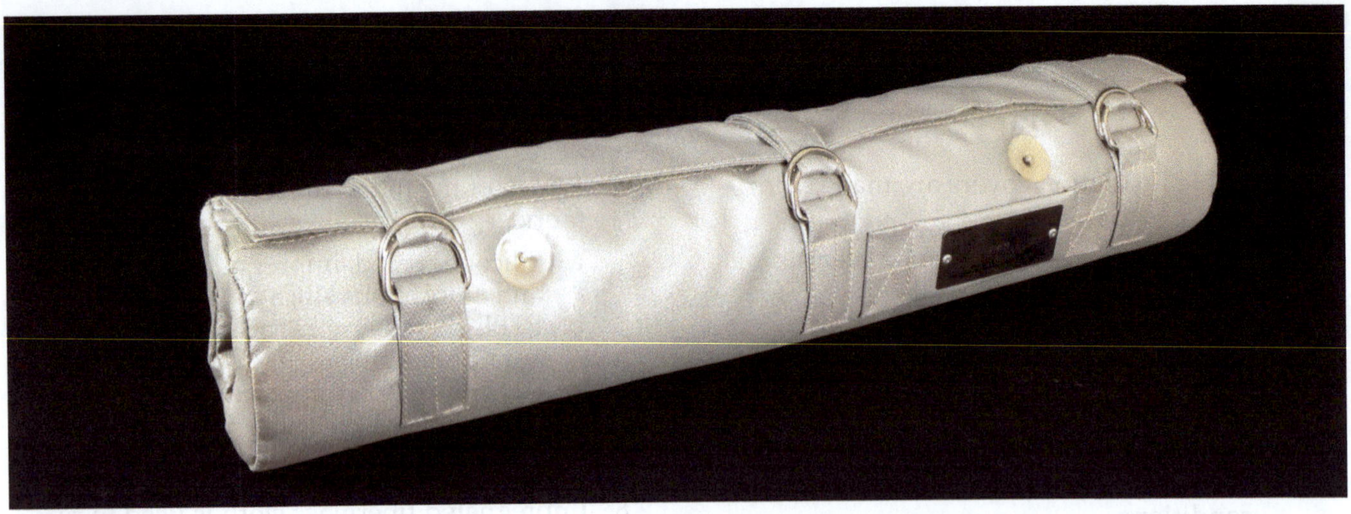

Figure RQ02

10. The primary method of securement in *Figure RQ02* is _____.
    a. D-rings
    b. hook-and-loop
    c. lacing hooks
    d. tie-down anchor straps

Trade Terms Introduced in This Module

Amorphous silica: A term identifying especially pure, noncrystalline (that is, amorphous) artificial mineral wool fibers made from molten silica.

E-glass: One of several chemical categories of glass. E-glass is most commonly used in fiberglass and glass-reinforced products.

Encasement: The fabric that encloses the insulation in a reusable insulation cover.

Hog rings: Open, c-shaped rings with both ends beveled so that they interlock when clamped into place. Special pliers or pneumatic tools are used to install and remove them. Also called *C-rings*.

Hook-and-loop: A material used to temporarily and repeatedly attach fabrics and other kinds of objects together. One surface is covered with a material containing tiny hooks and the other surface has a material containing tiny loops. When brought together, the hooks and loops engage. Originally manufactured under the Velcro® brand name.

Needled: A term describing a manufacturing method for nonwoven fabrics with randomly oriented fibers. These fabrics are created by rapidly and repeatedly inserting numerous barbed needles into a mass of fibers, during which they are pulled into a tightly tangled, cloth-like mass.

Noise abatement: Noise reduction; in this case, by use of insulating material.

Additional Resources

This module presents thorough resources for task training. The following reference material is recommended for further study.

Mechanical Insulation Design Guide, National Institute of Building Sciences. Available at **www.wbdg.org/guides-specifications/mechanical-insulation-design-guide**

National Industrial and Commercial Insulation Standards Manual, Midwest Insulation Contractors Association (MICA).

The following websites offer resources for products and training:

National Insulation Association (NIA). **www.insulation.org**

Firwin Corporation *(a manufacturer of custom reusable insulation products).* **www.firwinthermowrap.com**

Auburn Manufacturing, Inc. *(a manufacturer of modular insulation blanket wrap kits).* **www.auburnmfg.com**

Figure Credits

Auburn Manufacturing, Figures 10–14,

Coverflex Manufacturing, Inc. (US), Module opener, Figures 1-3, 5,6, 8, Review Question Figures 1, 2, Exam Figures 1-5

Brock Services, Ltd., Figure 9

Section Review Answer Key

Answer	Section Reference	Objective
Section One		
1. d	1.1.0	1a
2. c	1.2.0	1b

NCCER CURRICULA — USER UPDATE

NCCER makes every effort to keep its textbooks up-to-date and free of technical errors. We appreciate your help in this process. If you find an error, a typographical mistake, or an inaccuracy in NCCER's curricula, please fill out this form (or a photocopy), or complete the online form at **www.nccer.org/olf**. Be sure to include the exact module ID number, page number, a detailed description, and your recommended correction. Your input will be brought to the attention of the Authoring Team. Thank you for your assistance.

Instructors – If you have an idea for improving this textbook, or have found that additional materials were necessary to teach this module effectively, please let us know so that we may present your suggestions to the Authoring Team.

NCCER Product Development and Revision

13614 Progress Blvd., Alachua, FL 32615

Email: curriculum@nccer.org
Online: www.nccer.org/olf

❑ Trainee Guide ❑ Lesson Plans ❑ Exam ❑ PowerPoints Other ____________________

Craft / Level: Copyright Date:

Module ID Number / Title:

Section Number(s):

Description:

Recommended Correction:

Your Name:

Address:

Email: Phone:

Specialized Insulation Systems

OVERVIEW

There are some insulation systems the average insulation worker will not encounter on a day-to-day basis. Some are less-common applications in conventional insulation work, while others are highly specialized and would rarely be seen in typical mechanical insulation jobs. However, a competent mechanical insulator should have some knowledge of all types of insulation systems.

Module 19308

Trainees with successful module completions may be eligible for credentialing through the NCCER Registry. To learn more, go to **www.nccer.org** or contact us at 1.888.622.3720. Our website has information on the latest product releases and training, as well as online versions of our *Cornerstone* magazine and Pearson's product catalog.

Your feedback is welcome. You may email your comments to **curriculum@nccer.org**, send general comments and inquiries to **info@nccer.org**, or fill in the User Update form at the back of this module.

19308 V2

19308

Specialized Insulation Systems

Objectives

When you have completed this module, you will be able to do the following:

1. Describe various specialized insulation systems and their functions.
   a. Describe the construction and use of low-temperature panels, freezer rooms, and cryogenic insulation systems.
   b. Explain how refractory insulation is used.
   c. Describe spray-on and foamed-in-place polyurethane insulation.
   d. Describe fire-rated construction and the application of firestopping materials.
   e. Describe the use of insulation for sound control.
   f. Describe poured-in-place insulation and pre-insulated piping systems for underground pipe.

Performance Tasks

This is a knowledge-based module. There are no performance tasks.

Trade Terms

Cryogenic
Endothermic
Firestopping
Footings
Ganister
Intumescent
Loadbearing
Non-conducting
Refractory insulation
Sound isolation
Ventilation

Industry Recognized Credentials

If you are training through an NCCER-accredited sponsor, you may be eligible for credentials from NCCER's Registry. The ID number for this module is 19308. Note that this module may have been used in other NCCER curricula and may apply to other level completions. Contact NCCER's Registry at 888.622.3720 or go to **www.nccer.org** for more information.

Contents

Figures

Section One

1.0.0 Specialized Insulation Systems and Their Functions

Objective

Describe various specialized insulation systems and their functions.

a. Describe the construction and use of low-temperature panels, freezer rooms, and cryogenic insulation systems.
b. Explain how refractory insulation is used.
c. Describe spray-on and foamed-in-place polyurethane insulation.
d. Describe fire-rated construction and the application of firestopping materials.
e. Describe the use of insulation for sound control.
f. Describe poured-in-place insulation and pre-insulated piping systems for underground pipe.

Trade Terms

Cryogenic: For mechanical insulating purposes, any application or system operating at temperatures below –50°F (–46°C). However, the National Institute of Standards and Technology defines the upper end of the cryogenic band at –292°F (–180°C) for scientific purposes.

Endothermic: Any process that gains thermal energy (absorbs heat) from it surroundings.

Firestopping: A material or construction method that interferes with the entry or movement of fire through a fire-rated barrier.

Footings: The underlying supports for a column or wall.

Ganister: A hard, fine-grained quartzite rock (naturally-cemented quartz sandstone) used in high-temperature insulation.

Intumescent: The property of a material that swells when heated, increasing its volume and decreasing its density, usually by the production of gas voids inside the material. The insulating property of the voids slows the transmission of heat through the material.

Loadbearing: The function of any structural member, such as a wall, that supports part of the structure's weight.

Non-conducting: The property of a substance that resists or prevents the movement of heat or electricity.

Refractory insulation: Temperature-resistant insulation that can retain its properties above 1,000°F (538°C). Typically installed inside of furnaces and other high-temperature apparatus.

Sound isolation: Any material or construction method that interferes with the movement of sound (noise) from its source to somewhere else.

Ventilation: The natural or artificial entry and exchange of fresh air within a structure.

In most instances, specialized insulation systems are installed by specifically-trained insulation contractors. In this module you will learn to identify some of the more important systems, understand their functions, and become familiar with their materials and general installation methods. You will not be required to demonstrate practical skills in their application.

1.1.0 Low-Temperature Systems

Any system operating at temperatures below ambient require extra measures to ensure an effective insulation system. Refrigeration or cold-service systems operate within the temperature range of –50°F to 75°F (–46°C to 24°C). Service temperatures below –50°F (–46°C) are in the cryogenic range. In order to maintain the low refrigeration temperatures required to cool and freeze food in storage or processing facilities, the containment systems must be well insulated and sealed against leakage. Although mechanical systems establish and maintain the required temperatures, they would not be very effective without the insulation that keeps heat from penetrating the containment.

Most modern walk-in coolers and freezers are built with manufactured preinsulated panels. The panels have a core of polymer foam insulation (polystyrene or polyisocyanurate) and outside skins of metal, plastic, or laminates. The exterior layer is the vapor retarder, which must be sealed to resist water vapor diffusion. Grocery stores and restaurants have small freezers and coolers built from panels. These may be located indoors or outdoors using the exterior of the panels as the exterior walls and roof. Larger installations such as refrigerated warehouses (*Figure 1*) and food-processing plants may be built entirely of insulated panels.

Low-temperature panels have special interlocking hardware and gaskets to allow the panels to be attached to one another. The gaskets form vaportight, non-conducting joints. Both sidewall

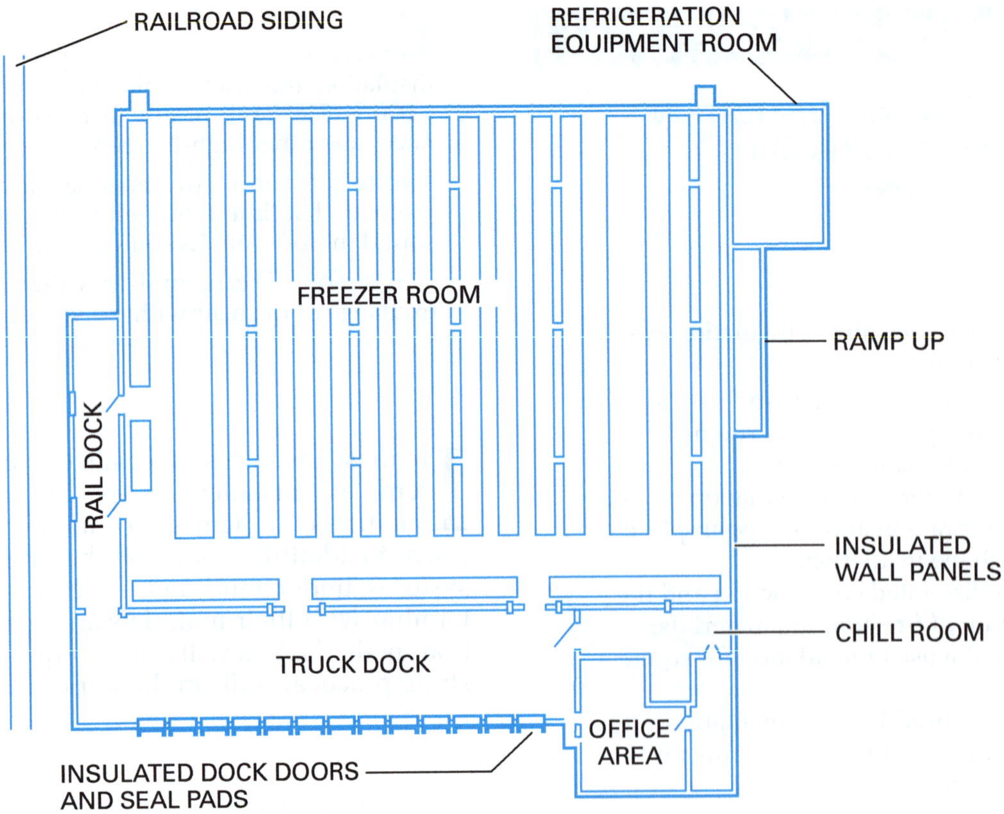

Figure 1 Layout of a refrigerated warehouse.

and roof panels are available. Sidewall panels may be **loadbearing** for small-room construction. This feature allows the roof panels to be installed directly on top of the walls without additional wall construction. Larger rooms may be built over a steel framework, permitting the use of nonbearing panels. The panels are manufactured in a range of sizes so that no major cutting is required for erection. Cutouts for doors and pipe penetrations may be required and can be made with conventional tools.

Another panel type has standing seams on the outside skin joints. These panels are erected and the standing seams are closed with an interlocking seam machine. This type of panel is used in applications where the panels are exposed to the weather. The exterior panel joints must be sealed vaportight. This is accomplished by applying a vapor retarder compound in the seams before they are closed.

Low-temperature rooms may be coolers, storage freezers, or blast freezers. Coolers are usually maintained above freezing at 32°F to 45°F (0°C to 7°C). Freezers generally operate at temperatures of 20°F to –20°F (–7°C to –29°C). Blast freezers operate below –40°F (–40°C); the lower the temperature, the thicker the panel. Cooler panels may be 2" (5 cm) thick and freezer panels may be 4" to 6" (10–15 cm) thick, with colder applications even thicker. The thicker the panel, the thicker the insulation and the lower the temperature that can be maintained for a given refrigeration plant.

Panels are usually erected by contractors employing crews of carpenters who specialize in cold-storage work. Regular insulation workers are usually not involved with this type of construction except for occasional repair jobs. Insulation workers may be required to insulate pipe that passes through panels, so it is beneficial to have a general understanding of cold storage panel construction.

Industry standards recommend the use of cellular glass, closed-cell phenolic, flexible elastomeric, polyisocyanurate, and polystyrene for refrigeration applications. These all are closed-cell foam materials that have low water vapor transmission and water absorption characteristics.

1.1.1 Freezer Rooms

Insulation must be installed in floors of freezer rooms and is usually installed in larger cooler room floors. A **ventilation** or heating system may be required beneath larger freezers to prevent under-floor freezing. A vapor retarder membrane of polyethylene sheet or hot asphalt must be installed first and will probably be installed over a rough floor.

Insulation board is installed over the vapor retarder. The board is usually installed in two layers with the joints staggered. If there are to be interior support columns, the floor vapor retarder and insulation must continue under the column **footings** or extend up around the column to above the finish floor level, as shown in *Figure 2*.

The floor insulation and vapor retarder must also be turned up the walls to a point above the finished floor (*Figure 3*). A concrete floor is usually poured over the insulation board. The floor vapor retarder and insulation must connect with the wall panels, and the vapor seal between the floor and walls must be complete. An anchor plate for the sidewall panels may be placed on top of the floor insulation.

Moisture may enter a freezer room in several ways, such as when the doors are opened, from the products stored within the rooms, and through ruptures in the vapor retarder. The moisture forms ice on the freezer unit coils. These units periodically run a defrost cycle to melt ice from the coils. The defrost drain must be heat-traced and insulated inside the freezer room to prevent it from freezing in the pipe before it exits the room.

Electrical conduit, sprinkler pipe, and other items penetrating through freezer room insulation must be insulated on the warm side. This is usually about 18 inches (46 cm) to prevent condensation forming on cold surfaces (*Figure 4*).

Other cold storage rooms are built by applying a vapor retarder, insulation, and interior finish to walls and ceiling of rooms built in a conventional manner. This type of built-up system is often found in older buildings.

Regardless of the type of low-temperature construction (panel or built-up) it is important to know the following:

- For any low-temperature insulation, a vapor retarder must be complete on the warm side of the insulation.
- Any localized ice formation on the inside of a freezer room is probably caused by a rupture in the adjacent vapor retarder on the warm side of the insulation.
- Any vapor retarder leak in a freezer panel allows ice to form in the insulation or insulation joints, causing separation. If not corrected, the ice will severely damage the insulation.
- Freezer defrost drains must be heat traced and insulated inside the room to prevent the condensate from freezing before it exits the space.

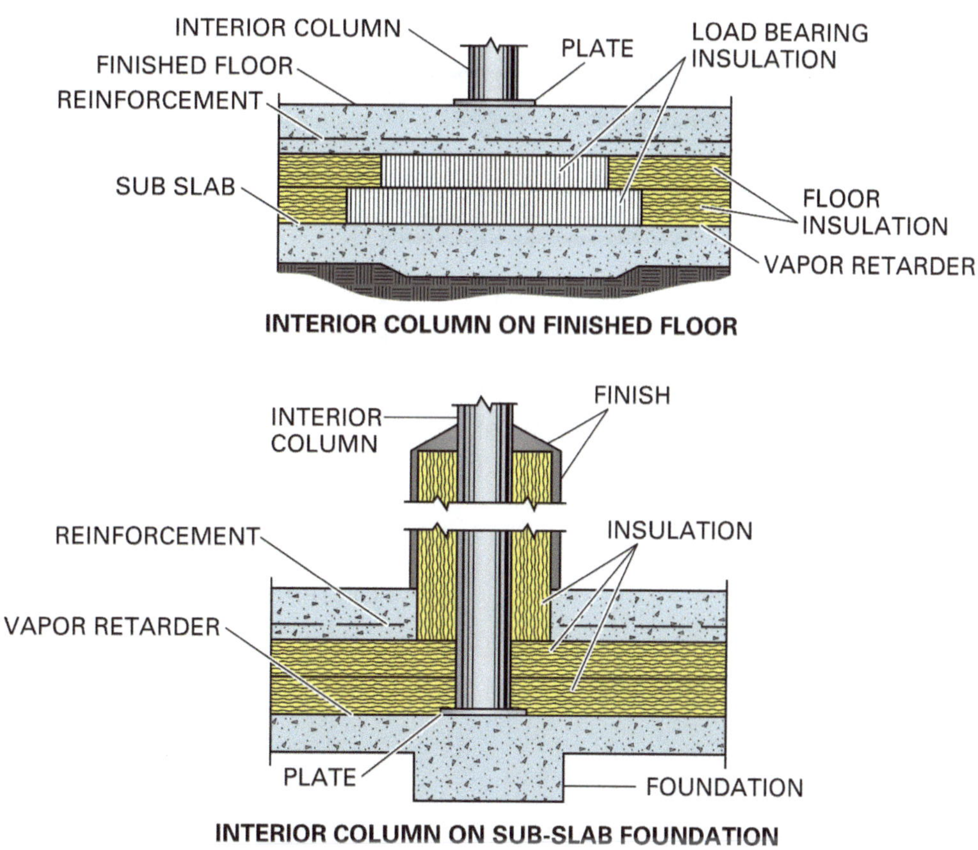

Figure 2 Insulation board installed over a vapor retarder.

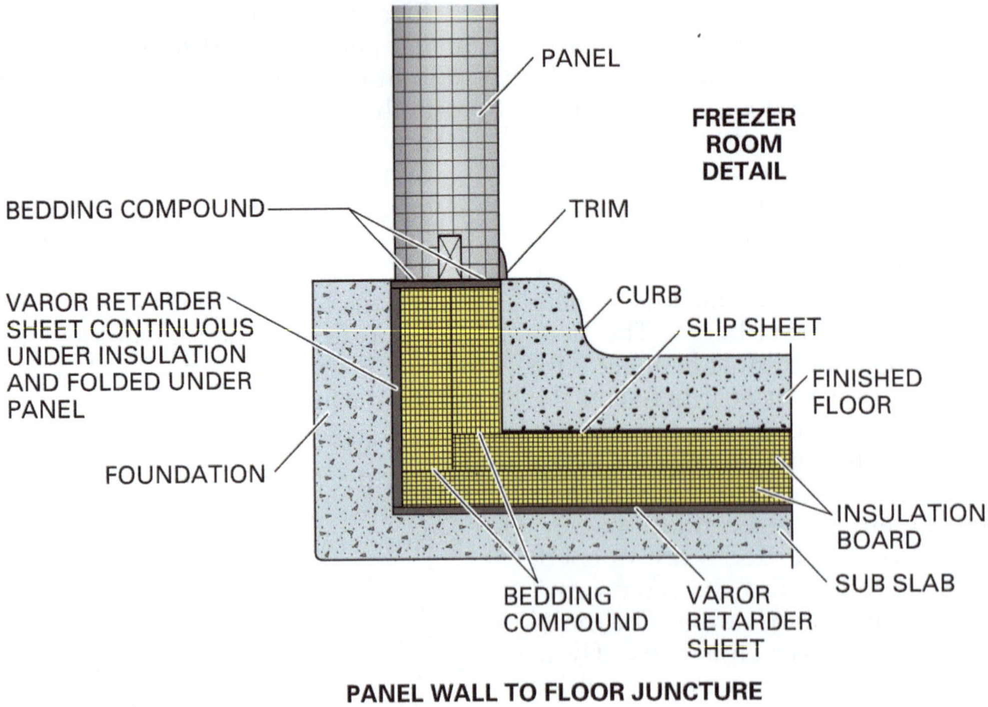

Figure 3 Installation of a vapor retarder sheet.

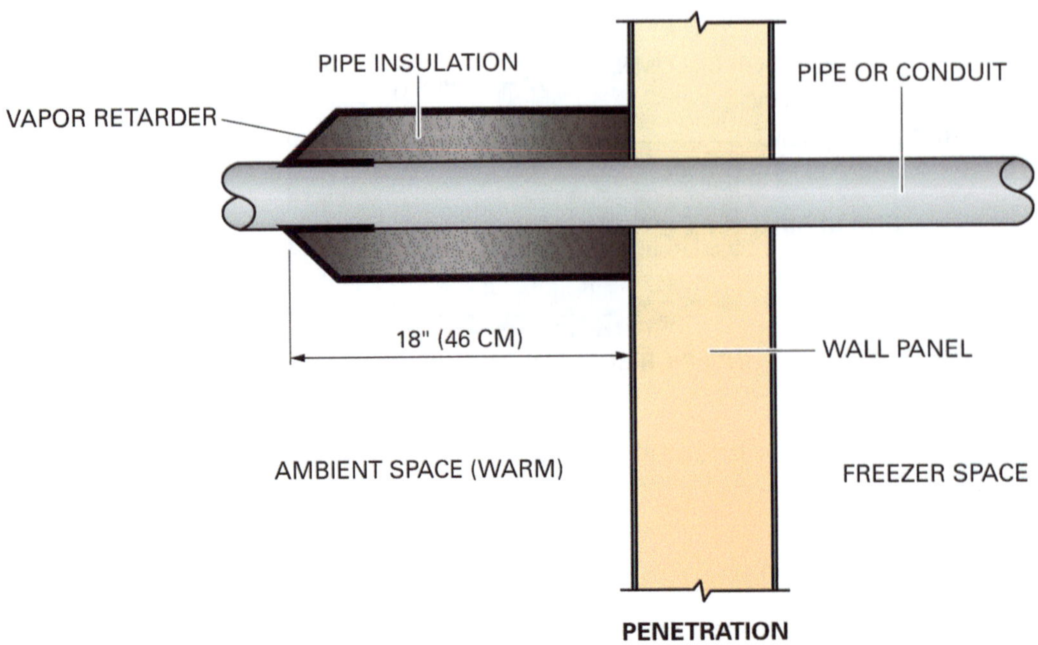

Figure 4 Insulating to prevent condensation on surfaces.

Other prefabricated, low-temperature panels are available for insulation of large air conditioning units, vessels, and other apparatus. While these are specialized, the basics are the same as for low-temperature room panels.

1.1.2 Cryogenic Systems

In addition to food storage and processing, cryogenic systems are used in liquefying gases such as oxygen, nitrogen, and natural gas. When gases of the elements and natural compounds such as ammonia and methane are subject to extreme compression and low temperatures, they change into liquid form. Oxygen, for example, liquefies at –297.3°F (–182.9°C). It is beneficial to convert gases to liquid form because a much greater mass of gas can be transported and stored in a given volume as a liquid than in gaseous form. These liquids must be kept cold and under pressure.

Liquid oxygen is an example of liquefied gas stored in tanks at hospitals and other locations where large amounts of oxygen are used. Another example is natural petroleum gas condensed to a liquid form (called liquefied natural gas, or LNG) so it can be carried from production facilities to distribution points in special LNG tanker ships and tank trucks. Most cryogenic systems are double-walled vacuum systems. Larger bulk-storage tanks and systems may be insulated with conventional insulation, typically in layers separated by a vacuum, or have an outer shell that is filled with dry air or nitrogen.

Cryogenic systems are also used to flash-freeze food. A system used to flash-freeze food at cryogenic temperatures is shown in *Figure 5*. This unit is a large, built-in structure that would have thick, multiple layers of insulation and vapor retarders to reduce water vapor entry into the insulation system.

Figure 5 Large cryogenic freeze-drying unit.

Did You Know?

Breathing Produce

Fresh produce such as apples, peaches, and tomatoes continue to ripen long after they are harvested. The biological processes associated with ripening produce heat and ethylene gas. Cold storage removes the heat, but if proper airflow is not maintained, the trapped ethylene gas will speed up the ripening process, especially in ethylene-sensitive produce, such as carrots, cucumbers, and potatoes. Too-rapid ripening can cause premature produce spoilage.

Insulation workers may be involved in insulating pipe, vessels, and equipment that operate at cryogenic temperatures. Typical insulation found in cryogenic applications would be double- or multilayer polyisocyanurate and cellular glass board as well as polystyrene and polyurethane rigid foam. Tightly-sealed, close-fitting joints in material are extremely important when applying insulation to these systems. The vapor-retardant membrane must be complete, as any moisture entering this type of insulation will have an immediate and damaging effect. This means sealing all seams, butt joints, and termination points according to cryogenic work standards. Cryogenic insulation work normally requires additional training and even certification for some industries, such as maritime cargo ship construction and maintenance.

NOTE

Polystyrene board insulation is not used in cryogenic systems due to temperature limits of the insulation material (–65°F to 165°F or –54°C to 74°C).

1.2.0 Refractory and High-Temperature Insulation

Refractory insulation is high-temperature insulation that is generally able to withstand direct exposure to flame or high-temperature combustion products. Refractory insulation is installed on the inside of furnaces, boilers, and other high-temperature (1,000°F/538°C and above) apparatus. One common use is to line the interiors of the furnaces for the huge boilers that are used in electrical power generation plants.

Brick, refractory, insulation, and lagging (BRIL) materials are required for boiler operation to minimize hotspots, prevent loss of heat, protect piping from the elements, and prevent injury to operators

Aerogel Insulations

Low-density insulations such as aerogels are now occasionally being used in modern cryogenic applications. Aerogels are insulation materials technically considered solids, but with a structure similar to kitchen gelatin after the liquid portion has been removed. What is left is a material consisting of a rigid, microscopically open network of a solid. Aerogels are very lightweight and extremely effective both for thermal insulation and sound control. The thermal resistance of aerogels is better than that of other types of insulating materials. Aerogel insulation is normally available in particulate and blanket forms. Many companies are working to develop block and board aerogel insulation in economical quantities. The blanket form is the most common and is the type found in cryogenic applications.

and maintenance people. For the most part, the brick and refractory materials are used inside the furnace area, and the insulation and lagging materials are on the outside of the brick and refractory materials. There may be a metal casing separating the lagging materials from the brick and refractory inside the furnace. The various parts of the wall of a small boiler system are shown in *Figure 6*.

Refractory materials include brick, refractory cement, and ceramic fiber. Most refractory materials are installed by masonry contractors specializing in this type of work. Refractory insulation is available in many types, sizes, and materials. It is also available in two forms, brick and castable. The difference between the two is in the way they are made. Bricks are pressed and castables are poured into a mold and allowed to set until they harden, a process referred to as *casting*.

1.2.1 Bricks

The most common type of refractory material is brick (*Figure 7*). Bricks can be made of the following materials:

- *Mineral wool* – Formed by taking molten slag, glass, or rock, and firing a jet of air or steam through it to create fibers that can be woven. Mineral wool is mixed with clay and pressure-fired to form bricks. It is used to insulate tube walls and boiler casings.
- *Calcium silicate* – After the raw materials are processed, the moist calcium silicate mixture is pressed into forms. It is used on enclosures and piping with temperatures less than 1,200°F (650°C).
- *High alumina* – Made from crushed **ganister** rocks that are mixed with lime and pressed to form bricks. The bricks are then baked in a kiln

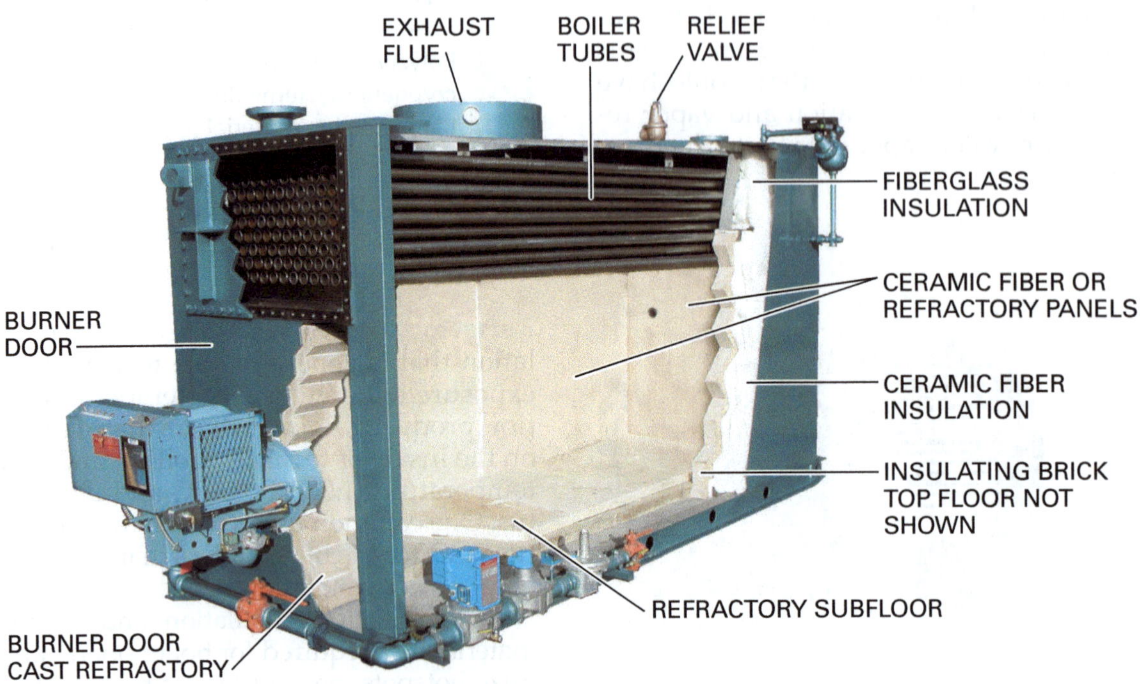

Figure 6 Parts of a boiler system furnace wall.

Figure 7 Examples of refractory bricks.

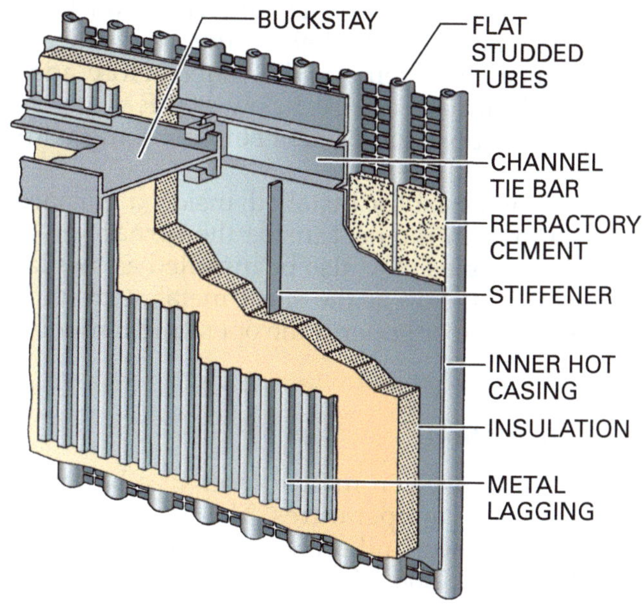

Figure 8 Furnace wall with metal lagging.

for several days to dry them out and stabilize their size. They are mainly used for the high-temperature conditions that exist in lime kilns.

- *Magnesite* – Made from magnesium oxide, the bricks are formed and sent through a tunnel kiln for drying. Much of the raw material for making this type of brick comes from sea water.
- *Chromite* – Formed the same way as magnesite bricks but made from igneous chromite (chromium oxide) ores. A typical use for this type of brick is in sulfite recovery furnaces.
- *Insulating firebrick* – Made from fire clay. One method of manufacturing is to add organic materials to the clay and then bake it so the organic material is baked out, leaving holes in the brick. This type of brick is frequently used on the hot side of the furnace due to its ability to expand and contract, thus shortening boiler warmup times. It is also good for oil-fired boilers. This brick cannot withstand high-lime slags or fluid coal slags that collect at the bottoms of industrial boiler furnaces.
- *Graphite* – Made from almost pure carbon, this material is formed by pressing the carbon into brick shapes. These bricks are particularly well-suited for sound-deadening walls.

Boilers frequently have brick insulation or insulation blankets on the outside of water walls and other furnace structures, with sheet metal or lagging installed over it to protect the insulating materials. The use of these external materials is for insulating purposes and to protect operators and workers from the high temperatures inside the furnace.

Brick can be used in just about every situation found in a boiler due to the large variety of types available. *Figure 8* shows the location of insulation brick in a furnace with wall tubes tied together by membrane. In this case, the tubes would be closer to the heat source and the insulating brick would be toward the outside of the surface of the furnace/boiler. The metal lagging sub-girts help keep the insulation brick in place, and lagging installation separates the heated insulation materials from the outside surface of the furnace/boiler.

1.2.2 Castable Refractory

Castable refractories are manufactured by taking portland cement and mixing it with alumina (aluminum oxide) in various percentages. The mixture is then poured into molds. It is available in the following three types:

- *Standard*—Contains 40 percent alumina.
- *Intermediate*—Contains 50–55 percent alumina.
- *Very pure*—Contains 70–80 percent alumina.

GOING GREEN

Hazardous Fumes

Furnaces burn fossil fuels that give off fumes or flue gases. These combustion gases contain chemicals that can cause disease and breathing problems in humans and animals, and are believed to be contributing to global climate change. Ensuring that the bricks and refractory materials are not leaking those fumes or gases out of the facility before they enter pollution control stages of the plant helps protect people and the natural environment.

To produce castables that can withstand temperatures above 800°F (430°C), the portland cement is heated to remove the water before adding the alumina. For use below 800°F, water is added to the portland cement before mixing with alumina.

Bricks or castables installed inside a furnace area help retain the heat inside the furnace area. Some refractories may also be installed as insulation on the outside of the sheet metal surrounding the furnace or boiler if the operating temperature warrants it.

1.2.3 Insulation

Insulation is used in many places in a boiler to reduce surface temperatures. This is done to reduce radiated heat and protect personnel. See NCCER Module 19306, "Industrial Boiler Systems" from *Mechanical Insulating Level Two,* for a more complete description of the insulation of industrial boiler and process plants. Insulation is also used on cool surfaces to prevent heat transfer and sweating of cold piping. *Figure 9* shows a few examples of insulation other than refractory that may be used on boiler furnaces.

1.3.0 Spray-On and Foamed-in-Place Polyurethane Insulation

Spray-on insulations include polyurethane expanding foam, and mineral wool and cellulose fibers in a foam carrier. Spray-on (SPF) and foam-in-place (FIP) polyurethane is used for low temperature coatings, roughing-in, and tank insulation coatings for ease of application. SPF is suitable for service temperatures less than 300°F (150°C). Spray-on insulations that contain cellulose and mineral wool are commonly required for fire-rated structures, which will be discussed in Section 1.4.0.

SPF insulation is used to insulate large surfaces such as roofs, irregularly-shaped equipment, and cavities. It is installed with special spray equipment. Kits of small pressurized cylinders of polyurethane are used for small jobs and to fill cavities. SPF is a two-part component consisting of a resin and an activator, which are mixed in the spray gun or dispenser. It is discharged as a liquid, but immediately foams and becomes rigid. Most mechanical insulation workers will not apply SPF, since its use and application is considered structural or architectural insulation. Those who do must always follow the manufacturer's guidelines for safe use when working with SPF insulation.

SPF can be installed on any surface that operates between –22°F and 300°F (–30°C to 150°C), depending on the formulation of the product. Surfaces must be dry and clean for application. Outside SPF systems must include a UV-protective coating because it will disintegrate when exposed to direct or reflected sunlight.

NOTE

Polyurethane foam is made with flame-retardant additives, but it will produce toxic gases when exposed to fire or flame. These materials will burn and contribute smoke when exposed to a direct flame. It is considered safe when installed in a fire-retardant structure such as a gypsum board wall.

FIP kits may be used to fill voids or insulate irregular surfaces. A kit consists of two pressurized cylinders with plastic hoses and a plastic dispenser. Both hoses connect to the chemicals which mix in the dispenser nozzle. Several dispenser nozzles may be furnished with the kit since nozzles cannot be reused.

CAUTION

Do not install FIP polyurethane when the temperature of the surface to which it will be applied is above 180°F (82°C) during application.

1.4.0 Fire-Rated Construction

Every wall, floor, and ceiling in a building is rated for its fire resistance, as established by building codes. A fire rating is stated in terms of hours, such as a one-, two-, three-, or four-hour wall. Other parts of the structure are also rated, such as floors, ceilings, and the roof. The rating denotes the length of time an assembly can withstand fire and provide protection from it, as determined under laboratory conditions. The greater the fire rating, the thicker the wall is likely to be, and the more fire-resistant the materials and construction methods. Two examples of fire-rated wall assemblies are shown in *Figure 10.*

In multifamily residential construction, such as apartments and town houses, the walls and ceilings dividing the adjoining occupied spaces must meet special fire and soundproofing requirements. The code requirements will vary from one location to another and may even vary within areas of a jurisdiction. For example, dwellings in high-risk areas may have stricter standards than those in other areas of the same city or county. In some cases, the code may

(A) CELLULAR GLASS INSULATION

(B) FOAMED IN INSULATION

(C) LOOSE FILL INSULATION

Figure 9 Examples of furnace insulation.

require a masonry wall between dwellings. This masonry wall may even be required to penetrate the roof of the building so that if a fire occurs, it is contained within the unit in which it started because it is unable to travel through the walls or across the attic space.

There are many different construction methods for these so-called party walls between apartments. Each is designed to meet different fire and soundproofing standards. The wall is likely to be more than 3 inches (8 cm) thick and contain several layers of drywall and insulation. A fire-rated wall may abut a nonrated partition or wall. In this case, the rated wall must be carried through to maintain the fire rating. Most conventional insulation material must terminate at fire walls or floors. An approved firestop material must be placed in any opening in the wall or floor in order to maintain the fire rating. Voids around uninsulated pipe or duct must be filled with a firestop material.

Spray-on insulation consisting of mineral wool or cellulose fiber is commonly used as fire retardants to improve the fire-rating of structures. The cellulose type of spray-on insulation is used for protection of metal support systems to prevent the intense heat from compromising the building structure. In many commercial structures, the steel used for beams and girders must be encased in fire-retardant material in order to meet the building code requirements. In *Figure 11*, fireproofing material is being sprayed onto a beam.

1.4.1 Firestopping

Firestopping is the combination of materials and methods used to close holes and penetrations made in fire-rated walls and floors for cabling, electrical conduits, and pipes. The firestopping in combination with the penetrating object restores the barrier's design fire rating. If such openings are not sealed, fire will travel through the openings, supplying oxygen to the fire. Penetrations and breaches in fire-rated barriers also include

Polyurethane Spray Foam Application

The SS *Coastal Progress* is a seafood storage and processing ship that receives and processes product from fishing vessels. The on-board storage space must be maintained at –30°F (–34°C) in order to preserve the seafood while the ship is at sea. In 2011, the ship was put into dry dock for a major overhaul and expansion. The ship was taken apart and her size was increased by 45 feet (14 m), expanding her storage capacity by 35 percent. Part of that expansion involved removing all the old foam insulation and replacing it with a minimum depth of 2 inches (5 cm) of polyurethane spray foam. One of the reasons for choosing urethane spray foam is that fact that it requires no fasteners that could affect the integrity of the ship's hull.

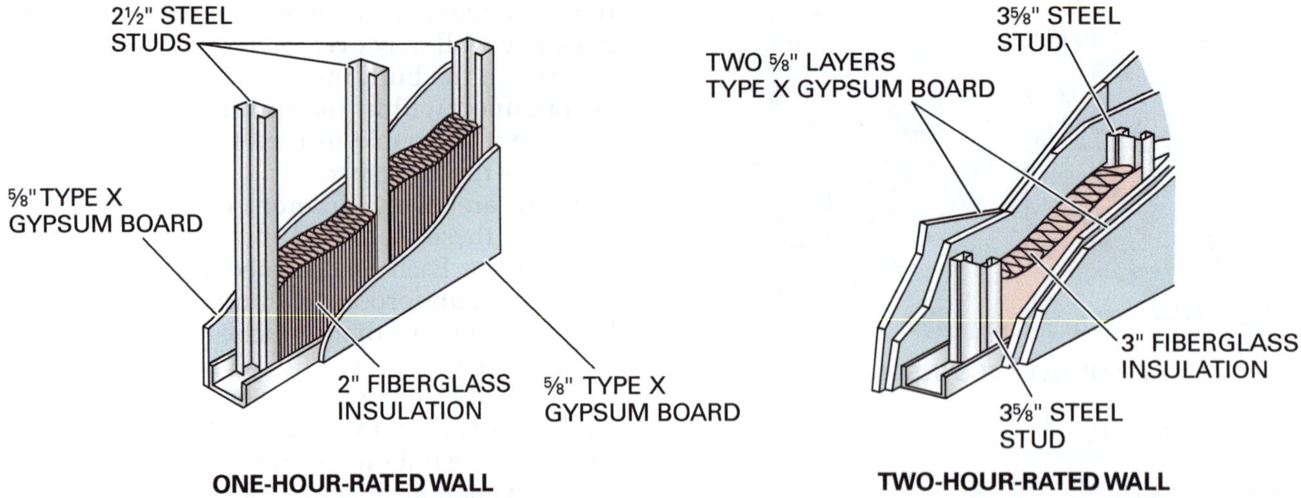

Figure 10 Examples of fire-rated wall assemblies.

Figure 11 Spray-on fireproofing application.

air conditioning ducts, reserved spaces for future penetrations, and spaces at the heads of nonbearing walls. Other types of assemblages, such as fire and smoke dampers, as well as structural modifications are required to address those conditions and are not considered firestopping. Various firestopping methods are shown in *Figure 12*.

In order to meet the fire-rating standards established by the building and fire codes, the openings must be sealed. The firestopping methods used for this purpose are classified as *mechanical* and *non-mechanical*. Mechanical firestops are prefabricated solid or flexible devices such as the one shown in *Figure 12* (*D*) that are designed to physically fill up the opening. Non-mechanical firestops are field-applied fire-resistant materials, such as caulks, sprays, and putties that are used to fill the space around the conduit or piping. Firestopping materials are typically applied around all types of piping, electrical conduit, ductwork, electrical and

Did You Know?

Party Walls

A party wall is a fire-rated wall that separates dwellings, such as town houses and multi-occupant commercial structures. The codes for such a wall require that the structure on one side of the wall is protected from a fully involved fire on the opposite side. The photo shown here illustrates why so much emphasis is placed on party walls.

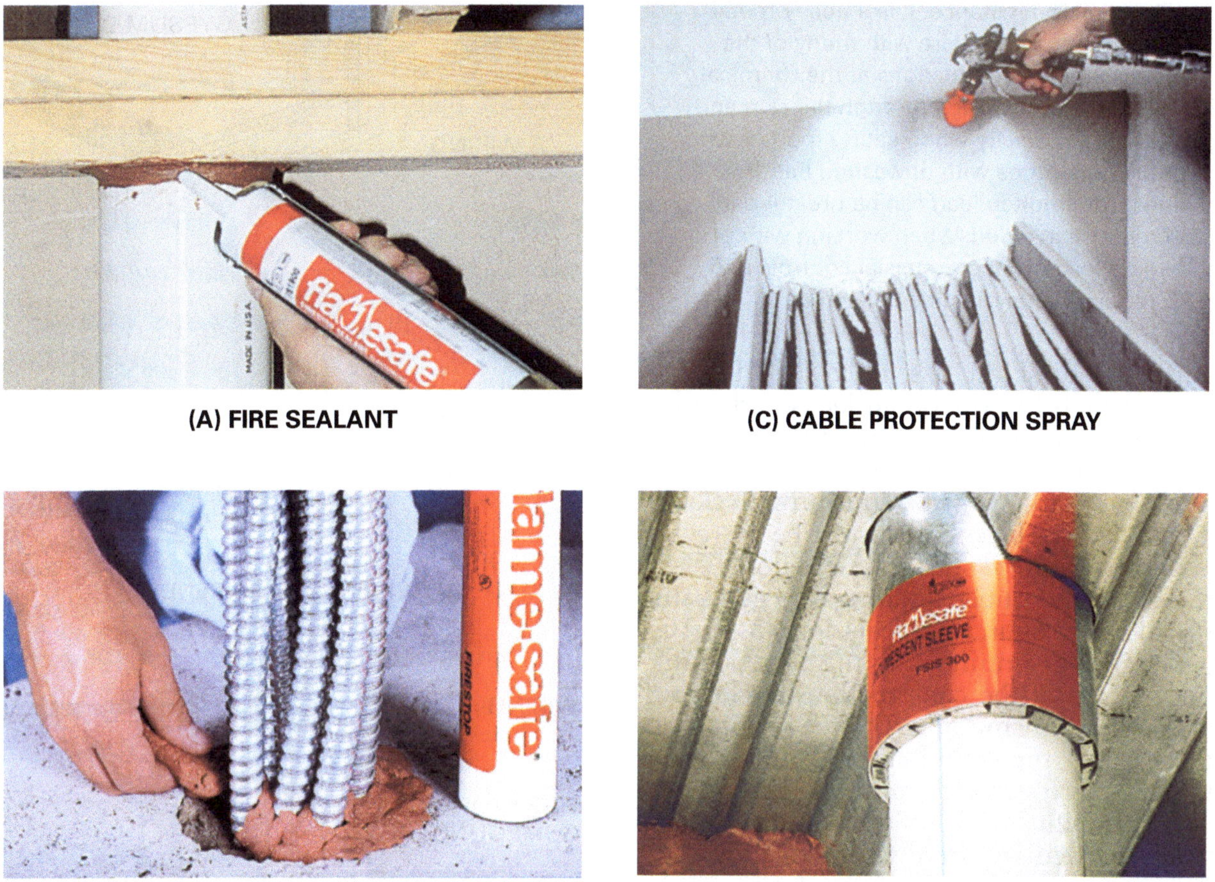

(A) FIRE SEALANT

(C) CABLE PROTECTION SPRAY

(B) INTUMESCENT PUTTY

(D) INTUMESCENT PIPE SLEEVE

Figure 12 Firestopping applications.

communication cables (*Figure 13*), and similar devices that run through openings in floor slabs, walls, and other fire-rated building partitions and assemblies.

Non-mechanical firestopping materials that act by themselves are classified as **intumescent** or **endothermic**. Both are made to help control the spread of fire before, during, and after exposure to open flames. Intumescent materials expand, increasing their insulating abilities, and can fill up holes when the penetrating materials are burned up, limiting spread of the fire. Endothermic materials react with heat, absorbing it and lowering the temperatures of the fire-rated structure in contact with it. Other materials chemically change when heated, forming an incombustible ash coating that flakes off as it protects the material.

These firestopping materials are made in a variety of forms, including composite sheeting, caulks, silicone sealants, foams, moldable putty, wrap strips, and spray coatings. They come in both one-part and two-part formulations. The installation of these materials must always be done in accordance with the applicable building codes and the manufacturer's instructions for the product being used.

1.5.0 Insulation for Sound Control

Sound control applications consist of conventional mineral or glass fiber insulations, special coating compounds, lead sheets, lead-laminated jackets, and equivalent-but-safer mass-loaded vinyl (MLV) jackets. The lead-laminated pipe and duct jackets may be regular aluminum jacket with a light lead sheet laminated to the inside. The mastics are similar to vapor retarder or weather-barrier mastics. Other laminates may contain lead as a major component. Products that do not contain lead but work in the same way may be used to replace lead products. Application techniques are similar to those used to apply conventional insulation and weatherproofing. Other applications of sound control material include sprayed-on cellulose and mineral fibers.

WARNING!

Lead is a toxic substance. It is a heavy metal element that can interfere with many of the body's systems and functions at the chemical level. It can be absorbed through the skin and by mouth, especially when eating food or when smoking cigarettes with unwashed hands. Fumes from molten lead can be breathed in and must be avoided. When working with or handling lead products, special controls and PPE requirements must be observed.

The first step of an assembly for airborne sound isolation is to close off air leaks and flanking paths, such as routes sound can travel over or under, around walls, through the windows and doors, through air ducts, and through floors and crawl spaces. These paths must be correctly treated in order to limit sound penetration.

Commercial and public buildings are generally required to meet a sound transmission class (STC) rating. The STC is a numeric rating representing the effectiveness of the construction in isolating airborne sound transmission. The higher the STC rating, the better the sound absorption. Hairline cracks and other openings can have an adverse effect on the ability of a building to achieve its STC rating, particularly in higher-rated construction. Where a very high STC performance is needed, air conditioning, heating, and ventilating ducts should not be included as parts of the structure. Failure to observe special construction and design details can destroy the effectiveness of the best assembly. Improved sound isolation is obtained by the following methods:

- Separate framing for the two sides of a wall
- Resilient-channel mounting for the drywall
- Using sound-absorbing materials in wall cavities
- Using adhesive-applied drywall of varying thicknesses in multilayer construction
- Caulking the perimeters of drywall partitions, openings in walls and ceilings, partition/mullion intersections, and electrical receptacle box openings
- Locating recessed wall fixtures in different stud cavities

The entire perimeter of sound-isolating partitions should be caulked around the drywall edges to make it airtight. The caulking should be a non-hardening, non-shrinking, non-bleeding, non-staining resilient sealant. Sound-control sealing must be covered in the specifications, understood by all those working with it, supervised by the appropriate party, and inspected carefully as the construction progresses.

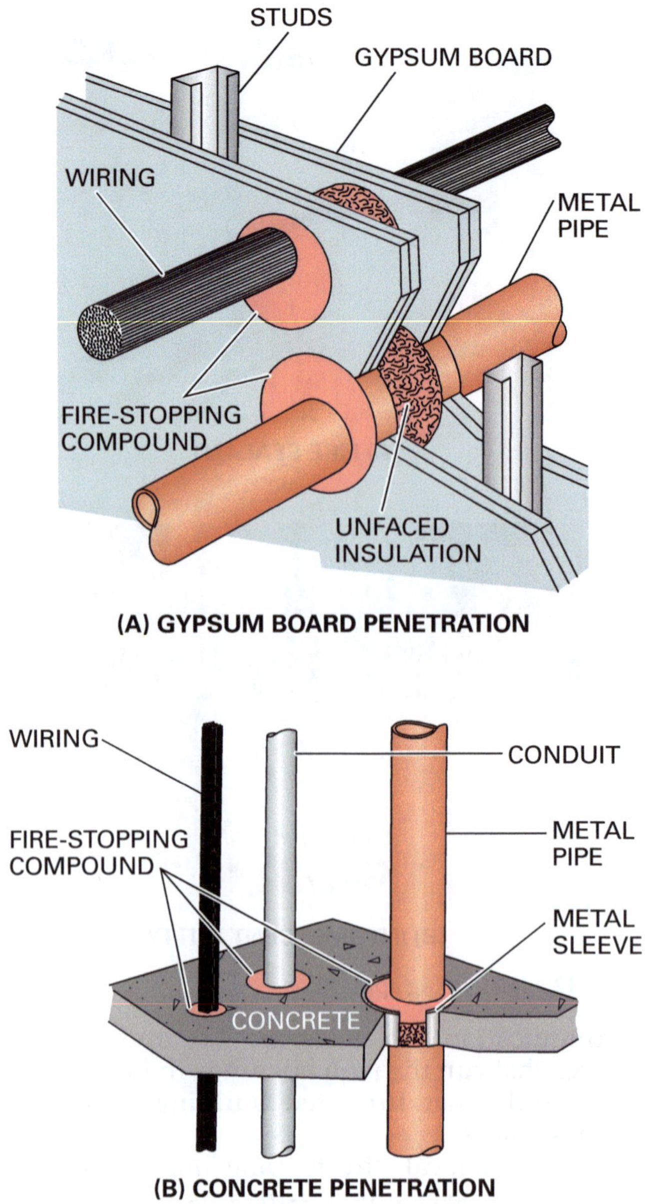

Figure 13 Fire-rated openings.

1.5.1 Sound-Isolating Materials

A variety of materials are used for sound control. These sound-isolating materials include the following:

- Mineral fiber (including glass) blankets and batts used in wood stud assemblies
- Semi-rigid mineral or glass-fiber blankets for use with steel studs and laminated gypsum partitions
- Mineral (including glass) fiberboard
- Gypsum-core sound-insulating board used behind drywall, applied with adhesive or mechanically fastened
- Rigid plastic foam furring systems
- Lead or other special shielding materials
- Sprayed-on cellulose

- Mass-loaded vinyl products

Mineral wool or glass fiber insulating batts and blankets may be used in assembly cavities to absorb airborne sound within the cavity. They should be placed in the cavity and carefully fitted behind electrical outlets and around any cutouts necessary for plumbing lines. Insulating batts and blankets may be faced with paper or a vapor barrier, and they may have flanges or be of the unfaced friction-fitted type.

1.6.0 Poured-In-Place Insulation and Preinsulated Piping Systems

It is common for piping runs to be buried under ground or under foundation slabs. Examples include hot water, chilled water, and steam pipes. Some buried piping must be insulated in order to maintain the temperature of the fluid being carried. Cellular glass foam insulation, mineral wool, and fiberglass pipe wrap are among the methods used to insulate such pipes. Other methods include poured-in-place insulation and preinsulated pipe systems.

1.6.1 Poured-in-Place Insulation

Poured-in-place insulation is typically a granular substance designed to be poured over underground piping and ductwork once they are placed in a trench. These insulation materials are made from calcium carbonate or calcium silicate combined with other materials.

These materials will not absorb water. They are poured into a form and then consolidated and compacted using a vibrator of the type used to consolidate concrete. When compacted, the insulation becomes solid and is designed to hold the weight of a person without leaving more than a 1-inch (2.5 cm) deep impression.

1.6.2 Preinsulated Pipe Systems

Preinsulated pipe systems are widely used for underground chilled water pipe, steam pipe, and other hot-pipe systems. Most preinsulated chilled water pipe systems and moderate-to-warm temperature (below 200°F or 93°C) pipe systems consist of the pipe carrying the fluid and an outer shell of PVC or fiberglass reinforced plastic. The space between the pipe and the outer shell is filled with polyurethane foam. Joint covers may be furnished with the pipe system and are installed after the pipe sections are connected and tested.

Valves on pipe systems of this type are usually installed in valve pits. The pipe in the pits may be conventional pipe and require field-applied insulation. This type of pipe system is rarely installed above ground.

Preinsulated underground steam pipe and other high-temperature, hot-pipe systems (above 200°F or 93°C) consist of the steam pipe, insulation, and an outer steel shell (*Figure 14*). More than one pipe can be inside the same shell. The shell is similar to a steel culvert. These pipe systems stop at valve pits and manholes. Conventional pipe and insulation are installed in the pits and manholes.

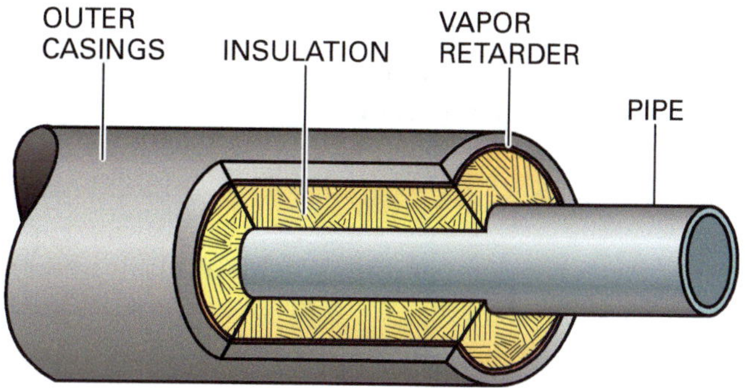

(A)

(B)

Figure 14 Preinsulated pipe.

Additional Resources

Mechanical Insulation Design Guide, National Institute of Building Sciences. Available at **www.wbdg.org/guides-specifications/mechanical-insulation-design-guide**

National Industrial and Commercial Insulation Standards Manual, Midwest Insulation Contractors Association (MICA).

National Insulation Association (NIA) website offers resources for products and training. **www.insulation.org**

Gypsum Construction Handbook, US Gypsum. Current Edition. Chicago, IL: RSMeans.

"Dritherm™ Installation Sheet," Dritherm International, Inc., copyright © 2007. **www.dritherm.com**

"Tech Data Sheet 3.0—Installation Sequence," Gilsulate International Inc., copyright ©2004. **www.gilsulate.com**

"How to Insulate for Sound Control," Chris Deziel, San Francisco Chronicle/SFGATE. **homeguides.sfgate.com**

1.0.0 Section Review

1. At what temperature might a blast freezer operate?
   a. –50°F (–46°C)
   b. –10°F (–23°C)
   c. 0°F (–18°C)
   d. 20°F (–7°C)

2. What material is mixed with mineral wool to make mineral wool refractory bricks?
   a. Asbestos
   b. Alumina
   c. Clay
   d. Graphite

3. Which of these is a common use for spray-on polyurethane foam insulation?
   a. Low-temperature work
   b. Fireproofing
   c. Insulating boilers
   d. Poured-in-place insulation

4. Which of the following items is a type of mechanical firestopping method?
   a. Steel, self-closing room door
   b. Field-applied, void-filling foam
   c. Prefabricated solid and flexible devices
   d. Field-applied, fire-retardant caulking

5. An STC rating is related to _____.
   a. fire retardant capability
   b. thermal conductivity
   c. moisture control
   d. sound isolation

6. Poured-in-place underground pipe insulation is typically a _____.
   a. granular substance
   b. liquid substance
   c. block material
   d. loose fill

SUMMARY

There are specialized insulation systems that the average mechanical insulation worker will not regularly encounter. These include the following:

- Low-temperature systems
- Refractory insulations
- Spray-on polyurethane insulations
- Fire-rated construction
- Sound transmission control
- Poured-in-place insulation and underground preinsulated piping

Some of these materials are used as part of conventional insulation work, while others are highly specialized and rarely encountered. However, a competent insulation mechanic should have some knowledge of all types of insulation systems. Generally, these systems are installed by specialty contractors. This module covers the most common materials and applications to enable the insulation mechanic to become familiar with these systems, know what they consist of, and understand their functions.

Review Questions

1. Preinsulated panels used on refrigeration units typically have a core of polystyrene or _____.
   a. mineral wool
   b. refractory material
   c. polyisocyanurate
   d. polyurethane foam

2. A bulk storage tank that has an outer shell filled with dry nitrogen gas is typically used in _____.
   a. cryogenic applications
   b. sound transmission control
   c. fireproofing
   d. below-ambient refrigeration

3. Insulating material made from magnesium oxide is likely to be used in _____.
   a. low-temperature applications
   b. firestopping putty
   c. sound transmission control
   d. refractory insulation

4. An aluminum oxide castable refractory that contains 70–80 percent alumina is classified as _____.
   a. standard
   b. intermediate
   c. very pure
   d. inadequate

5. A material commonly used as a spray-on fire retardant substance is _____.
   a. mineral wool
   b. refractory
   c. glass fiber
   d. polyurethane foam

6. The structure that separates adjoining dwellings for fire protection is often called a(n) _____.
   a. firestop
   b. party wall
   c. endothermic wall
   d. non-mechanical fire barrier

7. Firestopping applies to _____.
   a. treating penetrations to restore a fire-rated barrier to its full standard
   b. fireproof finishes on walls to provide a greater than 3-hour rating
   c. major openings such as doorways and ventilation plenums
   d. protective coatings on sprinkler system piping

8. Mass-loaded vinyl jackets are used in _____.
   a. firestopping
   b. refractory materials
   c. sound control
   d. low-temperature systems

9. Which of the following materials is *not* used in sound control?
   a. Mineral fiber
   b. Rigid foam
   c. Fiberglass blanket
   d. Refractory brick

10. Which temperature range is suitable for pre-insulated pipe?
    a. Cryogenic (less than –50°F or –46°C)
    b. Above 47°F or 7°C (chilled water temperature) to less than 200°F (93°C)
    c. Above 47°F (7°C) to greater than 200°F (93°C)
    d. Greater than 200°F (93°C)

Trade Terms Introduced in This Module

Cryogenic: For mechanical insulating purposes, any application or system operating at temperatures below –50°F (–46°C). However, the National Institute of Standards and Technology defines the upper end of the cryogenic band at –292°F (–180°C) for scientific purposes.

Endothermic: Any process that gains thermal energy (absorbs heat) from it surroundings.

Firestopping: A material or construction method that interferes with the entry or movement of fire through a fire-rated barrier.

Footings: The underlying supports for a column or wall.

Ganister: A hard, fine-grained quartzite rock (naturally-cemented quartz sandstone) used in high-temperature insulation.

Intumescent: The property of a material that swells when heated, increasing its volume and decreasing its density, usually by the production of gas voids inside the material. The insulating property of the voids slows the transmission of heat through the material.

Loadbearing: The function of any structural member, such as a wall, that supports part of the structure's weight.

Non-conducting: The property of a substance that resists or prevents the movement of heat or electricity.

Refractory insulation: Temperature-resistant insulation that can retain its properties above 1,000°F (538°C). Typically installed inside of furnaces and other high-temperature apparatus.

Sound isolation: Any material or construction method that interferes with the movement of sound (noise) from its source to somewhere else.

Ventilation: The natural or artificial entry and exchange of fresh air within a structure.

Additional Resources

This module presents thorough resources for task training. The following reference material is recommended for further study.

Mechanical Insulation Design Guide, National Institute of Building Sciences. Available at **www.wbdg.org/guides-specifications/mechanical-insulation-design-guide**

National Industrial and Commercial Insulation Standards Manual, Midwest Insulation Contractors Association (MICA).

National Insulation Association (NIA) website offers resources for products and training. **www.insulation.org**

Gypsum Construction Handbook, US Gypsum. Current Edition. Chicago, IL: RSMeans.

"Dritherm™ Installation Sheet," Dritherm International, Inc., copyright © 2007. **www.dritherm.com**

"Tech Data Sheet 3.0—Installation Sequence," Gilsulate International Inc., copyright ©2004. **www.gilsulate.com**

"How to Insulate for Sound Control," Chris Deziel, San Francisco Chronicle/SFGATE. **homeguides.sfgate.com**

Figure Credits

GEA Process Engineering, Inc. - www.niroinc.com, Figure 5

Rite Engineering & Manufacturing Corporation, Figure 6

Photo courtesy of Joe Moore & Company, Inc., Figure 7

Courtesy of The Babcock and Wilcox Company, Figure 8

Pittsburgh Corning Corporation, Figure 9A

Unifrax I LLC, Figures 9B, 9C

RectorSeal, Figure 12

National Gypsum Company, SA01

Insul-Tek® Piping Systems Inc., Figure 14B

Section Review Answer Key

Answer	Section Reference	Objective
Section One		
1. a	1.1.0	1a
2. c	1.2.1	1b
3. a	1.3.0	1c
4. c	1.4.1	1d
5. d	1.5.0	1e
6. a	1.6.1	1f

NCCER CURRICULA — USER UPDATE

NCCER makes every effort to keep its textbooks up-to-date and free of technical errors. We appreciate your help in this process. If you find an error, a typographical mistake, or an inaccuracy in NCCER's curricula, please fill out this form (or a photocopy), or complete the online form at **www.nccer.org/olf**. Be sure to include the exact module ID number, page number, a detailed description, and your recommended correction. Your input will be brought to the attention of the Authoring Team. Thank you for your assistance.

Instructors – If you have an idea for improving this textbook, or have found that additional materials were necessary to teach this module effectively, please let us know so that we may present your suggestions to the Authoring Team.

NCCER Product Development and Revision

13614 Progress Blvd., Alachua, FL 32615

Email: curriculum@nccer.org
Online: www.nccer.org/olf

❑ Trainee Guide ❑ Lesson Plans ❑ Exam ❑ PowerPoints Other ____________

Craft / Level: Copyright Date:

Module ID Number / Title:

Section Number(s):

Description:

Recommended Correction:

Your Name:

Address:

Email: Phone:

Fundamentals of Crew Leadership

Overview

When a crew is assembled to complete a job, one person is appointed the leader. This person is usually an experienced craft professional who has demonstrated leadership qualities. While having natural leadership qualities helps in becoming an effective leader, it is more true that "leaders are made, not born." Whether you are a crew leader or want to become one, this module will help you learn more about the requirements and skills needed to succeed.

Module 46101

NCCER

President: Don Whyte
Vice President: Steve Greene
Chief Operations Officer: Katrina Kersch
Director of Product Development: Daniele Dixon
Fundamentals of Crew Leadership Project Manager: Mark Thomas
Senior Development Manager: Mark Thomas
Senior Production Manager: Tim Davis
Quality Assurance Coordinator: Karyn Payne
Desktop Publishing Coordinator: James McKay
Permissions Specialist: Adrienne Payne
Production Specialist: Adrienne Payne
Editor: Graham Hack

Writing and development services provided by Topaz Publications, Liverpool, NY

Lead Writer/Project Manager: Thomas Burke
Desktop Publisher: Joanne Hart
Art Director: Alison Richmond
Permissions Editor: Andrea LaBarge
Writer: Thomas Burke

Pearson

Director of Alliance/Partnership Management: Andrew Taylor
Editorial Assistant: Collin Lamothe
Program Manager: Alexandrina B. Wolf
Digital Studio Project Managers: Heather Darby, Tanika Henderson, Jose Carchi
Director of Marketing: Leigh Ann Simms
Senior Marketing Manager: Brian Hoehl
Composition: NCCER
Printer/Binder: RR Donnelley
Cover Printer: RR Donnelley
Text Fonts: Palatino, Minion Pro, and Univers

Trainees with successful module completions may be eligible for credentialing through the NCCER Registry. To learn more, go to **www.nccer.org** or contact us at 1.888.622.3720. Our website has information on the latest product releases and training, as well as online versions of our *Cornerstone* magazine and Pearson's product catalog.

Your feedback is welcome. You may email your comments to **curriculum@nccer.org**, send general comments and inquiries to **info@nccer.org**, or fill in the User Update form at the back of this module.

This information is general in nature and intended for training purposes only. Actual performance of activities described in this manual requires compliance with all applicable operating, service, maintenance, and safety procedures under the direction of qualified personnel. References in this manual to patented or proprietary devices don't constitute a recommendation of their use.

46101 V3

46101
Fundamentals of Crew Leadership

Objectives

When you have completed this module, you will be able to do the following:

1. Describe current issues and organizational structures in industry today.
   a. Describe the leadership issues facing the construction industry.
   b. Explain how gender and cultural issues affect the construction industry.
   c. Explain the organization of construction businesses and the need for policies and procedures.
2. Explain how to incorporate leadership skills into work habits, including communications, motivation, team-building, problem-solving, and decision-making skills.
   a. Describe the role of a leader on a construction crew.
   b. Explain the importance of written and oral communication skills.
   c. Describe methods for motivating team members.
   d. Explain the importance of teamwork to a construction project.
   e. Identify effective problem-solving and decision-making methods.
3. Identify a crew leader's typical safety responsibilities with respect to common safety issues, including awareness of safety regulations and the cost of accidents.
   a. Explain how a strong safety program can enhance a company's success.
   b. Explain the purpose of OSHA and describe the role of OSHA in administering worker safety.
   c. Describe the role of employers in establishing and administering safety programs.
   d. Explain how crew leaders are involved in administering safety policies and procedures.
4. Demonstrate a basic understanding of the planning process, scheduling, and cost and resource control.
   a. Describe how construction contracts are structured.
   b. Describe the project planning and scheduling processes.
   c. Explain how to implement cost controls on a construction project.
   d. Explain the crew leader's role in controlling project resources and productivity.

Performance Tasks

Under the supervision of your instructor, you should be able to do the following:

1. Develop and present a look-ahead schedule.
2. Develop an estimate for a given work activity.

Trade Terms

Autonomy
Bias
Cloud-based applications
Craft professionals
Crew leader
Critical path
Demographics
Ethics
Infer
Intangible
Job description
Job diary
Legend
Lethargy
Letter of instruction (LOI)
Local area networks (LAN)
Lockout/tagout (LOTO)
Look-ahead schedule
Negligence
Organizational chart
Paraphrase
Pragmatic
Proactive
Project manager
Return on investment (ROI)
Safety data sheets (SDS)
Sexual harassment
Smartphone
Superintendent
Synergy
Textspeak
Wide area networks (WAN)
Wi-Fi
Work breakdown structure (WBS)

Industry Recognized Credentials

If you are training through an NCCER-accredited sponsor, you may be eligible for credentials from NCCER's Registry. The ID number for this module is 46101. Note that this module may have been used in other NCCER curricula and may apply to other level completions. Contact NCCER's Registry at 888.622.3720 or go to **www.nccer.org** for more information.

Contents

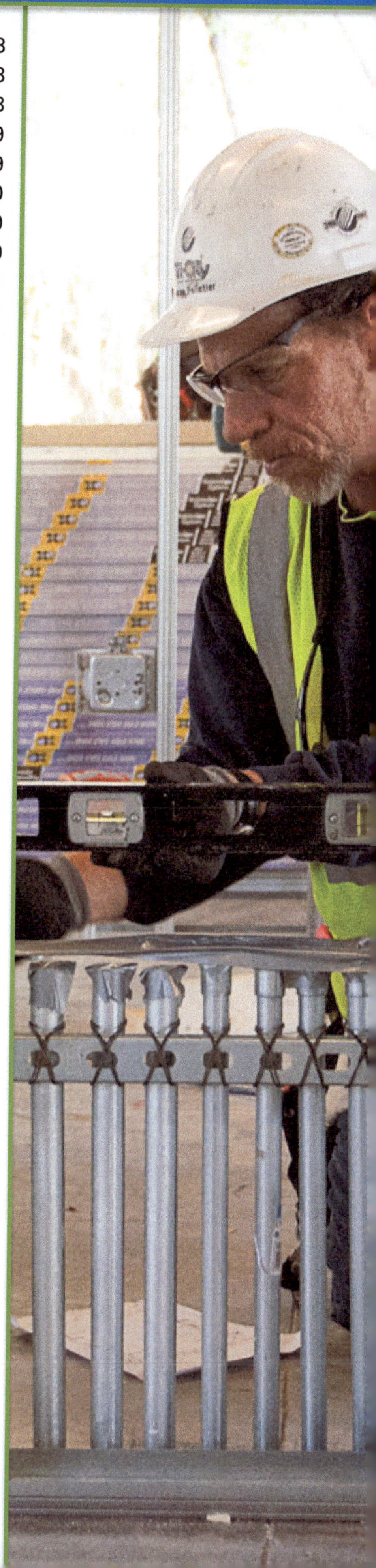

Figures and Tables

SECTION ONE

1.0.0 BUSINESS STRUCTURES AND ISSUES IN THE INDUSTRY

Objective

Describe current issues and organizational structures in industry today.

a. Describe the leadership issues facing the construction industry.
b. Explain how gender and cultural issues affect the construction industry.
c. Explain the organization of construction businesses and the need for policies and procedures.

Trade Terms

Cloud-based applications: Mobile and desktop digital programs that can connect to files and data located in distributed storage locations on the Internet ("the Cloud"). Such applications make it possible for authorized users to create, edit, and distribute content from any location where Internet access is possible.

Craft professionals: Workers who are properly trained and work in a particular construction trade or craft.

Crew leader: The immediate supervisor of a crew or team of craft professionals and other assigned persons.

Demographics: Social characteristics and other factors, such as language, economics, education, culture, and age, that define a statistical group of individuals. An individual can be a member of more than one demographic.

Job description: A description of the scope and responsibilities of a worker's job so that the individual and others understand what the job entails.

Letter of instruction (LOI): A written communication from a supervisor to a subordinate that informs the latter of some inadequacy in the individual's performance, and provides a list of actions that the individual must satisfactorily complete to remediate the problem. Usually a key step in a series of disciplinary actions.

Local area networks (LAN): Communication networks that link computers, printers, and servers within a small, defined location, such as a building or office, via hard-wired or wireless connections.

Organizational chart: A diagram that shows how the various management and operational responsibilities relate to each other within an organization. Named positions appear in ranked levels, top to bottom, from those functional units with the most and broadest authority to those with the least, with lines connecting positions indicating chains of authority and other relationships.

Paraphrase: Rewording a written or verbal statement in one's own words in such a way that the intent of the original statement is retained.

Pragmatic: Sensible, practical, and realistic.

Sexual harassment: Any unwelcome verbal or nonverbal form of communication or action construed by an individual to be of a sexual or gender-related nature.

Smartphone: The most common type of cellular telephone that combines many other digital functions within a single mobile device. Most important of these, besides the wireless telephone, are high-resolution still and video cameras, text and email messaging, and GPS-enabled features.

Synergy: Any type of cooperation between organizations, individuals, or other entities where the combined effect is greater than the sum of the individual efforts.

Wide area networks (WAN): Dedicated communication networks of computers and related hardware that serve a given geographic area, such as a work site, campus, city, or a larger but distinct area. Connectivity is by wired and wireless means, and may use the Internet as well.

Wi-Fi: The technology allowing communications via radio signals over a LAN or WAN equipped with a wireless access point or the Internet. (Wi-Fi stands for wireless fidelity.) Many types of mobile, portable, and desktop devices can communicate via Wi-Fi connections.

Today's managers, supervisors, and lower-level managers face challenges different from those of previous generations. To be a **crew leader** today, it is essential to be well prepared. Crew leaders must understand how to use various types of new technology. In addition, they must have the knowledge and skills needed to manage, train, and communicate with a culturally-diverse workforce whose attitudes toward work and job expectations may differ from those of earlier generations and cultures.

A summary of the changes in the workforce, in the work environment, and of industry needs include the following:

- A shrinking workforce
- The growth of construction, communication, and scheduling technology
- Changes in the attitudes and values of **craft professionals**
- The rapidly-changing **demographics** of gender, gender-identity, and foreign-born workers
- Increased emphasis on workplace safety and health
- Greater need for more education and training

1.1.0 Leadership Issues and Training Strategies

Effective craft training programs are necessary if the industry is to meet the forecasted worker demands. Many skilled, knowledgeable craft professionals, crew leaders, and managers have reached retirement age. In 2015, the generation of workers called *"Baby Boomers,"* who were born between 1946 and 1964, represented 50 percent of the workforce. Their departure creates a large demand for craftworkers across the industry. The US Department of Labor (DOL) concludes that the best way for industry to reduce shortages of skilled workers is to create more education and training opportunities. The DOL suggests that companies and community groups form partnerships and create apprenticeship programs. Such programs could provide younger workers, including women and minorities, with the opportunity to develop job skills by giving them hands-on experience.

When training workers, it is important to understand that people learn in different ways. Some people learn by doing, some people learn by watching or reading, and others need step-by-step instructions as they are shown the process. Most people learn best through a combination of styles. While you may have the tendency to teach in the style that you learn best, you must always stay in tune with what kind of learner you are teaching. Have you ever tried to teach somebody and failed, and then another person successfully teaches the same thing in a different way? A person who acts as a mentor or trainer needs to be able to determine what kind of learner they are addressing, and teach according to those needs.

"The mediocre teacher tells.
The good teacher explains.
The superior teacher demonstrates.
The great teacher inspires."
—William Arthur Ward (1921)

The need for training isn't limited to craft professionals. There must be supervisory training to ensure there are qualified leaders in the industry to supervise the craft professionals.

1.1.1 Motivation

As a supervisor or crew leader, it is important to understand what motivates your crew. Money is often considered a good motivator, but it is sometimes only a temporary solution. Once a person has reached a level of financial security, other factors come into play. Studies show that environment and conditions motivate many people. For those people, a great workplace may mean more to them than better pay.

If you give someone a raise, they tend to work harder for a period of time. Then the satisfaction wanes and they may want another raise. A sense of accomplishment is what motivates most people. That is why setting and working toward recognizable goals tends to make employees more productive. A person with a feeling of involvement or a sense of achievement is likely to be better motivated and help to motivate others.

1.1.2 Understanding Workers

Many older workers grew up in an environment where they learned to work hard with the same employer until retirement. They expected to stay with a company for a long time, and the structure of the companies created a family-type environment.

Times have changed. Younger workers have grown up in a highly mobile society and expect frequent rewards and rapid advancement. Some might perceive this generation of workers as lazy, narcissistic, and unmotivated, but in reality, they simply have a different perspective on life, work, and priorities. For such workers, it may be better to give them small projects or break up large projects into smaller pieces so that they are rewarded more often by successfully achieving short-term goals. The following strategies are important for keeping young workers motivated and engaged:

- *Goal setting* – Set short-term and long-term goals, including tasks to be done and expected time frames. Things can happen to change, delay, or upset the short-term goals. This is one reason to set long-term goals as well. Don't set workers up for failure, as this leads to frustration, and frustration can lead to reduced productivity.
- *Feedback* – Timely feedback is important. For example, telling someone they did a good job last year, or criticizing them for a job they did a month ago, is meaningless. Simple recognition isn't always enough. Some type of reward should accompany positive feedback, even if it is simply recognizing the employee in a public way. Constructive criticism or reprimands should always be given in private. You can also provide some positive action, such as one-on-one training, to correct a problem.

1.1.3 Craft Training

Craft training is often informal, taking place on the job site, outside of a traditional training classroom. According to the American Society for Training and Development (ASTD), a qualified co-worker or a supervisor conducts craft training generally through on-the-job instruction. The Society of Human Resources Management (SHRM) offers the following tips to supervisors in charge of training their employees:

- *Help crew members establish career goals* – Once career goals are established, you can readily identify the training required to meet the goals.
- *Determine what kind of training to give* – Training can be on the job under the supervision of a co-worker. It can be one-on-one with the supervisor. It can involve cross training to teach a new trade or skill, or it can involve delegating new or additional responsibilities.
- *Determine the trainee's preferred method of learning* – Some people learn best by watching, others from verbal instructions, and others by doing. These are categorized as visual learners, auditory learners, and tactile learners, respectively. When training more than one person at a time, try to use a mix of all three methods.

Communication is a critical component of training employees. SHRM advises that supervisors do the following when training their employees:

- Explain the task, why it is important, and how to do it. Confirm that the worker trainees understand these three areas by asking questions. Allow them to ask questions as well.
- Demonstrate the task. Break the task down into manageable parts and cover one part at a time.
- Ask your trainees to do the task while you observe them. Try not to interrupt them while they are doing the task unless they are doing something that is unsafe or potentially harmful.
- Give the trainees feedback. Be specific about what they did and mention any areas where they need to improve.

1.1.4 Supervisory Training

Because of the need for skilled craft professionals and qualified supervisory personnel, some companies offer training to their employees through in-house classes, or by subsidizing outside training programs. However, for a variety of reasons, many contractors don't offer training at all. Some common reasons include the following:

- Lack of money to train
- Lack of time to train
- Lack of knowledge about the benefits of training programs
- High rate of employee turnover
- Workforce is too small
- Past training involvement was ineffective
- The company hires only trained workers
- Lack of interest from workers

For craft professionals to move up into supervisory and managerial positions, they must continue their education and training. Those who are willing to acquire and develop new skills have the best chance of finding stable employment. This makes it critical for them to take advantage of training opportunities, and for companies to incorporate training into their business culture.

If your company has recognized the need for training, your participation in a leadership training program such as this will begin to fill the gap between craft and supervisory training.

1.1.5 Impact of Technology

Many industries, including the construction industry, have embraced technology to remain competitive. Benefits of technology include increased productivity and speed, improved quality of documents, greater access to common data, and better financial controls and communication. As technology becomes a greater part of supervision, crew leaders must be able to use it properly.

Cellphones (in particular the smartphone) have made it easy to keep in touch through numerous forms of media. As of 2016, 95 percent of all Americans own a cellphone of some kind

Did You Know?

Learning Strategies

There is a lot more to training than simply telling a group of people what you know about a subject or talking your way through a series of slides. Studies have shown that the information and understanding that learners retain depends on how they receive the information and use it. These learning processes are as follows, listed from least effective to most effective:

1. Reading about a process
2. Hearing a description of a process
3. Observing a process
4. Observing and hearing a description of a process
5. Observing, hearing, and responding to instruction
6. Observing, hearing, and doing a process

Although some companies may attempt to assign percentage effectiveness to these instructional approaches, the results are very arbitrary and inconclusive. It is more important to understand which are the most effective methods for teaching and training.

and 77 percent own a smartphone, according to the Pew Research Center. They are particularly useful communication tools for contractors or crew leaders who are on a job site, away from their offices, or constantly on the go. Workers use smartphones at any time for phone calls, emails, text messages, and voicemail, as well as to share photos and videos. The hundreds of thousands of mobile computing apps available allow smartphones to perform numerous other functions on the go. However, the number of accidents due to inattention while focusing on a smartphone or other mobile device is also rapidly rising. Always check the company's policy regarding cell phone use on the job.

As a crew leader, you should be aware that smartphones and tablets (*Figure 1*) allow supervisors to plan their calendars, schedule meetings, manage projects, and access their company email from remote locations. These devices are far more powerful than the computers that took us to the moon, and they can hold years of information from various projects. **Cloud-based applications** now permit remote access to and updating of files, plans, and data from any place with wireless connectivity. In fact, it is becoming common for work sites to set up **local area networks (LAN)**, **wide area networks (WAN)**, and dedicated **Wi-Fi** services to support mobile communications among workers on the job.

In all forms of electronic communication (verbal, written, and visual), it is important to keep messages brief, factual, and legal. Text-based communications can be easily misunderstood because there are no visual or auditory cues to indicate the sender's intent. In other words, it is more difficult to tell if someone is just joking via email because you can't see the sender's expression or hear their tone of voice.

1.2.0 Gender and Cultural Issues

In the past several years, the construction industry in the United States has experienced a shift in worker expectations and diversity. These two issues are converging at a rapid pace.

The generation of learners is also a factor in the learning process and in the workplace. The various generations include Baby Boomers, Generation X, Millennials (Generation Y), and Generation Z. The ranges of birth years for these generations are as follows (note that there may be

Figure 1 Digital tablets are a convenient management and scheduling tool for supervisors.

some overlap, and opinions on the generational names as well as the range of years for each generation may vary):

- *Baby Boomers* – born 1946 through 1964
- *Generation X* – born 1965 through 1979
- *Millennials* (*Generation Y*) – born 1980 through 2000
- *Generation Z* – born 2000 through 2015

Each generation has been studied to some degree to determine their different interpretations of and approaches to learning and training. Family norms, religious values and morals, educational methods, music, movies, politics, and global events are all influential factors that define a generation. Remember that individuals within a generation can vary widely in their expectations and behaviors.

This trend, combined with industry diversity initiatives, has created a climate in which companies recognize the need to embrace a diverse workforce that crosses generational, gender, and ethnic boundaries (*Figure 2*). To do this effectively, they are using their own resources, as well as relying on associations with the government and trade organizations.

All current research indicates that industry will be more dependent on the critical skills of a diverse workforce—but a workforce that must be both culturally and ethnically fused. Across the United States, construction and other industries are aggressively seeking to bring new workers into their ranks, including women, racial, and ethnic minorities. Social and political issues are no longer the main factors driving workplace diversity, but by consumers and citizens who need more hospitals, malls, bridges, power plants, refineries, and many other commercial and residential structures. The construction industry needs more workers.

Figure 2 The modern workforce is diverse.

There are some potential issues relating to a diverse workforce that may be encountered on the job site. These issues include different communication styles of men and women, language barriers associated with cultural differences, **sexual harassment**, and gender or racial discrimination.

1.2.1 Communication Styles of Men and Women

As more and more women enter construction workforce, it becomes increasingly important to break down communication barriers between men and women and to understand differences in behaviors so that men and women can work together more effectively. The Jamestown, Area Labor Management Committee (JALMC) in New York offers the following explanations and tips (*emphasized text* is from *Exchange and Deception: A Feminist Perspective*, ed. by Caroline Gerschlager, Monika Mokre, 2002. Springer Science & Business Media):

- *Women tend to ask more questions than men do* – Men are more likely to proceed with a job and figure it out as they go along, while women are more likely to ask questions first.
- *Men tend to offer solutions before empathy; women tend to do the opposite* – Both men and women should say what they want up front, whether it's the solution to a problem, or simply a sympathetic ear. That way, both genders will feel understood and supported.
- *Women are more likely to ask for help when they need it* – Women are generally more **pragmatic** when it comes to completing a task. If they need help, they will ask for it. Men are more likely to attempt to complete a task by themselves, even when they need assistance.
- *Men tend to communicate more competitively, and women tend to communicate more cooperatively* – Both parties need to hear one another out without interruption.

This doesn't mean that any one method is better or worse than the other; it simply means that men and women inherently use different approaches to achieve the same result. It's either genetic or cultural, but awareness can overcome these tendencies.

1.2.2 Language Barriers

Language barriers are a real workplace challenge for crew leaders. Millions of American workers speak languages other than English. Bilingual job sites are increasingly common. As the makeup of the immigrant population continues to change with the influx of refugees from around the world, the number of non-English speakers in the American workforce is rising and diversifying dramatically.

In addition, there are many different dialects of the English language in America, which can present some communication challenges. For example, some workers may speak a dialect commonly called *Ebonics* (referred to by linguists as *African-American Vernacular English*). The dialect called *Rural White Southern English* can also be difficult to understand by those not born in the American South. Many find the American New England accent and dialect difficult to comprehend. Some sources list as many as 16 distinct American dialects.

Companies have the following options to overcome the language challenge, mainly for foreign or English-as-a-second-language (ESL) workers:

- Offer English classes either at the work site or through school districts and community colleges.
- Offer incentives for workers to learn English.

Communication with ESL workers, and varying dialects in general, becomes even more critical as the workforce grows more diverse. The following tips will help when communicating across language barriers:

- Be patient. Give workers time to process the information in a way they can comprehend.
- Avoid humor. Humor is easily misunderstood. The worker may misinterpret what you say as a joke at the worker's expense.
- Don't assume workers are unintelligent simply because they don't understand what you are saying. Explaining something in multiple ways, and having trainees **paraphrase** what they heard, is an excellent way to ensure mutual understanding and prevent miscommunication.
- If a worker is not fluent in English, ask the worker to demonstrate his or her understanding through other means.
- Speak slowly and clearly, and avoid the tendency to raise your voice.
- Use face-to-face communication whenever possible. Over-the-phone communication is often more difficult when a language barrier is involved.
- Use pictures or sketches to get your point across.

1.2.3 Cultural Differences

As workers from a multitude of backgrounds and cultures work together, there are bound to be differences and conflicts in the workplace. To overcome cultural conflicts, the SHRM suggests the following approach to resolving cultural conflicts between individuals:

- *Define the problem from both points of view* – How does each person involved view the conflict? What does each person think is wrong? This involves moving beyond traditional thought processes to consider alternate ways of thinking.
- *Uncover cultural interpretations* – What assumptions may the parties involved be making based on cultural programming? This is particularly true for certain gestures, symbols, and words that mean different things in different cultures. By doing this, the supervisor may realize what motivated an employee to act in a specific way.
- *Create cultural* **synergy** – Devise a solution that works for both parties involved. The purpose is to recognize and respect other's cultural values, and work out mutually acceptable alternatives.

1.2.4 Sexual Harassment

In today's business world, men and women are working side-by-side in careers of all kinds, creating increased opportunity for sexual harassment to occur. Sexual harassment can be defined as any unwelcome behavior that makes someone feel uncomfortable in the workplace by focusing attention on their gender or gender identity. Activities that might qualify as sexual harassment include, but are not limited to, the following:

- Telling an offensive, sexually-oriented joke
- Displaying a poster of a man or woman in a revealing swimsuit
- Wearing a patch or article of clothing blatantly promoting or degrading a specific gender
- Making verbal or physical advances (*Figure 3*)
- Speaking abusively about a specific gender

Historically, many have thought of sexual harassment as an act perpetrated by men against women, especially those in subordinate positions. However, the nature of sexual harassment cases over the years have shown that the perpetrator can be of any gender.

Figure 3 Any unwelcome physical contact can be considered harassment.

Sexual harassment can occur in various circumstances, including the following:

- The victim and harasser may be either male or female. The victim does not have to be of the opposite gender.
- The harasser can be the victim's supervisor, an agent of the employer, a supervisor in another area, a co-worker, a subordinate, or a nonemployee.
- Legally, the victim doesn't have to be the person harassed, but could be anyone offended by the relevant conduct.
- Unlawful sexual harassment may occur without economic injury to or the discharge of the victim.
- The harasser's conduct must be unwelcome.

The Equal Employment Opportunity Commission (EEOC) enforces sexual harassment laws within industries. When investigating allegations of sexual harassment, the EEOC looks at the whole record, including the circumstances and the context in which the alleged incidents occurred. A decision on the allegations is made from the facts on a case-by-case basis. Supervision can hold the crew leader responsible who is aware of sexual harassment and does nothing to stop it. The crew leader therefore should not only take action to stop sexual harassment, but should serve as a good example for the rest of the crew.

Prevention is the best tool to eliminate sexual harassment in the workplace. The EEOC encourages employers to take steps to prevent sexual harassment from occurring. Employers should clearly communicate to employees that they and their company won't tolerate sexual harassment. They do so by developing a policy on sexual harassment, establishing an effective complaint or grievance process, and taking immediate and appropriate action when an employee complains.

Controversial or sexually-related remarks, jokes, and swearing are not only offensive to coworkers, but also tarnish a worker's character. Crew leaders need to emphasize that abrasive or crude behavior may affect opportunities for advancement. If disciplinary action becomes necessary, written company policies should spell out the type of actions warranted by such behavior. A typical approach is a three-step process in which the perpetrator is first given a verbal reprimand. In the event of further violations, a written reprimand, often including a **letter of instruction (LOI)** and a warning are given. Dismissal typically accompanies any subsequent violations.

1.2.5 Gender and Minority Discrimination

Employers are giving more attention to fair recruitment, equal pay for equal work, and promotions for women and minorities in the workplace. Consequently, many businesses are analyzing their practices for equity, including the treatment of employees, the organization's hiring and promotional practices, and compensation.

The construction industry, which was once male-dominated, is moving away from this image and actively recruiting and training women, younger workers, people from other cultures, and disabled workers. This means that organizations hire the best person for the job, without regard for race, sex, religion, age, etc.

Did You Know?

Respectful Workplace Training

Some companies employ a tool called *sensitivity training* in cases where individuals or groups have trouble adapting to a multi-cultural, multi-gender workforce. Sensitivity training is a psychological technique using group discussion, role playing, and other methods to allow participants to develop an awareness of themselves and how they interact with others. Critics of earlier forms of sensitivity training noted that the methods used were not very different from brainwashing. Modern forms of cultural-awareness and gender-issue training are more ethical, and go by the term *respectful workplace training*.

To prevent discrimination cases, employers must have valid job-related criteria for hiring, compensation, and promotion. They must apply these measures consistently for every applicant interview, employee performance appraisal, and hiring or promotion decision. Therefore, employers must train all workers responsible for recruitment, selection, supervision of employees, and evaluating job performance on how to use the job-related criteria legally and effectively.

1.3.0 Business Organization

An organization is the relationship among the people within the company or project. The crew leader needs to be aware of two types of organizations: formal organizations and informal organizations.

A formal organization exists when the people within a work group led by someone direct their activities toward achieving a common goal. An example of a formal organization is a work crew consisting of four carpenters and two laborers led by a crew leader, all working together to accomplish a project.

An organizational chart typically illustrates a formal organization. It outlines all the positions that make up an organization and shows how those positions are related. Some organizational charts even identify the person within each position and the person to whom that person reports, as well as the people that the person supervises. *Figure 4* and *Figure 5* show examples of what an organization chart might look like for a construction company and an industrial company. Note that each organizational position represents an opportunity for advancement in the construction industry that a crew leader can eventually achieve.

An informal organization allows for communication among its members so they can perform as a group. It also establishes patterns of behavior that help them to work as a group, such as agreeing to use a specific training or certification program.

An example of an informal organization is a trade association such as Associated Builders and Contractors (ABC), Associated General Contractors (AGC), and the National Association of Women in Construction (NAWIC). Those, along with the thousands of other trade associations in the United States, provide forums in which members with common concerns can share information, work on issues, and develop standards for their industry.

Both formal and informal organizations establish the foundation for how communication flows. The formal structure is the means used to delegate authority and responsibility and to exchange information. The informal structure provides means for the exchange of information.

Members in an organization perform best when they can do the following:

- Know the job and how it will be done
- Communicate effectively with others in the group
- Understand their role in the organization
- Recognize who has the authority and responsibility

"Leadership is not about titles, positions, or flow charts. It is about one life influencing another."
—John C. Maxwell

1.3.1 Division of Responsibility

The conduct of a business involves certain functions. In a small organization, one or two people may share responsibilities. However, in a larger organization with many different and complex activities, distinct activity groups may lie under the responsibility of separate department managers. In either case, the following major department functions exist in most companies:

- *Executive* – This office represents top management. It is responsible for the success of the company through short-range and long-range planning.
- *Human Resources (HR)* – This office is responsible for recruiting and screening prospective employees, managing employee benefits programs, advising management on pay and benefits, and developing and enforcing procedures related to hiring practices.
- *Accounting* – This office is responsible for all recordkeeping and financial transactions, including payroll, taxes, insurance, and audits.
- *Contract Administration* – This office prepares and executes contractual documents with owners, subcontractors, and suppliers.
- *Purchasing* – This office obtains material prices and then issues purchase orders. The purchasing office also obtains rental and leasing rates on equipment and tools.
- *Estimating* – This office is responsible for recording the quantity of material on the jobs, the takeoff, pricing labor and material, analyzing subcontractor bids, and bidding on projects.
- *Operations* – This office plans, controls, and supervises all project-related activities.

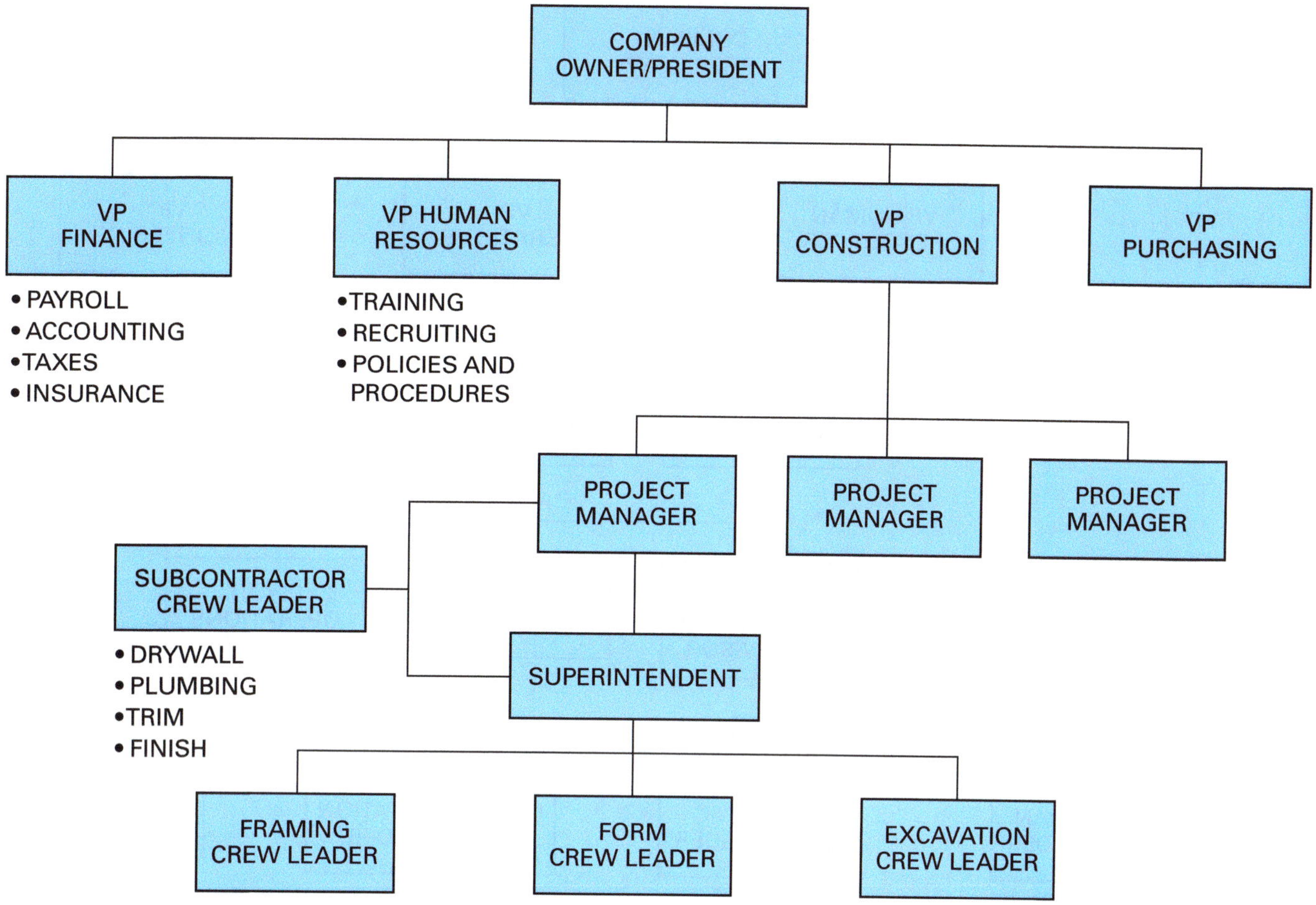

Figure 4 Sample organization chart for a construction company.

Other divisions of responsibility a company may create involve architectural and engineering design functions. These divisions usually become separate departments.

1.3.2 Responsibility, Authority, and Accountability

As an organization grows, the manager must ask others to perform many duties so that the manager can concentrate on management tasks. Managers typically assign (delegate) activities to their subordinates. When delegating activities, the crew leader assigns others the responsibility to perform the designated tasks.

Responsibility is the obligation to perform the duties. Along with responsibility comes authority. *Authority* is the right to act or make decisions in carrying out an assignment. The type and amount of authority a supervisor or worker has depends on the employee's company. There must be a balance between the authority and responsibility workers have so that they can carry out their assigned tasks. In addition, the crew leader must delegate sufficient authority to make a worker accountable to the crew leader for the results.

Accountability is holding an employee responsible for completing the assigned activities. Even though the crew leader may delegate authority and responsibility to crew members, the overall responsibility—the accountability—for the tasks assigned to the crew always rests with the crew leader.

1.3.3 Job Descriptions

Many companies furnish each employee with a written **job description** that explains the job in detail. Job descriptions set a standard for the employee. They make judging performance easier, clarify the tasks each person should handle, and simplify the training of new employees.
Each new employee should understand all the duties and responsibilities of the job after reviewing the job description. Thus, the time is shorter for the employee to make the transition from being a new and uninformed employee to a more experienced member of a crew.

A job description need not be long, but it should be detailed enough to ensure there is no misunderstanding of the duties and responsibilities of the position. The job description should

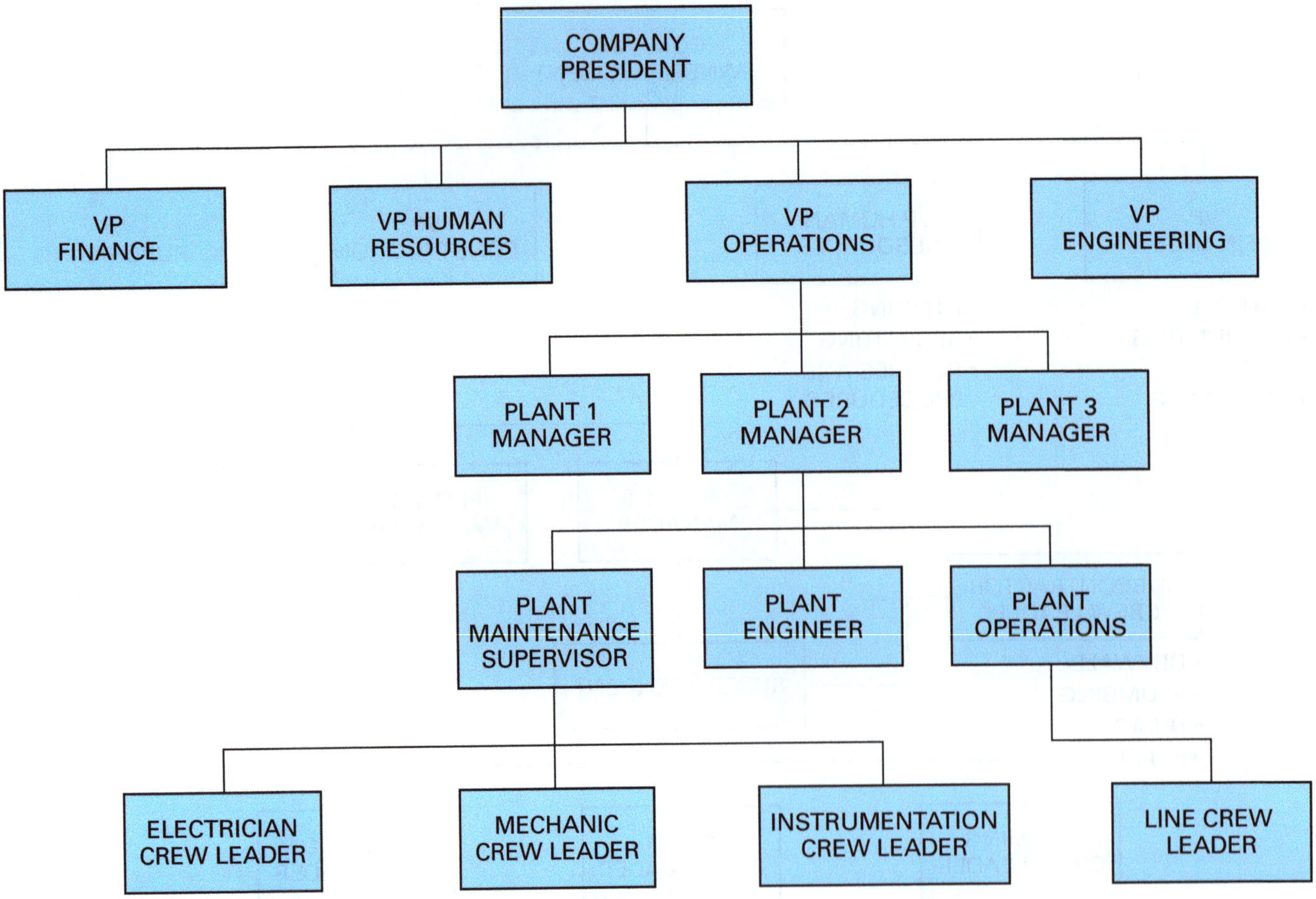

Figure 5 Sample organization chart for an industrial company.

contain all the information necessary to evaluate the employee's performance and hold the employee accountable.

A job description should contain, at minimum, the following:

- Job title
- General description of the position
- Minimum qualifications for the job
- Specific duties and responsibilities
- The supervisor to whom the position reports
- Other requirements, such as qualifications, certifications, and licenses

Figure 6 is an example of a job description.

1.3.4 Policies and Procedures

Most companies have formal policies and procedures established to help crew leaders carry out their duties. A policy is a general statement establishing guidelines for a specific activity. Examples include policies on vacations, breaks, workplace safety, and checking out tools. Procedures are formal instructions to carry out and meet policies. For example, a procedure written to implement a policy on workplace safety would include guidelines expected of all employees for reporting accidents and to follow general safety practices.

A crew leader must be familiar with the company policies and procedures, especially regarding safety practices. When OSHA inspectors visit a jobsite, they often question employees and crew leaders about the company policies related to safety. If they are investigating an accident, they will want to verify that the responsible crew leader knew the applicable company policy and followed it.

Position:
Crew Leader

General Summary:
First line of supervision on a construction crew installing concrete formwork.

Reports To:
Job Superintendent

Physical and Mental Responsibilities:

- Ability to stand for long periods
- Ability to solve basic math and geometry problems

Duties and Responsibilities:

- Oversee crew
- Provide instruction and training in construction tasks as needed
- Make sure proper materials and tools are on the site to accomplish tasks
- Keep project on schedule
- Enforce safety policies and procedures

Knowledge, Skills, and Experience Required:

- Extensive travel throughout the Eastern United States, home base in Atlanta
- Ability to operate a backhoe and trencher
- Valid commercial driver's license with no DUI violations
- Ability to work under deadlines with the knowledge and ability to foresee problem areas and develop a plan of action to solve the situation

Figure 6 A sample job description.

Did You Know?

The ADA

The Americans with Disabilities Act (ADA) is a law that prohibits employers from discriminating against disabled persons in their hiring practices. While the ADA doesn't require written job descriptions, investigators use existing job descriptions as evidence in determining whether discrimination occurred. This means that job descriptions must specifically define the job duties and the abilities needed to perform them. In making a determination of discrimination under the ADA, investigators give consideration to whether a reasonable accommodation could have qualified the disabled person to fill a job.

Additional Resources

Construction Workforce Development Professional, NCCER. 2016. New York, NY: Pearson Education, Inc.

Mentoring for Craft Professionals, NCCER. 2016. New York, NY: Pearson Education, Inc.

Generational Cohorts and their Attitudes Toward Work Related Issues in Central Kentucky, Frank Fletcher, et al. 2009. Midway College, Midway, KY. **www.kentucky.com**

The Young Person's Guide to Wisdom, Power, and Life Success: Making Smart Choices. Brian Gahran, PhD. 2014. San Diego, CA: Young Persons Press. **www.WPGBlog.com**

The following websites offer resources for products and training:

Aging Workforce News, **www.agingworkforcenews.com**

American Society for Training and Development (ASTD), **www.astd.org**

Equal Employment Opportunity Commission (EEOC), **www.eeoc.gov**

National Association of Women in Construction (NAWIC), **www.nawic.org**

Society for Human Resources Management (SHRM), **www.shrm.org**

United States Census Bureau, **www.census.gov**

United States Department of Labor, **www.dol.gov**

Wi-Fi® is a registered trademark of the Wi-Fi Alliance, **www.wi-fi.org**

1.0.0 Section Review

1. According to the US Department of Labor, the best way for the construction industry to reduce skilled-worker shortages is to _____.
   a. create training opportunities
   b. avoid discrimination lawsuits
   c. update the skills of older workers who are retiring at a later age than they previously did
   d. implement better policies and procedures

2. Which of the following is *not* helpful when dealing with language diversity in the workplace?
   a. Speaking slowly and clearly in an even tone
   b. Using humor when communicating
   c. Using sketches or diagrams to explain what needs to be done.
   d. Being patient and giving bilingual workers time to process your instructions.

3. Members tend to function best within an organization when they _____.
   a. are allowed to select their own uniform for each project
   b. understand their role within the organization
   c. don't disagree with the statements of other workers or supervisors
   d. are able to work without supervision

SECTION TWO

2.0.0 LEADERSHIP SKILLS

Objective

Explain how to incorporate leadership skills into work habits, including communications, motivation, team-building, problem-solving, and decision-making skills.

a. Describe the role of a leader on a construction crew.
b. Explain the importance of written and oral communication skills.
c. Describe methods for motivating team members.
d. Explain the importance of teamwork to a construction project.
e. Identify effective problem-solving and decision-making methods.

Trade Terms

Autonomy: The condition of having complete control over one's actions, and being free from the control of another. To be independent.

Bias: A preconceived inclination against or in favor of something.

Ethics: The moral principles that guides an individual's or organization's actions when dealing with others. Also refers to the study of moral principles.

Infer: To reach a conclusion using a method of reasoning that starts with an assumption and considers a set of logically-related events, conditions, or statements.

Legend: In maps, plans, and diagrams, an explanatory table defining all symbolic information contained in the document.

Lockout/tagout (LOTO): A system of safety procedures for securing electrical and mechanical equipment during repairs or construction, consisting of warning tags and physical locking or restraining devices applied to controls to prevent the accidental or purposeful operation of the equipment, endangering workers, equipment, or facilities.

Proactive: To anticipate and take action in the present to deal with potential future events or outcomes based on what one knows about current events or conditions.

Project manager: In construction, the individual who has overall responsibility for one or more construction projects; also called the general superintendent.

Superintendent: In construction, the individual who is the on-site supervisor in charge of a given construction project.

Textspeak: A form of written language characteristic of text messaging on mobile devices and text-based social media, which usually consists of acronyms, abbreviations, and minimal punctuation.

It is important to define some of the supervisory positions discussed throughout this module. You are already familiar with the roles of the craft professional and crew leader. A **superintendent** is essentially an on-site supervisor who is responsible for one or more crew leaders or front-line supervisors. A **project manager** or general superintendent may be responsible for managing one or more projects. This training will concentrate primarily on the supervisory role of the crew leader.

Craftworkers and crew leaders differ in that the crew leader manages the activities that the craft professionals perform. To manage a crew of craft professionals, a crew leader must have first-hand knowledge and experience in their activities. Additionally, the crew leader must be able to act directly in organizing and directing the activities of the various crew members.

This section explains the importance of developing leadership skills as a new crew leader. It will cover effective ways to communicate with coworkers and employees at all levels, build teams, motivate crew members, make decisions, and resolve problems.

Crew leaders are generally promoted up from a work crew. A worker's ability to accomplish tasks, get along with others, meet schedules, and stay within the budget have a significant influence on the selection process. The crew leader must lead the team to work safely and provide a quality product.

Making the transition from crew member to a crew leader can be difficult, especially when the new position involves overseeing a group of former peers. For example, some of the crew may try to take advantage of their friendship by seeking special favors. They may also want to be privy to supervisory information that is normally closely held. When you become a crew leader, you are no longer responsible for your work alone. Crew leaders are accountable for the work of an entire crew of people with varying skill levels, personalities, work styles, and cultural and educational backgrounds.

Crew leaders must learn to put personal relationships aside and work for the common goals of the entire crew. The crew leader can overcome these problems by working with the crew to set mutual performance goals and by freely communicating with them within permitted limits. Use their knowledge and strengths along with your own so that they feel like they are key players on the team.

As employees move from being a craftworker into the role of a crew leader and above, they will begin to spend more hours supervising the work of others (supervisory work) than practicing their own craft skills (technical work). *Figure 7* illustrates the relative amounts of time craft professionals, crew leaders, superintendents, and project managers spend on technical and supervisory work as their management responsibilities increase.

There are many ways to define a leader. One simple definition of a leader is a person who influences other people to achieve a goal. Some people may have innate leadership qualities that developed during their upbringing, or they may have worked to develop the traits that motivate others to follow and perform. Research shows that people who possess such talents are likely to succeed as leaders.

"Leadership is the art of getting someone else to do something you want done because he wants to do it."
—Dwight D. Eisenhower

2.1.0 The Qualities and Role of a Leader

Leadership traits are similar to the skills that a crew leader needs to be effective. Although the characteristics of leadership are many, there are some definite commonalities among effective leaders.

First and foremost, effective leaders lead by example. They work and live by the standards that they establish for their crew members or followers, making sure they set a positive example.

Effective leaders also tend to have a high level of drive and determination, as well as a persistent attitude. When faced with obstacles, effective leaders don't get discouraged. Instead, they identify the potential problems, make plans to overcome them, and work toward achieving the intended goal. In the event of failure, effective leaders learn from their mistakes and apply that knowledge to future situations. They also learn from their successes.

Effective leaders are typically good communicators who clearly lay out the goals of a project to their crew members. Accomplishing this may require that the leader overcome issues such as language barriers, gender **bias**, or differences in personalities to ensure that each member of the crew understands the established goals of the project.

Effective leaders can motivate their crew members to work to their full potential and become useful members of the team. Crew leaders try to develop crew-member skills and encourage them to improve and learn so they can contribute more to the team effort. Effective leaders strive for excellence from themselves and their team, so they work hard to provide the skills and leadership necessary to do so.

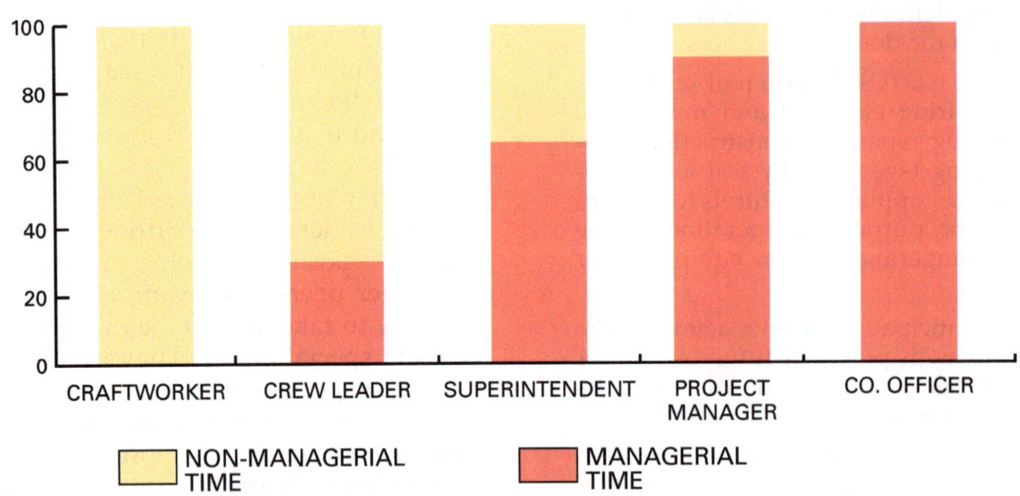

Figure 7 Percentages of technical and supervisory work by role.

In addition, effective leaders must possess organizational skills. They know what needs to be accomplished, and they use their resources to make it happen. Because they can't do it alone, leaders require the help of their team members to share in the workload. Effective leaders delegate work to their crew members, and they implement company policies and procedures to complete the work safely, effectively, and efficiently.

Finally, effective leaders have the experience, authority, and self-confidence that allows them to make decisions and solve problems. To accomplish their goals, leaders must be able to calculate risks, take in and interpret information, assess courses of action, make decisions, and assume the responsibility for those decisions.

2.1.1 Functions of a Leader

The functions of a leader will vary with the environment, the group of individuals they lead, and the tasks they must perform. However, there are certain functions common to all situations that the leader must fulfill. Some of the major functions are as follows:

- Accept responsibility for the successes and failures of the group's performance
- Be sensitive to the differences of a diverse workforce
- Ensure that all group members understand and abide by company policies and procedures
- Give group members the confidence to make decisions and take responsibility for their work
- Maintain a cohesive group by resolving tensions and differences among its members and between the group and those outside the group
- Organize, plan, staff, direct, and control work
- Represent the group

2.1.2 Leadership Traits

There are many traits and skills that help create an effective leader. Some of these include the following:

- Ability to advocate an idea
- Ability to motivate
- Ability to plan and organize
- Ability to teach others
- Enthusiasm
- Fairness
- Good communication skills
- Initiative
- Loyalty to their company and crew
- Willingness to learn from others

2.1.3 Expected Leadership Behavior

Followers have expectations of their leaders. They look to their leaders to do the following:

- Abide by company policies and procedures
- Be a loyal member of the team
- Communicate effectively
- Have the necessary technical knowledge
- Lead by example
- Make decisions and assume responsibility
- Plan and organize the work
- Suggest and direct
- Trust the team members

2.1.4 Leadership Styles

Through history, there have been many terms used to describe different leadership styles and the amount of **autonomy** they permit. The amount of crew autonomy directly relates to the leadership style used by the crew leader. *Figure 8* illustrates the linear relationship between leader authority and worker autonomy.

There are three main styles of leadership. At one extreme is the *controller style* of leadership, where the crew leader makes all of the decisions independently, without seeking the opinions of crew members. (Very little autonomy exists for workers under this type of leader.) At the other extreme is the *advisor style*, where the crew leader empowers the employees to make decisions (high autonomy). In between these extremes is the *directive style*, where the crew leader seeks crew member opinions and makes the appropriate decisions based on their input.

The following are some characteristics of each of the three leadership styles:

Controller style (high authority, low autonomy):

- Expect crew members to work without questioning procedures
- Seldom seek advice from crew members
- Insist on solving problems alone
- Seldom permit crew members to assist each other
- Praise and criticize on a personal basis
- Have no sincere interest in creatively improving methods of operation or production

Directive style (equal authority and autonomy):

- Discuss problems with their crew members
- Listen to suggestions from crew members
- Explain and instruct
- Give crew members a feeling of accomplishment by commending them when they do a job well

- Are friendly and available to discuss personal and job-related problems

Advisor style (low authority, high autonomy):

- Believe no supervision is best
- Rarely give orders
- Worry about whether they are liked by their crew members

Effective leadership takes many forms. The correct style for a particular situation or operation depends on the nature of the crew as well as the work it has to accomplish. For example, if the crew does not have enough experience for the job ahead, then a controller style may be appropriate. The controller style of leadership is also effective when jobs involve repetitive operations that require little decision-making.

However, if a worker's attitude is an issue, a directive style may be appropriate. In this case, providing the missing motivational factors may increase performance and result in the improvement of the worker's attitude. The directive style of leadership is also used when the work is of a creative nature, because brainstorming and exchanging ideas with such crew members can be beneficial.

The advisor style is effective with an experienced crew on a well-defined project. The company must give a crew leader sufficient authority to do the job. This authority must be commensurate with responsibility, and it must be made known to crew members when they are hired so that they understand who is in charge.

A crew leader must have an expert knowledge of the activities to be supervised in order to be effective. This is important because the crew members need to know that they have someone to turn to when they have a question or a problem, when they need some guidance, or when modifications or changes are warranted by the job.

Respect is probably the most useful element of authority. Respect usually derives from being fair to employees by listening to their complaints and suggestions, and by using incentives and rewards appropriately to motivate crew members. In addition, crew leaders who have a positive attitude and a favorable personality tend to gain the respect of their crew members as well as their peers. Along with respect comes a positive attitude from the crew members.

2.1.5 Ethics in Leadership

Crew leaders should maintain the highest standards of honesty and legality. Every day, the crew leader must make decisions that may have fair-

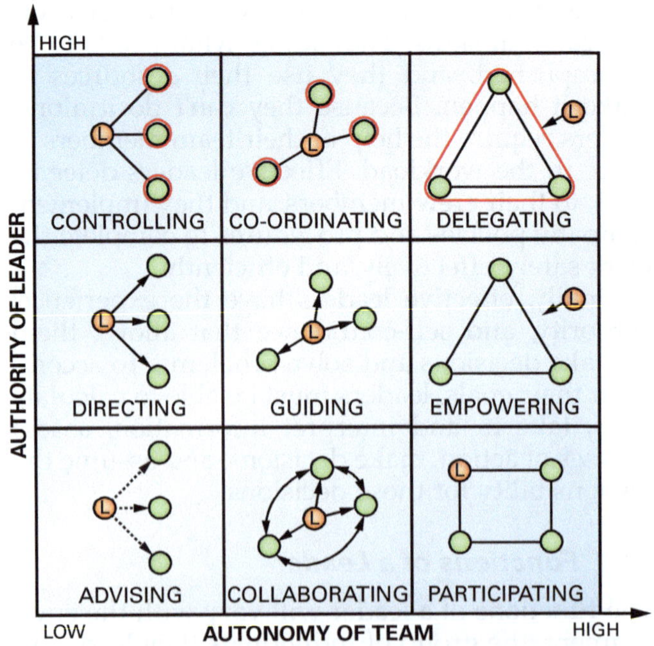

Figure 8 The authority-autonomy leadership style matrix.

ness and moral implications. When you make a dishonorable or unethical decision, it not only reflects on you, but also impacts other workers, peers, and the company as a whole.

There are three basic types of **ethics**:

- *Business or legal ethics* – Business, or legal, ethics concerns adhering to all laws and published regulations related to business relationships or activities.
- *Professional/balanced ethics* – Professional, or balanced, ethics relates to carrying out all activities in such a manner as to be honest and fair to everyone under one's authority.
- *Situational ethics* – Situational ethics pertains to specific activities or events that may initially appear to be a gray area. For example, you may ask yourself, "How will I feel about myself if my actions were going to be published in the newspaper or if I need to justify my actions to my family, friends, and colleagues? Would I still do the same thing?"

You will often find yourself in a situation where you will need to assess the ethical consequences of an impending decision. For example, if one of your crew members is showing symptoms of heat exhaustion, should you keep him working just because the superintendent says the project is behind schedule? If you are the only one aware that your crew did not properly erect the reinforcing steel, should you stop the pour and correct the situation? If supervision ever asks a crew leader to carry through on an unethical decision, it is up to that individual to inform the next-higher level of authority of the unethical nature of the issue. It is a sign of good character to refuse to carry out an unethical act.

> *"The right way is not always the popular and easy way. Standing for right when it is unpopular is a true test of character."*
> *—Margaret Chase Smith*

NOTE

Anyone who is aware of or knowingly a party to an illegal activity can face legal consequences simply due to inactivity, even if the activities are instigated by one's supervisor. However, a person of good character will never consider weighing the legal consequences against the ethics of one's actions.

2.2.0 Communication

Successful crew leaders learn to communicate effectively with people at all levels of the organization. In doing so, they develop an understanding of human behavior and acquire communication skills that enable them to understand and influence others.

There are many definitions of communication. Communication is the act of accurately and effectively conveying to or exchanging facts, feelings, and opinions with another person. Simply stated, communication is the process of exchanging information and ideas. Just as there are many definitions of communication, it also comes in many forms, including verbal, written, and nonverbal.

2.2.1 The Communication Process and Verbal Communication

There are two basic steps to clear communication, as illustrated in *Figure 9*. First, a sender sends a message (either verbal or written) to a receiver. When the receiver gets the message, he or she figures out what it means by listening or reading carefully. If anything is unclear, the receiver gives

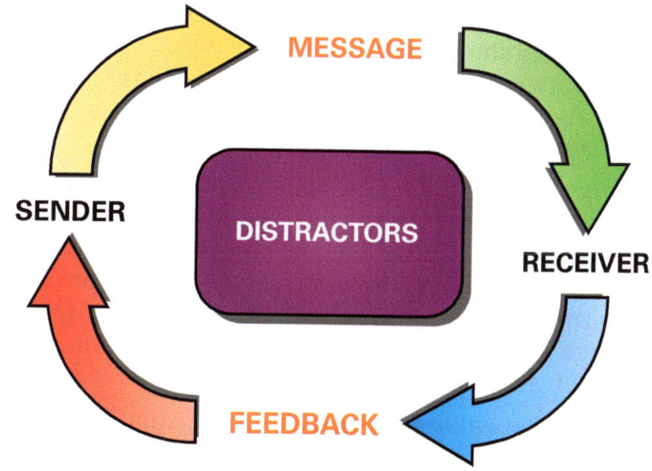

Figure 9 The communication process.

the sender feedback by asking the sender for more information.

This process is called *two-way communication*, and it is the most effective way to make sure that everyone understands what's going on. This process sounds simple, so you may ask why good communication is so hard to achieve? When we try to communicate, a lot of things can get in the way. These communication obstacles and distractors are called *noise*.

The communication process illustrated in *Figure 9* consists of the following major components:

- *The Sender* – The sender is the person who creates and transmits the message. In verbal communication, the sender speaks the message aloud to the person(s) receiving the message. The sender must be sure to speak in a clear and concise manner easily understood by others. This isn't a natural skill; it takes practice. Some basic speaking tips are as follows:
  - Avoid talking with anything in your mouth (food, gum, etc.).
  - Avoid swearing, or other crude language, and acronyms.
  - Don't speak too quickly or too slowly. In extreme cases, people tend to focus on the rate of speech rather than the words themselves.

Did You Know?

Supervisor's Communication Breakdown

No, this isn't about failure of communication, but how a supervisor's daily communications break down into the various types, and how much time the supervisor spends doing each. Research shows that about 80 percent of the typical supervisor's day is spent communicating through writing, speaking, listening, or using body language. Of that time, studies suggest that approximately 20 percent of communication is in the written form, and 80 percent involves speaking or listening.

- Pronounce words carefully to prevent misunderstandings.
- Speak pleasantly and with enthusiasm. Avoid speaking in a harsh voice or in a monotone.

- *The Message* – The message is what the sender is attempting to communicate to the receiver(s). A message can be a set of directions, an opinion, or dealing with a personnel matter (praise, reprimand, etc.). Whatever its function, a message is an idea or fact that the sender wants the receiver to know.

  Before speaking, determine what must be communicated, and then take the time to organize what to say. Ensure that the message is logical and complete. Take the time to clarify your thoughts to avoid confusing the receiver. This also permits the sender to get to the point quickly.

  In delivering the message, the sender should assess the audience. It is important not to talk down to them. Remember that everyone, whether in a senior or junior position, deserves respect and courtesy. Therefore, the sender should use words and phrases that the receiver can understand and avoid technical language or slang if the receiver is unfamiliar with such terms. In addition, the sender should use short sentences, which gives the audience time to understand and digest one point or fact at a time.

- *The Receiver* – The receiver is the person who takes in the message. For the verbal communication process to be successful, it is important that the receiver understands the message as the speaker intended. Therefore, the receiver must avoid things that interfere with the delivery of the message. There are many barriers to effective listening, particularly on a busy construction job site. Some of these obstacles include the following:
  - Noise, visitors, cell phones, or other distractions
  - Preoccupation, being under pressure, or daydreaming
  - Reacting emotionally to what is being communicated
  - Thinking about how to respond instead of listening
  - Giving an answer before the message is complete
  - Personal bias/prejudice against the sender or the sender's communication style
  - Finishing the sender's sentence

  Some tips for overcoming these barriers include the following:
  - Take steps to minimize or remove distractions; learn to tune out your surroundings
  - Listen for key points
  - Take notes
  - Be aware of your personal biases, and try to stay open-minded; focus on the message, not the speaker
  - Allow yourself time to process your thoughts before responding
  - Let the sender communicate the message without interruption

  There are many ways for a receiver to show that he or she is actively listening to the message. The receiver can even accomplish this without saying a word. Examples include maintaining eye contact (*Figure 10*), nodding your head, and taking notes. Feedback can also provide an important type of response to a message.

- *Feedback* – Feedback is the receiver's communication back to the sender in response to the message. Feedback is a very important part of the communication process because it shows the sender how the receiver interpreted the message and that the receiver understood it as intended. In other words, feedback is a checkpoint to make sure the receiver and sender are on the same page.

  The receiver can paraphrase the message as a form of feedback to the sender. When paraphrasing, you use your own words to repeat the message. That way, you can show the sender that you interpreted the message correctly and could explain it to others if needed.

Figure 10 Eye contact is important when communicating face-to-face.

When providing feedback, the receiver can also request clarification or additional information, generally by asking questions.

One opportunity a crew leader can take to provide feedback is in the performance of crew evaluations. Many companies use formal evaluations on a yearly basis to assess workers' performance for pay increases and eligibility for advancement. These evaluations should not come as a once-a-year surprise to workers. An effective crew leader provides frequent performance feedback, and should ultimately summarize these communications in the annual performance evaluation. It is also important to stress the importance of self-evaluation with your crew.

2.2.2 Written or Visual Communications

Some communication must be written or visual. Written or visual communication includes messages or information documented on paper, or transmitted electronically using words, sketches, or other types of images.

Many messages on a job are in text form. Examples include weekly reports, requests for changes, purchase orders, and correspondence on a specific subject. These items are in writing because they can form a permanent record for business and historical purposes. Some communications on the job must be in the form of images. People are visual beings, and sketches, diagrams, graphs, and photos or videos can more effectively communicate some things. Examples include the plans or drawings used on a job, or a video of a work site accident scene.

"One picture is worth a thousand words."
—San Antonio Express-News (1918)

When writing or creating a visual message, it is best to assess the receiver (the reader) or the audience before beginning. The receiver must be able to read the message and understand the content; otherwise, the communication process will be unsuccessful. Therefore, the sender (the writer) should consider the actual meaning of words or diagrams and how others might interpret them. In addition, the writer should make sure that handwriting, if applicable, is legible.

Here are some basic tips for writing:

- Avoid emotion-packed words or phrases.
- Avoid making judgments unless asked to do so.
- Avoid using unfamiliar acronyms and textspeak (*Figure 11*).

Figure 11 Avoid textspeak for on-the-job written communications.

- Avoid using technical language or jargon unless the writer knows the receiver is familiar with the terms used.
- Be positive whenever possible.
- Be prepared to provide a verbal or visual explanation, if needed.
- Make sure that the document is legible.
- Present the information in a logical manner.
- Proofread your work; check for spelling, typographical, and grammatical errors—especially if relying on a spell-checker or an autocorrect typing feature.
- Provide an adequate level of detail.
- State the purpose of the message clearly.
- Stick to the facts.

Entire books are available that explain how to create effective diagrams and graphs. There are many factors that contribute to clear, concise, unambiguous images. The following are some basic tips for creating effective visual communications:

- Avoid making complex graphics; simplicity is better.
- Be prepared to provide a written or verbal explanation of the diagram, if needed.
- Ensure that the diagram is large enough to be useful.
- Graphs are better for showing numerical relationships than columns of numbers.
- Present the information in a logical order.
- Provide a legend if you include symbolic information.
- Provide an adequate level of detail and accuracy.

2.2.3 Nonverbal Communication

Unlike verbal or written communication, nonverbal communication doesn't involve spoken or written words, or images. Rather, nonverbal communication refers to things the speaker and the receiver see and hear in each other while communicating face to face. Examples include facial expressions, body movements, hand gestures, tone of voice, and eye contact.

Nonverbal communication can provide an external signal of an individual's inner emotions. It occurs at the same time as verbal communication. In most cases, the transmission of the sender and receiver nonverbal cues is totally unconscious.

People observe physical, nonverbal cues when communicating, making them just as important as words. Often, nonverbal signals influence people more than spoken words. Therefore, it's important to be conscious of nonverbal cues to avoid miscommunication based on your posture or expression. After all, these things may have nothing to do with the communication exchange; instead, they may be carrying over from something else going on in your day.

2.2.4 Communication Issues

It is important to note that everyone communicates differently; that is what makes us unique as individuals. As the diversity of the workforce changes, communication becomes even more challenging because the audience may include individuals from many different backgrounds. Therefore, it is necessary to assess the audience to determine how to communicate effectively with each individual.

The key to effective communication is to acknowledge that people are different, and to be able to adjust the communication style to meet the needs of the audience or the person on the receiving end of your message. This involves relaying the message in the simplest way possible, and avoiding the use of words that people may find confusing. Be aware of how you use technical language, slang, jargon, and words that have multiple meanings. Present the information in a clear, concise manner. Avoid rambling and always speak clearly, using good grammar.

> **NOTE**
>
> When working with bilingual workers, it is sometimes helpful to learn words and phrases commonly used in their language(s). They will respect you more for attempting to learn their language as they learn yours, and the effort will generate a closer comradery.

In addition, be prepared to communicate the message in multiple ways or adjust your level of detail or terminology to ensure that everyone understands the meaning as intended. For instance, a visual learner may need a map to comprehend directions. It may be necessary to overcome language barriers on the job site by using graphics or visual aids to relay the message. *Figure 12* shows how to tailor the message to the audience.

2.3.0 Motivation

The ability to motivate others is a key skill that leaders must develop. To motivate is to influence others to put forth more effort to accomplish something. For example, a crew member who skips breaks and lunch to complete a job on time is thought to be highly motivated, but crew members who do the bare minimum or just enough to keep their jobs are considered unmotivated.

A leader can infer an employee's motivation by observing various factors in the employee's performance. Examples of factors that may indicate a crew's motivation include the level or rate of unexcused absenteeism, the percentage of employee turnover, and the number of complaints, as well as the quality and quantity of work produced.

Different things motivate different people in different ways. Consequently, there is no single best approach to motivating crew members. It is important to recognize that what motivates one crew member may not motivate another. In addition, what works to motivate a crew member once may not motivate that same person again in the future.

Often, the needs and factors that motivate individuals are the same as those that create job satisfaction. These include the following:

- Accomplishment
- Change
- Job importance
- Opportunity for advancement
- Personal growth
- Recognition and praise
- Rewards

A crew leader's ability to satisfy these needs increases the likelihood of high morale within a crew. Morale refers to an individual's emotional outlook toward work and the level of satisfaction gained while performing the jobs assigned. High morale means that employees will be motivated to work hard, and they will have a positive attitude about coming to work and doing their jobs.

2.3.1 Accomplishment

Accomplishment refers to a worker's need to set challenging goals and achieve them. There is nothing quite like the feeling of achieving a goal, particularly a goal one never expected to accomplish in the first place.

Participant Exercise A, Part I

Read the following verbal conversations, and identify any problems:

Conversation I:

Judy: Hey, José...
José: What's up?
Judy: Has the site been prepared for the job trailer yet?
José: Job trailer?
Judy: The job trailer—it's coming in today. What time will the job site be prepared?
José: The trailer will be here about 1:00 PM.
Judy: The job site! What time will the job site be prepared?

____

____

____

____

____

____

Conversation II:

Jimar: Hey, Mike, I need your help.
Mike: What is it?
Jimar: You and Miguel go over and help Al's crew finish laying out the site.
Mike: Why me? I can't work with Miguel. He can't understand a word I say.
Jimar: Al's crew needs some help, and you and Miguel are the most qualified to do the job.
Mike: I told you, Jimar, I can't work with Miguel.

____

____

____

____

____

____

Conversation III:

Hiro: Hey, Jill.
Jill: Sir?
Hiro: Have you received the latest DOL, EEO requirement to be sure the OFCP administrator finds our records up to date when he reviews them in August?
Jill: DOL, EEO, and OFCP?
Hiro: Oh, and don't forget the MSHA, OSHA, and EPA reports are due this afternoon.
Jill: MSHA, OSHA, and EPA?

____

____

____

____

____

____

____

____

Conversation IV:

Lakeisha: Good morning, Roberto, would you do me a favor?
Roberto: Okay, Lakeisha. What is it?
Lakeisha: I was reading the concrete inspection report and found the concrete in Bays 4A, 3B, 6C, and 5D didn't meet the 3,000 psi strength requirements. Also, the concrete inspector on the job told me the two batches that came in today had to be refused because they didn't meet the slump requirements as noted on page 16 of the spec. I need to know if any placement problems happened on those bays, how long the ready mix trucks were waiting today, and what we plan to do to stop these problems in the future.

____

____

____

____

____

Participant Exercise A, Part II

Read the following written memos, and identify any problems:

Memo I:

Let's start with the transformer vault $285.00 due. For what you ask? Answer: practically nothing I admit, but here is the story. Paul the superintendent decided it was not the way good ole Comm Ed wanted it, we took out the ladder and part of the grading (as Paul instructed us to do) we brought it back here to change it. When Comm Ed the architect or DOE found out that everything would still work the way it was, Paul instructed us to reinstall the work. That is the whole story there is please add the $285.00 to my next payout.

____
____
____
____
____
____
____
____

Memo II:

Let's take rooms C 307-C-312 and C-313 we made the light track supports and took them to the job to erect them when we tried to put them in we found direct work in the way, my men spent all day trying to find out what to do so ask your Superintendent (Frank) he will verify seven hours pay for these men as he went back and forth while my men waited. Now the Architect has changed the system of hanging and has the gall to say that he has made my work easier, I can't see how. Anyway, we want an extra two (2) men for seven (7) hours for April 21 at $55.00 per hour or $385.00 on April 28th DOE Reference 197 finally resolved this problem. We will have no additional charges on DOE Reference 197, please note.

____
____
____
____
____
____

Crew leaders can help their crew members attain a sense of accomplishment by encouraging them to develop performance plans, such as goals for the year, the attaining of which the crew leader will consider later in performance evaluations. In addition, crew leaders can provide the support and tools (such as training and coaching) necessary to help their crew members achieve these goals.

2.3.2 Change

Change refers to an employee's need to have variety in work assignments. Change can keep things interesting or challenging, and prevent the boredom that results from doing the same task day after day with no variety. However, frequent or significant changes in work can actually have a negative impact on morale, since people often prefer some consistency and predictability in their lives.

2.3.3 Job Importance

Job importance refers to an employee's need to feel that his or her skills and abilities are valued and make a difference. Employees who don't feel valued tend to have performance and attendance issues. Crew leaders should attempt to make every crew member feel like an important part of the team, as if the job wouldn't be possible without their help.

2.3.4 Opportunity for Advancement

Opportunity for advancement refers to an employee's need to gain additional responsibility and develop new skills and abilities. It is important that employees know that they aren't limited to their current jobs. Let them know that they have a chance to grow with the company and to be promoted as recognition for excelling in their work when such opportunities occur.

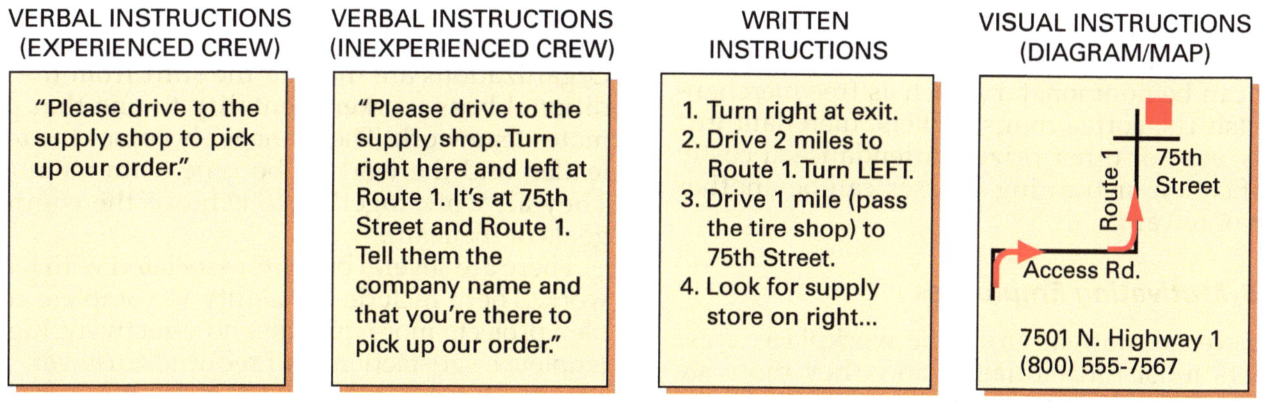

Different people learn in different ways. Be sure to communicate so you can be understood.

Figure 12 Tailor your message.

Effective leaders encourage each of their crew members to work to his or her full potential. In addition, they share information and skills with their employees to help them advance within the organization.

2.3.5 Recognition

Recognition and praise refer to the need to have good work appreciated, applauded, and acknowledged by others. You can accomplish this by simply thanking employees for helping on a project, or it can entail more formal praise, such as an award for Employee of the Month, or tangible or monetary awards. Some tips for giving recognition and praise include the following:

- Be available on the job site so that you have the opportunity to witness good work.
- Look for and recognize good work, and look for ways to praise it.
- Give recognition and praise only when truly deserved; people can quickly recognize insincere or artificial praise, and you will lose respect as a result.
- Acknowledge satisfactory performance, and encourage improvement by showing confidence in the ability of the crew members to do above-average work.

2.3.6 Personal Growth

Personal growth refers to an employee's need to learn new skills, enhance abilities, and grow as a person. It can be very rewarding to master a new competency on the job. Personal growth prevents the boredom associated with contemplation of working the same job indefinitely with no increase of technical knowledge, skills, or responsibility.

Crew leaders should encourage the personal growth of their employees as well as themselves. Learning should be a two-way street on the job site; crew leaders should teach their crew members and learn from them as well. In addition, crew members should be encouraged to learn from each other

Personal growth also takes place outside the workplace. Crew leaders should encourage their employees to acquire formal education applicable to their vocation as well as in subjects that will expand their understanding of the larger world. Encourage them to develop outside interests and ways to express their innate creativity.

2.3.7 Rewards

Rewards are additional compensation for hard work. Rewards can include an increase in a crew member's wages, or go beyond that to include bo-

Did You Know?

The Importance of Continuing Education

Education doesn't stop the day a person receives a diploma or certificate. It is a lifelong activity. Employers have long recognized and promoted continuing education as a factor in advancement, but it is essential to simply remaining in place as well. Regardless of what you do, new construction materials, methods, standards, and processes are constantly emerging. Those who don't make the effort to keep up will fall behind.

Also, studies have shown that a worker's lifetime income improved by 7–10 percent per year of community college. Consider continuing and vocational technical education (CTE and VTE) through local colleges.

nuses or other incentives. They can be monetary in nature (salary raises, holiday bonuses, etc.), or they can be nonmonetary, such as free merchandise (shirts, coffee mugs, jackets, merchant gift cards, etc.), or other prizes. Attendance at costly certification or training courses can be another form of reward.

2.3.8 Motivating Employees

To increase motivation in the workplace, crew leaders must individualize how they motivate different crew members. It is important that crew leaders get to know their crew members and determine what motivates them as individuals. Once again, as diversity increases in the workforce, this becomes even more challenging; therefore, effective communication skills are essential. Some tips for motivating employees include the following:

- Keep jobs challenging and interesting.
- Communicate your expectations. People need clear goals in order to feel a sense of accomplishment when they achieve them.
- Involve the employees. Feeling that their opinions are valued leads to pride in ownership and active participation.
- Provide sufficient training. Give employees the skills and abilities they need to be motivated to perform.
- Mentor the employees. Coaching and supporting employees boosts their self-esteem, their self-confidence, and ultimately their motivation.
- Lead by example. Employees will be far more motivated to follow someone who is willing to do the same things they are required to do.
- Treat employees well. Be considerate, kind, caring, and respectful; treat employees the way that you want to be treated.
- Avoid using scare tactics. Approaching your leadership responsibilities by threatening employees with negative consequences can result in higher employee turnover instead of motivation.
- Reward your crew for doing their best by giving them easier tasks from time to time. It is tempting to give your best employees the hardest or dirtiest jobs because you know they will do the jobs correctly.
- Reward employees for a job well done.
- Maintain a sense of humor, especially toward your own failings as a human being. No one is perfect. Your employees will appreciate that and be more motivated to be an encouragement to you.

2.4.0 Team Building

Organizations are making the shift from the traditional boss-worker mentality to one that promotes teamwork. The manager becomes the team leader, and the workers become team members. They all work together to achieve the common goals of the team.

There are several benefits associated with teamwork. These include the ability to complete complex projects more quickly and effectively, higher employee satisfaction, and reduced turnover.

2.4.1 Successful Teams

Successful teams consist of individuals who are willing to share their time and talents to reach a common goal—the goal of the team. Members of successful teams possess an "Us" or "We" attitude rather than an "I" and "You" attitude; they consider what's best for the team and put their egos aside.

Some characteristics of successful teams include the following:

- Everyone participates and every team member counts.
- All team members understand the goals of the team's work and are committed to achieve those goals.
- There is a sense of mutual trust and interdependence.
- The organization gives team members the confidence and means to succeed.
- They communicate.
- They are creative and willing to take risks.
- The team leader has strong people skills and is committed to the team.

2.4.2 Building Successful Teams

To be successful in the team leadership role, the crew leader should contribute to a positive attitude within the team. There are several ways in which the team leader can accomplish this. First, he or she can work with the team members to create a vision or purpose of what the team is to achieve. It is important that every team member is committed to the purpose of the team, and the team leader is instrumental in making this happen.

Within the construction industry, the company typically assigns a crew to a crew leader. However, it can be beneficial for the team leader to be involved in selecting the team members. The willingness of people to work on the team and the resources that they bring should be the key factors for selecting them for the team.

Participant Exercise B

You are the crew leader of a masonry crew. Sam Williams is the person whom the company holds responsible for ensuring that equipment is operable and distributed tothe jobs in a timely manner.

Occasionally, disagreements with Sam have resulted in tools and equipment arriving late. Sam, who has been with the company 15 years, resents having been placed in the job and feels that he outranks all the crew leaders.

Sam figured it was about time he talked with someone about the abuse certain tools and other items of equipment were receiving on some of the jobs. Saws were coming back with guards broken and blades chewed up, bits were being sheared in half, motor housings were bent or cracked, and a large number of tools were being returned covered with mud. Sam was out on your job when he observed a mason carrying a portable saw by the cord. As he watched, he saw the mason bump the swinging saw into a steel column. When the man arrived at his workstation, he dropped the saw into the mud.

You are the worker's crew leader. Sam approached as you were coming out of the work trailer. He described the incident. He insisted, as crew leader, you are responsible for both the work of the crew and how its members use company property. Sam concluded, "You'd better take care of this issue as soon as possible! The company is sick and tired of having your people mess up all the tools!"

You are aware that some members of your crew have been mistreating the company equipment.

1. How would you respond to Sam's accusations?

____

____

____

____

____

2. What action would you take regarding the misuse of the tools?

____

____

____

____

____

3. How can you motivate the crew to take better care of their tools? Explain.

____

____

____

____

____

When forming a new team, team leaders should do the following:

- Explain the purpose of the team. Team members need to know what they will be doing, how long they will be doing it (if they are temporary or permanent), and why they are needed.
- Help the team establish goals or targets. Teams need a purpose, and they need to know what it is they are responsible for accomplishing.
- Define team member roles and expectations. Team members need to know how they fit into the team and what they can expect to accomplish as members of the team.
- Plan to transfer a sense of responsibility to the team as appropriate. Teams should feel responsible for their assigned tasks. However, never forget that management will still hold the crew leader responsible for the crew's work.

"My model for business is The Beatles. They were four guys who kept each other's kind of negative tendencies in check. They balanced each other and the total was greater than the sum of the parts. That's how I see business: Great things in business are never done by one person, they're done by a team of people.

—Steve Jobs

2.4.3 Delegating

Once the various activities that make up the job have been determined, the crew leader must identify the person or persons who will be responsible for completing each activity. This requires that the crew leader be aware of the skills and abilities of the people on the crew. Then, the crew leader must put this knowledge to work in matching the crew's skills and abilities to accomplish the specific tasks needed to complete the job.

After matching crew members to specific activities, the crew leader must then delegate the assignments to the responsible person(s). Generally, when delegating responsibilities, the crew leader verbally communicates directly with the person who will perform or complete the activity.

When delegating work, remember to do the following:

- Delegate work to a crew member who can do the job properly. If it becomes evident that the worker doesn't perform to the desired standard, either teach the crew member to do the work correctly or turn it over to someone else who can (without making a public spectacle of the transfer).
- Make sure crew members understand what to do and the level of responsibility. Be clear about the desired results, specify the boundaries and deadlines for accomplishing the results, and note the available resources.
- Identify the standards and methods of measurement for progress and accomplishment, along with the consequences of not achieving the desired results. Discuss the task with the crew member and check for understanding by asking questions. Allow the crew member to contribute feedback or make suggestions about how to perform the task in a safe and quality manner.
- Give the crew member the time and freedom to get started without feeling the pressure of too much supervision. When making the work assignment, be sure to tell the crew member how much time there is to complete it, and confirm that this time is consistent with the job schedule.
- Examine and evaluate the result once a task is complete. Then, give the crew member some feedback as to how well the worker did the task. Get the crew member's comments. The information obtained from this is valuable and will enable the crew leader to know what kind of work to assign that crew member in the future. It will also provide a means of measuring the crew leader's own effectiveness in delegating work.

Be aware that there may be times when someone else in the company will issue written or verbal instructions to a crew member or to the crew without going through the crew leader. This kind of situation requires an extra measure of maturity and discretion on the part of the crew leader to first understand the circumstances, then establish an understanding with the responsible individual, if possible, to avoid work orders that circumvent the crew leader in the future.

2.4.4 Implementing Policies and Procedures

Every company establishes policies and procedures that crew leaders are expected to implement, and employees are expected to follow. Company policies and procedures are essentially guidelines for how the organization does business. They can also reflect organizational philosophies, such as putting safety first or making the customer the top priority. Examples of policies and procedures include safety guidelines, credit standards, and billing processes.

The following tips can help you effectively implement policies and procedures:

- Learn the purpose of each policy. This will help you follow the policy and apply it appropriately and fairly.
- If you're not sure how to apply a company policy or procedure, check the company manual or ask your supervisor.

NOTE

Try to obtain a supervisor's policy interpretation in writing or print out the email response so that you can append the decision to your copy of the company manual for future reference.

- Always follow company policies and procedures. Remember that they combine what's best for the customer and the company. In addition, they provide direction on how to handle specific situations and answer questions.

Crew leaders may need to issue orders to their crew members. An order is a form of communication that initiates, changes, or stops an activity. Orders may be general or specific, written or oral, and formal or informal. The decision of how an order will be issued is up to the crew leader, but the policies and procedures of the company may govern the choice.

When issuing orders, do the following:

- Make them as specific as possible. Avoid being general or vague unless it is impossible to foresee all the circumstances that could occur in carrying out the order.
- Recognize that it isn't necessary to write orders for simple tasks unless the company requires that supervisors write all orders.
- Write orders for more complex tasks, tasks that will take considerable time to complete, or that are permanent (standing) orders.
- Consider what is being said, the audience to whom it applies, and the situation under which it will be implemented to determine the appropriate level of formality for the order.

2.5.0 Making Decisions and Solving Problems

Decision making and problem solving and are a large part of every crew leader's daily work. They are a part of life for all supervisors, especially in fast-paced, deadlineoriented industries.

2.5.1 Decision Making Versus Problem Solving

Sometimes, the difference between decision making and problem solving isn't clear. Decision making refers to simply initiating an action, stopping one, or choosing an alternative course, as appropriate for the situation. Problem solving involves recognizing the difference between the way things are and the way things should be, then taking action to move toward the desired condition. The two activities are related because, to make a decision, you may have to use problem-solving techniques, just as solving problems requires making decisions.

2.5.2 Types of Decisions

Some decisions are routine or simple, and can be made based on past experiences. An example would be deciding how to get to work. If you've worked at the same place for a long time, you are already aware of the options for traveling to work (take the bus, drive a car, carpool with a co-worker, take a taxi, etc.). Based on past experiences with the options identified, you can make a decision about how best to get to work.

Other decisions are more difficult to make. These decisions require more careful thought about how to carry out an activity by using problem-solving techniques. An example is planning a trip to a new vacation spot. If you're not sure how to get there, where to stay, what to see, etc., one option is to research the area to determine the possible routes, hotel accommodations, and attractions. Then, you can make a decision about which route to take, what hotel to choose, and what sites to visit, without the benefit of direct experience. The Internet makes these tasks much easier, but the research still requires prioritizing what is most important to you.

2.5.3 Problem Solving

The ability to solve problems is an important skill in any workplace. It's especially important for craft professionals, whose workday is often not predictable or routine. This section provides a five-step process for solving problems, which you can apply to both workplace and personal issues.

Review the following steps and then see how you can be apply them to a job-related problem. Keep in mind that you can't solve a problem until everyone involved acknowledges the problem.

Step 1 *Define the problem.* This isn't as easy as it sounds. Thinking through the problem often uncovers additional problems. Also, drilling down to the facts of the problem may mean setting aside your own biases or presumptions toward the situation or the individuals involved.

Step 2 *Think about different ways to solve the problem.* There is often more than one solution to a problem, so you must think through each possible solution and pick the best one. The best solution might be taking parts of two different solutions and combining them to create a new solution.

Step 3 *Choose the solution that seems best, and make an action plan.* It is best to receive input both from those most affected by the problem, those who must correct the problem, and from those who will be most affected by any potential solution.

Step 4 *Test the solution to determine whether it actually works.* Many solutions sound great in theory but in practice don't turn out to be effective. On the other hand, you might discover from trying to apply a solution that it is acceptable with a little modification. If a solution doesn't work, think about how you could improve it, and then test your new plan.

Step 5 *Evaluate the process.* Review the steps you took to discover and implement the solution. Could you have done anything better? If the solution turns out to be satisfactory, you can add the solution to your knowledge base.

These five steps can be applied in specific situations. Read the following example situation, and apply the five-step problem-solving process to come up with a solution.

Example:

You are part of a team of workers assigned to a new shopping mall project. The project will take about 18 months to complete. The only available parking is half a mile from the job site. The crew must carry heavy toolboxes and safety equipment from their cars and trucks to the work area at the start of the day, and then carry them back at the end of their shifts. The five-step problem-solving process can be applied as follows:

Step 1 *Define the problem.* Workers are wasting time and energy hauling all their equipment to and from the work site.

Step 2 *Think about different ways to solve the problem.* Several workers have proposed solutions:

- Install lockers for tools and equipment closer to the work site.
- Have workers drive up to the work site to drop off their tools and equipment before parking.
- Bring in another construction trailer where workers can store their tools and equipment for the duration of the project.
- Provide a round-trip shuttle service to ferry workers and their tools.

> **NOTE**
> Each solution will have pros and cons, so it's important to receive input from the workers affected by the problem. For example, workers will probably object to any plan (like the drop-off plan) that leaves their tools vulnerable to theft.

Step 3 *Choose the solution that seems best, and make an action plan.* The work site superintendent doesn't want an additional trailer in your crew's area. The workers decide that the shuttle service makes the most sense. It should solve the time and energy problem, and workers can keep their tools with them. To put the plan into effect, the project supervisor arranges for a large van and driver to provide the shuttle service.

Step 4 *Test the solution to determine whether it actually works.* The solution works, but there is another problem. The workers' schedule has them all starting and leaving at the same time. There isn't enough room in the van for all the workers and their equipment. To solve this problem, the supervisor schedules trips spaced 15 minutes apart. The supervisor also adjusts worker schedules to correspond with the trips. That way, all the workers won't try to get on the shuttle at the same time.

Step 5 *Evaluate the process.* This process gave both management and workers a chance to express an opinion and discuss the various solutions. Everyone feels pleased with the process and the solution.

2.5.4 Special Leadership Problems

Because they are responsible for leading others, it is inevitable that crew leaders will encounter problems and be forced to make decisions about how to respond to the problem. Some problems will be relatively simple to resolve, like covering for a sick crew member who has taken a day off from work. Other problems will be complex and much more difficult to handle.

Some complex problems that are relatively common include the following:

- Inability to work with others
- Absenteeism and turnover
- Failure to comply with company policies and procedures

Inability to Work with Others – Crew leaders will sometimes encounter situations where an employee has a difficult time working with others on the crew. This could be a result of personality differences, gender or gender-identity prejudices, an inability to communicate, or some other cause. Whatever the reason, the crew leader must address the issue and get the crew working as a team.

The best way to determine the reason for why individuals don't get along or work well together is to talk to the parties involved. The crew leader should speak openly with the employee, as well as the other individual(s) to uncover the source of the problem and discuss its resolution.

After uncovering the reason for the conflict, the crew leader can determine how to respond. There may be a way to resolve the problem and get the workers communicating and working as a team again. On the other hand, there may be nothing the crew leader can do that will lead to a harmonious solution. In this case, the crew leader would need to either transfer one of the involved employees to another crew or have the problem crew member terminated. Resorting to this latter option should be the last measure and taken after discussing the matter with one's immediate supervisor or the company's Human Resources Department.

Absenteeism and Turnover – Absenteeism and turnover in the industry can delay jobs and cause companies to lose money. Absenteeism refers to workers missing their scheduled work time on a job. It has many causes, some of which cannot be helped. Sickness, family emergencies, and funerals are examples of unavoidable causes of worker absence. However, there are some causes of unexcused absenteeism that crew leaders can prevent.

The most effective way to control absenteeism is to make the company's policy clear to all employees. The policy should be explained to all new employees. This explanation should include the number of absences allowed and acceptable reasons for taking sick or personal days. In addition, all workers should know how to inform their crew leaders when they miss work and understand the consequences of exceeding the number of sick or personal days allowed.

Once crew leaders explain the policy on absenteeism to employees, they must be sure to implement it consistently and fairly. This makes employees more likely to follow it. However, if enforcement of the policy is inconsistent and the crew leader gives some employees exceptions, it won't be effective. Thus, the rate of absenteeism is likely to increase.

Despite having a policy on absenteeism, there will always be employees who are chronically late or miss work. In cases where an employee abuses the absenteeism policy, the crew leader should discuss the situation directly with the employee, ensure that they understand the policy, and insist that they comply with it. If the employee's behavior does not improve, disciplinary action will be in order.

Turnover refers to the rate at which workers leave a company and are replaced by others. Like absenteeism, there are some causes of turnover that cannot be prevented and others that can. For instance, an employee may find a job elsewhere earning twice as much money. However, crew leaders can prevent some employee turnover situations. They can work to ensure safe working conditions for their crew, treat their workers fairly and consistently, and help promote good working conditions. The key is communication and promoting the motivational factors discussed earlier. Crew leaders need to know the problems if they are going to be able to successfully resolve them.

Some major causes of employee turnover include the following:

- Unfair/inconsistent treatment by the immediate supervisor
- Unsafe project sites
- Lack of job security or opportunities for advancement

For the most part, the actions described for absenteeism are also effective for reducing turnover. Past studies have shown that maintaining harmonious relationships on the job site goes a long way in reducing both turnover and absenteeism. This requires effective and **proactive** leadership on the part of the crew leader.

Failure to Comply with Company Policies and Procedures – Policies are rules that define the relationship between the company, its employees, its clients, and its subcontractors. Procedures include the instructions for carrying out the policies. Some companies have policies that dictate dress codes. The dress code may be designed partly to ensure safety, and partly to define the image a company wants to project to the outside world.

Companies develop procedures to ensure that everyone who performs a task does it safely and efficiently. Many procedures directly relate to safety. A **lockout/tagout (LOTO)** procedure is an example. In this procedure, the company defines who may perform a LOTO, how to properly complete and remove a LOTO, and who has the authority to remove it. Workers who fail to follow the procedure endanger themselves, as well as their co-workers.

Companies typically have a policy on disciplinary action, which defines steps to take if an employee violates company policies or procedures. The steps range from counseling by a supervisor for the first offense, to a written warning and/or LOI, to dismissal for repeat offenses. This will vary from one company to another. For example, some companies will fire an employee for the first violation of a safety procedure with potential for loss of life or serious injury.

The crew leader has the first-line responsibility for enforcing company policies and procedures. The crew leader should take the time with a new crew member to discuss the policies and procedures and show the crew member how to access them. If a crew member shows a tendency to neglect a policy or procedure, it is up to the crew leader to counsel that individual. If the crew member continues to violate a policy or procedure, the crew leader has no choice but to refer that individual to the appropriate authority within the company for disciplinary action.

NOTE

When a crew shows a pattern of consistent policy violations, management will scrutinize the crew leader and potentially take disciplinary action if the crew leader does not handle the violations appropriately. The crew leader is responsible for all aspects of the crew, not just its work accomplishment on the job.

Participant Exercise C

Case I:

On the way over to the job trailer, you look up and see a piece of falling scrap heading for one of the laborers. Before you can say anything, the scrap material hits the ground about five feet in front of the worker. You notice the scrap is a piece of conduit. You quickly pick it up, assuring the worker you will take care of this matter.

Looking up, you see your crew on the third floor in the area from which the material fell. You decide to have a talk with them. Once on the deck, you ask the crew if any of them dropped the scrap. The men look over at Bob, one of the electricians in your crew. Bob replies, "I guess it was mine. It slipped out of my hand."

It is a known fact that the Occupational Safety and Health Administration (OSHA) regulations state that an enclosed chute of wood shall be used for material waste transportation from heights of 20 feet or more. It is also known that Bob and the laborer who was almost hit have been seen arguing lately.

1. Assuming Bob's action was deliberate, what action would you take?

   ____

   ____

   ____

2. Assuming the conduit accidentally slipped from Bob's hand, how can you motivate him to be more careful?

   ____

   ____

3. What follow-up actions, if any, should be taken relative to the laborer who was almost hit?

   ____

   ____

   ____

4. Should you discuss the apparent OSHA violation with the crew? Why or why not?

   ____

   ____

   ____

5. What acts of leadership would be effective in this case? To what leadership traits are they related?

   ____

   ____

   ____

Case II:

The company just appointed Antonio crew leader of a tile-setting crew. Before his promotion into management, he had been a tile setter for five years. His work had been consistently of superior quality.

Except for a little good-natured kidding, Antonio's co-workers had wished him well in his new job. During the first two weeks, most of them had been cooperative while Antonio was adjusting to his supervisory role.

At the end of the second week, a disturbing incident took place. Having just completed some of his duties, Antonio stopped by the job-site wash station. There he saw Steve and Ron, two of his old friends who were also in his crew, washing.

"Hey, Ron, Steve, you shouldn't be cleaning up this soon. It's at least another thirty minutes until quitting time," said Antonio. "Get back to your work station, and I'll forget I saw you here."

"Come off it, Antonio," said Steve. "You used to slip up here early on Fridays. Just because you have a little rank now, don't think you can get tough with us." To this Antonio replied, "Things are different now. Both of you get back to work, or I'll make trouble." Steve and Ron said nothing more, and they both returned to their work stations.

From that time on, Antonio began to have trouble as a crew leader. Steve and Ron gave him the silent treatment. Antonio's crew seemed to forget how to do the most basic activities. The amount of rework for the crew seemed to be increasing. By the end of the month, Antonio's crew was behind schedule.

1. How do you think Antonio should've handled the confrontation with Ron and Steve?

   ____

   ____

2. What do you suggest Antonio could do about the silent treatment he got from Steve and Ron?

   ____

   ____

3. If you were Antonio, what would you do to get your crew back on schedule?

   ____

   ____

4. What acts of leadership could be used to get the crew's willing cooperation?

   ____

   ____

5. To which leadership traits do they correspond?

   ____

   ____

Additional Resources

Construction Workforce Development Professional, NCCER. 2016. New York, NY: Pearson Education, Inc.

Mentoring for Craft Professionals, NCCER. 2016. New York, NY: Pearson Education, Inc.

It's Your Ship: Management Techniques from the Best Damn Ship in the Navy, Captain D. Michael Abrashoff, USN. 2012. New York City, NY: Grand Central Publishing.

Survival of the Fittest, Mark Breslin. 2005. McNally International Press.

The Definitive Book of Body Language: The Hidden Meaning Behind People's Gestures and Expressions, Barbara Pease and Allan Pease. 2006. New York City, NY: Random House / Bantam Books.

2.0.0 Section Review

1. A crew leader differs from a craftworker in that a crew leader _____.
   a. does not need direct experience in the job duties a craft professional typically performs
   b. can expect to oversee one or more workers in addition to performing typical craft duties
   c. is exclusively in charge of overseeing, since performing technical work isn't part of this role
   d. has no responsibility to be present on the job site

2. Feedback is important in verbal communication because it _____.
   a. requires the sender to repeat the message
   b. involves the receiver repeating back the message word for word
   c. informs the sender of how the message was received
   d. consists of a written analysis of the message

3. Once achieved, setting challenging goals for workers gives them a sense of _____.
   a. accomplishment
   b. entitlement
   c. persecution
   d. failure

4. Which of the following is *not* a characteristic of a successful team?
   a. Everyone participates and everyone counts.
   b. There is a sense of mutual trust.
   c. Members minimize communication with each other.
   d. The team leader is committed to the team.

5. Problem solving differs from decision making because it _____.
   a. involves finding an answer
   b. expresses an opinion
   c. involves starting or stopping an action
   d. separates facts from non-facts

Section Three

3.0.0 Safety and Safety Leadership

Objective

Identify a crew leader's typical safety responsibilities with respect to common safety issues, including awareness of safety regulations and the cost of accidents.

a. Explain how a strong safety program can enhance a company's success.
b. Explain the purpose of OSHA and describe the role of OSHA in administering worker safety.
c. Describe the role of employers in establishing and administering safety programs.
d. Explain how crew leaders are involved in administering safety policies and procedures.

Trade Terms

Intangible: Not touchable, material, or measureable; lacking a physical presence.

Lethargy: Sluggishness, slow motion, lack of activity, or a lack of enthusiasm.

Negligence: Lack of appropriate care when doing something, or the failure to do something, usually resulting in injury to an individual or damage to equipment.

Safety data sheets (SDS): Documents listing information about a material or substance that includes common and proper names, chemical composition, physical forms and properties, hazards, flammability, handling, and emergency response in accordance with national and international hazard communication standards. Also called material safety data sheets (MSDS).

Businesses lose millions of dollars every year because of on-the-job accidents. Work-related injuries, sickness, and fatalities have caused untold suffering for workers and their families. Resulting project delays and budget overruns can cause huge losses for employers, and work-site accidents damage the overall morale of the crew.

Craft professionals routinely face hazards. Examples of these hazards include falls from heights, working on scaffolds, using cranes in the presence of power lines, operating heavy machinery, and working on electrically-powered or pressurized equipment. Despite these hazards, experts believe that applying preventive safety measures could drastically reduce the number of accidents.

> **Did You Know?**
>
> **The Fatal Four**
>
> When OSHA inspects a job site, they focus on the types of safety hazards that are most likely to cause serious and fatal injuries. These hazards result in the following most common categories of injuries:
>
> - Falls from elevations
> - Struck-by hazards
> - Caught-in/between hazards
> - Electrical-shock hazards

As a crew leader, one of your most important tasks is to enforce the company's safety program and make sure that all workers are performing their tasks safely. To be successful, the crew leader should do the following:

- Be aware of the human and monetary costs of accidents.
- Understand all federal, state, and local governmental safety regulations applicable to your work.
- Be the most visible example of the best safe work practices.
- Be involved in training workers in safe work methods.
- Conduct training sessions.
- Get involved in safety inspections, accident investigations, and fire protection and prevention.

"Example is not the main thing in influencing others. It is the only thing."
—Albert Schweitzer

Crew leaders are in the best position to ensure that their crew members perform all jobs safely. Providing employees with a safe working environment by preventing accidents and enforcing safety standards will go a long way towards maintaining the job schedule and enabling a job's completion on time and within budget.

3.1.0 The Impact of Accidents

Each day, workers in construction and industrial occupations face the risk of falls, machinery accidents, electrocutions, and other potentially fatal hazards. The National Institute of Occupational Safety and Health (NIOSH) statistics show that roughly 1,000 construction workers are killed on the job each year, which is more than any other industry. Falls are the leading cause of deaths in the construction industry through accident or **negligence**, accounting for over 60 percent of the fatalities in recent years. Nearly half of the fatal falls occurred from roofs, scaffolds, or ladders. Roofers, structural metal workers, and painters are at the greatest risk of fall fatalities (*Figure 13*).

In addition to the number of fatalities that occur each year, there are a staggering number of work-related injuries. In 2015, for example, almost 200,000 job-related injuries occurred in the construction industry. NIOSH estimates that the total cost of fatal and non-fatal injuries in the construction industry represents about 15 percent of the costs for all private industry. The main causes of injuries on construction sites include falls, electrocution, fires, and mishandling of machinery or equipment. According to NIOSH, back injuries are the leading health-related problem in workplaces.

3.1.1 Cost of Accidents

Occupational accidents cost roughly $250 billion or more every year. These costs affect the individual employee, the company, and the construction industry as a whole.

Organizations encounter both direct and indirect costs associated with workplace accidents. Direct costs are the money companies must pay out to workers' compensation claims and sick pay; indirect costs are all the other tangible and **intangible** things and costs a company must account for as the result of a worker's injury or death. To compete and survive, companies must control these as well as all other employment-related costs. There are many costs involved with workplace accidents. A company can insure some of these costs, but not others.

Figure 13 Falls are the leading cause of deaths and injuries in construction.

Insured costs – Insured costs are those costs either paid directly or reimbursed by insurance carriers. Insured costs related to injuries or deaths include the following:

- Compensation for lost earnings (known as *worker's comp*)
- Funeral charges
- Medical and hospital costs
- Monetary awards for permanent disabilities
- Pensions for dependents
- Rehabilitation costs

Insurance premiums or charges related to property damages include the following:

- Fire or other safety-related peril
- Structural loss; material and equipment loss or damage
- Loss of business use and occupancy
- Public liability
- Replacement cost of equipment, material, and structures

Uninsured costs – The relative direct and indirect costs of accidents are comparable to the visible and hidden portions of an iceberg, as shown in *Figure 14*. The tip of the iceberg represents direct costs, which are the visible costs. Not all of these are covered by insurance. The more numerous indirect costs aren't readily measurable, but they can represent a greater financial burden than the direct costs.

Uninsured costs from injuries or deaths include the following:

- First aid expenses
- Transportation costs
- Costs of investigations
- Costs of processing reports
- Down time on the job site
- Costs to train replacement workers

Did You Know?

The Costs of Negligence

If you receive an injury as the result of a workplace accident, and a completed investigation shows that your injuries were due to your own negligence, you can't sue your employer for damages. In addition, workers' compensation insurance companies may decline to pay claims for negligent employees who are injured or killed.

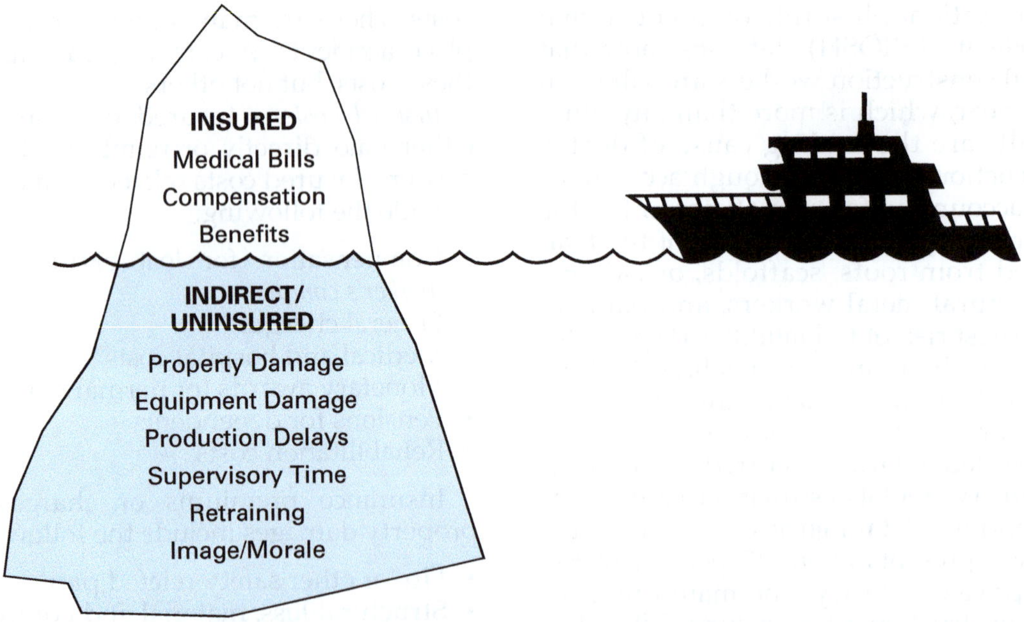

Figure 14 Costs associated with accidents.

Uninsured costs related to wage losses include the following:

- Idle time of workers whose work is interrupted
- Time spent cleaning the accident area
- Time spent repairing damaged equipment
- Time lost by workers receiving first aid
- Costs of training injured workers in a new career

Uninsured costs related to production losses include the following:

- Product spoiled by accident
- Loss of skill and experience; worker replacement
- Lowered production capacity
- Idle machine time due to lack of qualified operators

Associated costs may include the following:

- Difference between actual losses and amount recovered
- Costs of rental equipment used to replace damaged equipment
- Costs of inexperienced temp or permanent new workers used to replace injured workers
- Wages or other benefits paid to disabled workers
- Overhead costs while production is stopped
- Impact on schedule
- Loss of client bonus or payment of forfeiture for delays

Uninsured costs related to off-the-job activities include the following:

- Time spent on ensuring injured workers' welfare
- Loss of skill and experience of injured workers
- Costs of training replacement workers

Uninsured costs related to intangible factors include the following:

- Increased labor conflict
- Loss of bid opportunities because of poor safety records
- Loss of client goodwill
- Lowered employee morale
- Unfavorable public relations

3.2.0 OSHA

To reduce safety and health risks and the number of injuries and fatalities on the job, the federal government has enacted laws and regulations, including the Occupational Safety and Health Act of 1970 (OSH Act of 1970). This law created the Occupational Safety and Health Administration (OSHA), which is part of the US Department of Labor. OSHA also provides education and training for employers and workers. Through the administration of OSHA, the US Congress seeks "to assure so far as possible every working man and woman in the Nation safe and healthful working conditions and to preserve our human resources..." (OSH Act of 1970, Section 2[b]).

To promote a safe and healthy work environment, OSHA issues standards and rules for working conditions, facilities, equipment, tools, and work processes. It does extensive research into occupational accidents, illnesses, injuries, and deaths to reduce the number of occurrences and adverse effects. In addition, OSHA regulatory agencies conduct workplace inspections to ensure that companies follow the standards and rules.

To enforce OSHA regulations, the government has granted regulatory agencies the right to enter public and private properties to conduct workplace safety investigations. The agencies also have the right to take legal action if companies are not in compliance with the Act. These regulatory agencies employ OSHA Compliance Safety and Health Officers (CSHO), who are experts in the occupational safety and health field. The CSHOs are thoroughly familiar with OSHA standards and recognize safety and health hazards.

States with their own occupational safety and health programs conduct inspections by enlisting the services of qualified state CSHOs.

Companies are inspected for a multitude of reasons. They may be randomly selected, or they may be chosen as a result of employee complaints, a report of an imminent danger, or major accidents/fatalities that have occurred.

OSHA has established significant monetary fines for the violation of its regulations. *Table 1* lists the penalties as of 2016. In some cases, OSHA will hold a superintendents or crew leaders personally liable for repeat violations as well. In addition to the fines, there are possible criminal charges for willful violations resulting in death or serious injury. The attitude of the employer and their safety history can have a significant effect on the outcome of a case.

3.3.0 Employer Safety Responsibilities

Each employer must set up a safety and health program to manage workplace safety and health and to reduce work-related injuries, illnesses, and fatalities. The program must be appropriate for the conditions of the workplace. It should consider the number of workers employed and the hazards they face while at work.

To be successful, the safety and health program must have management, leadership, and employee participation. In addition, training and informational meetings play an important part in effective programs. Being consistent with safety policies is the key. Regardless of the employer's responsibility, however, the individual worker is ultimately responsible for his or her own safety.

Table 1 OSHA Penalties for violations established in 2016

Type of Violation	Maximum Penalty
Serious Other-Than-Serious Posting Requirements	$12,471 per violation
Failure to Abate	$12,471 per day
Willful or Repeated Violation	$124,709 per violation

Did You Know?

State OSHA Programs

Nearly half the states in the United States and some of its territories have state-run OSHA programs. These programs are set up under federal OSHA guidelines and must establish job health and safety standards that are at least as effective as the federal standards. The states/territories have the option of adopting more stringent standards or setting standards for hazards not addressed in the federal program. Of the 22 states and territories with non-federal OSHA programs, six of them limit their coverage to state and local government employees.

3.3.1 Safety Program

The crew leader plays a key role in the successful implementation of the safety program. The crew leader's attitude toward the program sets the standard for how crew members view safety. Therefore, the crew leader should follow all program guidelines and require crew members to do the same.

Safety programs should consist of the following:

- Safety policies and procedures
- Safety information and training
- Posting of safety notices
- Hazard identification, reporting, and assessment
- Safety record system
- Accident reporting and investigation procedures
- Appropriate discipline for not following safety procedures

3.3.2 Safety Policies and Procedures

Employers are responsible for following OSHA and state safety standards. Usually, they incorporate federal and state OSHA regulations into a safety policies and procedures manual. Employees receive such a manual when they are hired.

During orientation, appropriate company staff should guide the new employees through the general sections of the safety manual and the sections that have the greatest relevance to their job.

If the employee can't read, the employer should have someone read it to the employee and answer any questions that arise. The employee should then sign a form stating understanding of the information.

It isn't enough to tell employees about safety policies and procedures on the day they are hired and then never mention them again. Rather, crew leaders should constantly emphasize and reinforce the importance of adhering to all safety policies and procedures. In addition, employees should play an active role in determining job safety hazards and find ways to prevent and control hazards.

3.3.3 Hazard Identification and Assessment

Safety policies and procedures should be specific to the company. They should clearly present the hazards of the job and provide the means to report hazards to the proper level of management without prejudice to the individual doing the reporting. Crew leaders should also identify and assess hazards to which employees are exposed. They must also assess compliance with federal and state OSHA standards.

To identify and assess hazards, OSHA recommends that employers conduct periodic and random inspections of the workplace, monitor safety and health information logs, and evaluate new equipment, materials, and processes for potential hazards before they are used.

"You get what you inspect, not what you expect."
—Anonymous

Crew leaders and workers play important roles in identifying and reporting hazards. It is the crew leader's responsibility to determine what working conditions are unsafe and to inform employees of hazards and their locations. In addition, they should encourage their crew members to tell them about hazardous conditions. To accomplish this, crew leaders must be present and available on the job site.

The crew leader also needs to help the employee be aware of and avoid the built-in hazards to which craft professionals are exposed. Examples include working at elevations, working in confined spaces such as tunnels and underground vaults, on caissons, in excavations with earthen walls, and other naturally-dangerous projects. In addition, the crew leader can take safety measures, such as installing protective railings to prevent workers from falling from buildings, as well as scaffolds, platforms, and shoring.

3.3.4 Safety Information and Training

The employer must provide periodic information and training to new and long-term employees. This happens as often as necessary so that all employees receive adequate training. When safety and health information changes or workplace conditions create new hazards, the company must then provide special training and informational sessions. It is important to note that the company must present safety-related information in a manner that each employee will understand.

When a crew leader assigns an inexperienced employee a new task, the crew leader must ensure that the employee can do the work in a safe manner. The crew leader can accomplish this by providing safety information or training for groups or individuals.

When assigning an inexperienced employee a new task, do the following:

- Define the task.
- Explain how to do the task safely.
- Explain what tools and equipment to use and how to use them safely.
- Identify the necessary personal protective equipment and train the employee in its use.
- Explain the nature of the hazards in the work and how to recognize them.
- Stress the importance of personal safety and the safety of others.
- Hold regular safety training sessions with the crew's input.
- Review **safety data sheets (SDS)** that may be applicable.

3.3.5 Safety Record System

OSHA regulations (29 *CFR* 1904) require that employers keep records of hazards identified and document the severity of the hazard. The information should include the likelihood of employee exposure to the hazard, the seriousness of the harm associated with the hazard, and the number of exposed employees.

In addition, the employer must document the actions taken or plans for action to control the hazards. While it is best to take corrective action immediately, it is sometimes necessary to develop a plan to set priorities and deadlines and track progress in controlling hazards.

Employers who are subject to the recordkeeping requirements of the Occupational Safety and Health Act of 1970 must maintain records of all recordable occupational injuries and illnesses. The following are some OSHA forms that should be used for this recordkeeping:

- OSHA Form 300, *Log of Work-Related Injuries and Illnesses*
- OSHA Form 300A, *Summary of Work-Related Injuries and Illnesses*
- OSHA Form 301, *Injury and Illness Incident Report*

These three OSHA forms are included in the *Appendix* at the end of this module. Note that crew leaders directly handle the OSHA Form 301.

An SDS provides both workers and emergency personnel with the proper procedures for handling or working with a substance that may be dangerous. The document will include information such as physical data (melting point, boiling point, flash point, etc.), toxicity, health effects, first aid, reactivity, storage, disposal, protective equipment required for handling, and spill/leak procedures. These sheets are of particular use if a spill, fire, or other accident occurs.

Companies not exempted by OSHA must maintain required safety logs and retain them for 5 years following the end of the calendar year to which they relate. Logs must be available (normally at the company offices) for inspection and copying by representatives of the Department of Labor, the Department of Health and Human Services, or states given jurisdiction under the Act. Employees, former employees, and their representatives may also review these logs.

3.3.6 Accident Investigation

Employees must know from their training to immediately report any unusual event, accident, or injury. Policies and definitions of what these incidents consist of should be included in safety manuals. In the event of an accident, the employer is required to investigate the cause of the accident and determine how to avoid it in the future.

According to OSHA regulations, the employer must investigate each work-related death, serious injury or illness, or incident having the potential to cause death or serious physical harm. The employer should document any findings from the investigation, as well as the action plan to

Summaries of Work-Related Injuries and Illnesses

Most companies with 11 or more employees must post an OSHA Form 300A, *Summary of Work-Related Injuries and Illnesses*, between February 1 and April 30 of each year. Employees have the right to review this form. Check your company's policies regarding this and the related OSHA forms.

OSHA's Form 300A (Rev. 01/2004)

Summary of Work-Related Injuries and Illnesses

Year 20__ __

U.S. Department of Labor
Occupational Safety and Health Administration

Form approved OMB no. 1218-0176

All establishments covered by Part 1904 must complete this Summary page, even if no work-related injuries or illnesses occurred during the year. Remember to review the Log to verify that the entries are complete and accurate before completing this summary.

Using the Log, count the individual entries you made for each category. Then write the totals below, making sure you've added the entries from every page of the Log. If you had no cases, write "0."

Employees, former employees, and their representatives have the right to review the OSHA Form 300 in its entirety. They also have limited access to the OSHA Form 301 or its equivalent. See 29 CFR Part 1904.35, in OSHA's recordkeeping rule, for further details on the access provisions for these forms.

Number of Cases

Total number of deaths	Total number of cases with days away from work	Total number of cases with job transfer or restriction	Total number of other recordable cases
(G)	(H)	(I)	(J)

Number of Days

Total number of days away from work	Total number of days of job transfer or restriction
(K)	(L)

Injury and Illness Types

Total number of . . .
(M)

(1) Injuries	(4) Poisonings
(2) Skin disorders	(5) Hearing loss
(3) Respiratory conditions	(6) All other illnesses

Post this Summary page from February 1 to April 30 of the year following the year covered by the form.

Public reporting burden for this collection of information is estimated to average 58 minutes per response, including time to review the instructions, search and gather the data needed, and complete and review the collection of information. Persons are not required to respond to the collection of information unless it displays a currently valid OMB control number. If you have any comments about these estimates or any other aspects of this data collection, contact: US Department of Labor, OSHA Office of Statistical Analysis, Room N-3644, 200 Constitution Avenue, NW, Washington, DC 20210. Do not send the completed forms to this office.

Establishment information

Your establishment name ________

Street ________

City ________ State ____ ZIP ____

Industry description (*e.g., Manufacture of motor truck trailers*)

Standard Industrial Classification (SIC), if known (*e.g., 3715*)

OR

North American Industrial Classification (NAICS), if known (e.g., 336212)

Employment information *(If you don't have these figures, see the Worksheet on the back of this page to estimate.)*

Annual average number of employees ________

Total hours worked by all employees last year ________

Sign here

Knowingly falsifying this document may result in a fine.

I certify that I have examined this document and that to the best of my knowledge the entries are true, accurate, and complete.

Company executive — Title

( ) - Phone — / / Date

prevent future occurrences. The company should complete these actions immediately, with photos or video if possible. It's important that the investigation uncover the root cause of the accident to avoid similar incidents in the future. In many cases, the root cause was a flaw in the system that failed to recognize the unsafe condition or the potential for an unsafe act (*Figure 15*).

3.4.0 Leader Involvement in Safety

To be an effective, you must be actively involved in your company's safety program. Crew leader involvement includes conducting frequent safety training sessions and inspections, promoting first aid, and fire protection and prevention, preventing substance abuse on the job, and investigating accidents. Most importantly, crew leaders must practice safety at all times.

3.4.1 Safety Training Sessions

A safety training session may be a brief, informal gathering of a few employees or a formal meeting with instructional videos and talks by guest speakers. The size of the audience and the topics addressed determine the format of the meeting. You should plan to conduct small, informal safety sessions weekly.

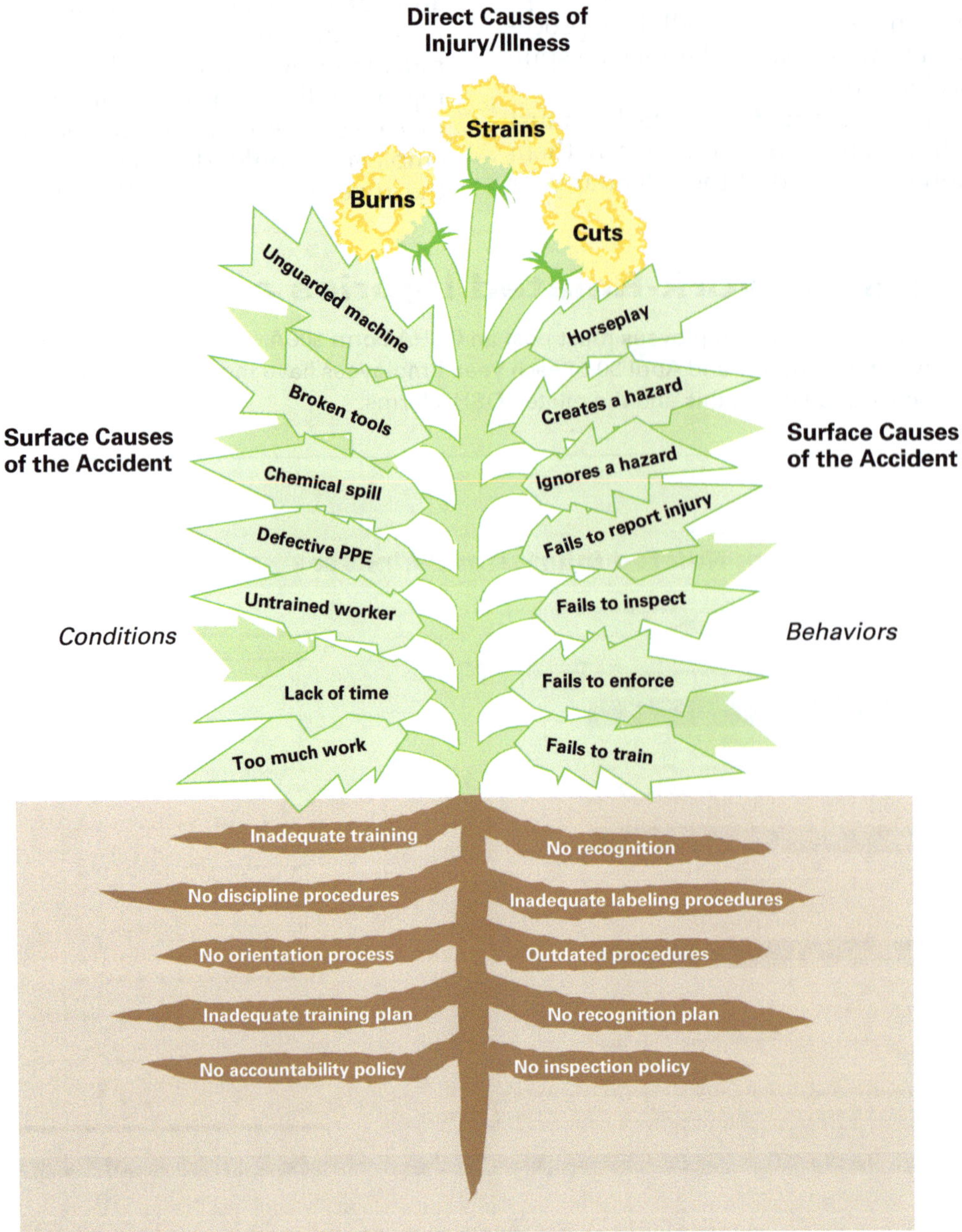

Figure 15 Root causes of accidents.

OSHA Accident Notification Requirements

There are urgent reporting requirements by the employer to OSHA for the following cases:

1. Within 8 hours: A work-related fatality.
2. Within 24 hours:
   - Work-related accident that resulted in an in-patient admission of one or more employees
   - Work-related amputation
   - Work-related loss of an eye

You should also plan safety training sessions in advance, and you should communicate the information to all affected employees. In addition, the topics covered in these training sessions should be timely and practical. Keep a log of each safety session signed by all attendees. It must be maintained as a record and available for inspection. It is advisable to attach a copy of a summary of the training session to the attendance record so you can keep track of what topics you covered and when.

3.4.2 Inspections

Crew leaders must make routine, frequent inspections to prevent accidents from happening. They must also take steps to avoid accidents. For that purpose, they need to inspect the job sites where their workers perform tasks. It's advisable to do these inspections before the start of work each day and during the day at random times.

Crew leaders must protect workers from existing or potential hazards in their work areas. Crews are sometimes required to work in areas controlled by other contractors. In these situations, the crew leader must maintain control over the safety exposure of the crew. If hazards exist, the crew leader should immediately bring the hazards to the attention of the contractor at fault, their superior, and the person responsible for the job site.

Crew leader inspections are only valuable if follow-up action corrects potential hazards. Therefore, crew leaders must be alert for unsafe and negligent acts on their work sites. When an employee performs an unsafe action, the crew leader must explain to the employee why the act was unsafe, tell that employee to not do it again, and request cooperation in promoting a safe working environment. The crew leader must document what happened and what the employee was told to do to correct the situation.

It is then important that crew leaders follow up to make certain the employee is complying with the safety procedures. Never allow a safety violation to go uncorrected. There are three courses of action that you, as a crew leader, can take in an unsafe work site situation:

- Get the appropriate party to correct the problem.
- Fix the problem yourself.
- Refuse to have the crew work in the area until the responsible party corrects the problem.

"As soon as you see a mistake and don't fix it, it becomes your mistake."

—Anonymous

3.4.3 First Aid

The primary purpose of first aid is to provide immediate and temporary medical care to employees involved in accidents, as well as employees experiencing non-work-related health emergencies, such as chest pain or breathing difficulty. To meet this objective, every crew leader should be aware of the location and contents of first aid kits available on the job site. Emergency numbers should be posted in the job trailer. In addition, OSHA requires that at least one person trained in first aid be present at the job site at all times. It's also advisable, but not required, that someone on site should have cardiopulmonary resuscitation (CPR) training.

NOTE

CPR certifications must be renewed every two years.

The victim of an accident or sudden illness at a job site may be harder to aid than elsewhere since the worker may be at a remote location. The site may be far from a rescue squad, fire department, or hospital, presenting a problem in the rescue and transportation of the victim to a hospital. The worker may also have received an injury from falling rock or other materials, so immediate special rescue equipment or first-aid techniques are often needed.

Employer benefits of having personnel trained in first aid at job sites include the following:

- The immediate and proper treatment of minor injuries may prevent them from developing into more serious conditions. Thus, these precautions can eliminate or reduce medical expenses, lost work time, and sick pay.

- It may be possible to determine if the injured person requires professional medical attention.
- Trained individuals can save valuable time preparing the inured for treatment for when professional medical care arrives. This service increases the likelihood of saving a life.

The American Red Cross, Medic First Aid, and the United States Bureau of Mines provide basic and advanced first aid courses at nominal costs. These courses include both first aid and CPR. The local area offices of these organizations can provide further details regarding the training available.

3.4.4 Fire Protection and Prevention

Fires and explosions kill and injure many workers each year, so it is important that crew leaders understand and practice fire-prevention techniques as required by company policy.

The need for protection and prevention is increasing as manufacturers introduce new building materials. Some building materials are highly flammable. They produce great amounts of smoke and gases, which cause difficulties for fire fighters, and can quickly overcome anyone present. Other materials melt when they burn and may puddle over floors, preventing fire-fighting personnel from entering areas where this occurs.

OSHA has specific standards for fire safety. Employers are required to provide proper exits, fire-fighting equipment, and employee training on fire prevention and safety. For more information, consult OSHA guidelines (available at **www.osha.gov**).

3.4.5 Substance Abuse

Substance abuse is a continuing problem in the workplace. Substance abuse is the inappropriate overuse of drugs and chemicals, whether they are legal or illegal. All substance abuse results in some form of mental, sensory, or physical impairment. Some people use illegal "street drugs", such as cocaine or crystal meth. Others use legal prescription drugs incorrectly by taking too many (or too few) pills, using other people's medications, or self-medicating. Alcohol can also be abused by consuming to the point of intoxication. Other substances that are legal in some states (e.g., marijuana) can cause prolonged impairment with heavy usage.

It is essential that crew leaders enforce company policies and procedures regarding substance abuse. Crew leaders must work with management to deal with suspected drug and alcohol abuse and should not deal with these situations themselves. The Human Resources department or a designated manager usually handles these cases.

There are legal consequences of substance abuse and the associated safety implications. If you observe an employee showing impaired behavior for any reason, immediately contact your supervisor and/or Human Resources department for assistance. You protect the business, and the employee's and other workers' safety by taking these actions. It is the crew leader's responsibility to maintain safe working conditions at all times. This may include removing workers from a work site where they may be endangering themselves or others.

For example, suppose several crew members go out and smoke marijuana or drink during lunch. Then, they return to work to erect scaffolding for a concrete pour in the afternoon. If you can smell marijuana on the crew member's clothing or alcohol on their breath, you must step in and take action. Otherwise, they might cause an accident that could delay the project, or cause serious injury or death to themselves or others.

> *"Concern for man himself and his safety must always form the chief interest of all technical endeavors."*
> —*Albert Einstein*

It is often difficult to detect drug and alcohol abuse because the effects can be subtle. The best way is to look for identifiable effects, such as those mentioned above, or sudden changes in behavior that aren't typical of the employee. Some examples of such behaviors include the following:

- Unscheduled absences; failure to report to work on time
- Significant changes in the quality of work
- Unusual activity or **lethargy**
- Sudden and irrational temper flare-ups
- Significant changes in personal appearance, cleanliness, or health

There are other more specific signs that should arouse suspicion, especially if more than one is visible:

- Slurring of speech or an inability to communicate effectively
- Shiftiness or sneaky behavior, such as an employee disappearing to wooded areas, storage areas, or other private locations
- Wearing sunglasses indoors or on overcast days to hide dilated or constricted pupils, conditions which impair vision (*Figure 16*)

Figure 16 An impaired worker is a dangerous one.

- Wearing long-sleeved garments, particularly on hot days, to cover marks from needles used to inject drugs
- Attempting to borrow money from co-workers
- The loss of an employee's tools or company equipment

3.4.6 Job-Related Accident Investigations

Crew leaders are sometimes involved with an accident investigation. When an accident, injury, or report of work-connected illness takes place. If present on site, the crew leader should proceed immediately to the accident location to ensure that the victim receives proper first aid. The crew leader will also want to make sure that responsible individuals take other safety and operational measures to prevent another incident.

If required by company policy, the crew leader will also need to make a formal investigation and submit a report after an incident (including the completion of an *OSHA Form 301* report). An investigation looks for the causes of the accident by examining the circumstances under which it occurred and talking to the people involved. Investigations are perhaps the most useful tool in the prevention of future accidents.

The following are four major parts to an accident investigation:

- Describing the accident and related events leading to the accident
- Determining the cause(s) of the accident
- Identifying the persons or things involved and the part played by each
- Determining how to prevent reoccurrences

3.4.7 Promoting Safety

The best way for crew leaders to encourage safety is through example. Crew leaders should be aware that their behavior sets standards for their crew members. If a crew leader cuts corners on safety, then the crew members may think that it is okay to do so as well.

CAUTION

Workers often "follow the leader" when it comes to unsafe work practices. It is common for supervisory personnel to engage in unsafe practices and take more risks because they are more experienced. However, inexperienced or careless workers who take the same risks won't be as successful avoiding injury. As a leader, you must follow all safety practices to encourage your crew to do the same.

"I cannot trust a man to control others who cannot control himself."

—Robert E. Lee

The key to effectively promote safety is good communication. It is important to plan and coordinate activities and to follow through with safety programs. The most successful safety promotions occur when employees actively participate in planning and carrying out activities.

Some activities used by organizations to help motivate employees on safety and help promote safety awareness include the following:

- Safety training sessions
- Contests
- Recognition and awards
- Publicity

Safety training sessions can help keep workers focused on safety and give them the opportunity to discuss safety concerns with the crew. A previous section addressed this topic.

Did You Know?

Substance Abuse

An employee who is involved in an accident while under the influence of drugs or alcohol may be denied workers compensation insurance benefits.

Case Study

For years, a prominent safety engineer was confused as to why sheet-metal workers fractured their toes frequently. The crew leader had not performed thorough accident investigations, and the injured workers were embarrassed to admit how the accidents really occurred. Further investigation discovered they used the metal-reinforced cap on their safety shoes as a "third hand" to hold the sheet metal vertically in place when they fastened it. The rigid and heavy metal sheet was inclined to slip and fall behind the safety cap onto the toes, causing fractures. The crew leader could have prevented several injuries by performing a proper investigation after the first accident.

3.4.8 Safety Contests

Contests are a great way to promote safety in the workplace. Examples of safety-related contests include the following:

- Sponsoring housekeeping contests for the cleanest job site or work area
- Challenging employees to come up with a safety slogan for the company or department
- Having a poster contest that involves employees or their children creating safety-related posters
- Recording the number of accident-free workdays or worker-hours
- Giving safety awards (hats, T-shirts, other promotional items or prizes)

One of the positive aspects of safety contests is their ability to encourage employee participation. It is important, however, to ensure that the contest has a valid purpose. For example, workers can display the posters or slogans created in a poster contest throughout the organization as safety reminders.

CAUTION

One mistake that some companies make when offering safety contests is providing tangible or monetary awards to departments or teams specifically for the lowest number of reported accidents or accident-free work hours. While well-intentioned, this approach appeals to the tendency of people to inflate their performance to win. Consequently, history shows this has the negative effect of encouraging the under-reporting of accidents and injuries, which defeats the purposes of safety contests.

3.4.9 Incentives and Rewards

Incentives and awards serve several purposes. Among them are acknowledging and encouraging good performance, building goodwill, reminding employees of safety issues, and publicizing the importance of practicing safety standards. There are countless ways to recognize and award safety. Examples include the following:

- Supplying food at the job site when a certain goal is achieved

- Providing a reserved parking space to acknowledge someone for a special achievement
- Giving gift items such as T-shirts or gift certificates to reward employees
- Giving awards to a department or an individual (*Figure 17*)
- Sending a letter of appreciation
- Publicly honoring a department or an individual for a job well done

You can use creativity to determine how to recognize and award good safety on the work site. The only precautionary measure is that the award should be meaningful and not perceived as a bribe. It should be representative of the accomplishment.

3.4.10 Publicity

Publicizing safety is the best way to get the message out to employees. An important aspect of publicity is to keep the message accurate and current. Safety posters that hang for years on end tend to lose effectiveness. It is important to keep ideas fresh.

Examples of promotional activities include posters or banners, advertisements or information on bulletin boards, payroll mailing stuffers, and employee newsletters. In addition, the company can purchase merchandise that promotes safety, including buttons, hats, T-shirts, and mugs.

Figure 17 An example of a safety award.

Participant Exercise D

Described here are three scenarios that reflect unsafe practices by craft workers. For each of these scenarios write down how you would deal with the situation, first as the crew leader of the craft worker, and then as the leader of another crew.

1. You observe a worker wearing his hard hat backwards and his safety glasses hanging around his neck. He is using a concrete saw.

___

___

___

___

___

___

___

2. As you are supervising your crew on the roof deck of a building under construction, you notice that a section of guard rail has been removed. Another contractor was responsible for installing the guard rail.

___

___

___

___

___

___

___

3. Your crew is part of plant shutdown at a power station. You observe that a worker is welding without a welding screen in an area where there are other workers.

___

___

___

___

___

___

___

Additional Resources

Construction Workforce Development Professional, NCCER. 2016. New York, NY: Pearson Education, Inc.

Mentoring for Craft Professionals, NCCER. 2016. New York, NY: Pearson Education, Inc.

The following websites offer resources for products and training:

National Census of Fatal Occupational Injuries (NCFOI), **www.bls.gov**

National Institute of Occupational Safety and Health (NIOSH), **www.cdc.gov/niosh**

National Safety Council, **www.nsc.org**.

Occupational Safety and Health Administration (OSHA), **www.osha.gov**

3.0.0 Section Review

1. One of a crew leader's most important responsibilities to the employer is to _____.
   a. enforce company safety policies
   b. estimate material costs for a project
   c. make recommendations for setting up a crew
   d. provide input for fixed-price contracts

2. What amount can OSHA fine a company for willfully committing a violation of an OSHA safety standard?
   a. $1,000
   b. $7,000
   c. $12,471
   d. $124,709

3. Who is ultimately responsible for a worker's safety?
   a. The individual worker
   b. The crew leader
   c. The project superintendent
   d. The company HR department head

4. A crew leader's safety responsibilities include all of the following *except* _____.
   a. conducting safety training sessions
   b. developing a company safety program
   c. performing safety inspections
   d. participating in accident investigations

Section Four

4.0.0 Project Planning

Objective

Demonstrate a basic understanding of the planning process, scheduling, and cost and resource control.

a. Describe how construction contracts are structured.
b. Describe the project planning and scheduling processes.
c. Explain how to implement cost controls on a construction project.
d. Explain the crew leader's role in controlling project resources and productivity.

Performance Tasks

1. Develop and present a look-ahead schedule.
2. Develop an estimate for a given work activity.

Trade Terms

Critical path: In manufacturing, construction, and other types of creative processes, the required sequence of tasks that directly controls the ultimate completion date of the project.

Job diary: A written record that a supervisor maintains periodically (usually daily) of the events, communications, observations, and decisions made during the course of a project.

Look-ahead schedule: A manual- or software-scheduling tool that looks several weeks into the future at the planned project events; used for anticipating material, labor, tool, and other resource requirements, as well as identifying potential schedule conflicts or other problems.

Return on investment (ROI): A measure of the gain or loss of money resulting from an investment; normally measured as a percentage of the original investment.

Work breakdown structure (WBS): A diagrammatic and conceptual method for subdividing a complex project, concept, or other thing into its various functional and organizational parts so that planners can analyze and plan for each part in detail.

This section describes methods of efficient project control. It examines estimating, planning and scheduling, and resource and cost control. All workers who participate in a job are responsible at some level for controlling cost and schedule performance, and for ensuring that they complete the project according to plans and specifications.

> **NOTE**
> This section mainly pertains to building-construction projects, but the project control principles described here apply generally to all types of projects.

The contractor, project manager, superintendent, and crew leader each have management responsibilities for their assigned jobs. For example, the contractor's responsibility begins with obtaining the contract, and it doesn't end until the client takes ownership of the project. The project manager is generally the person with overall responsibility for coordinating the project. Finally, the superintendent and crew leader are responsible for coordinating the work of one or more workers, one or more crews of workers within the company or, on occasion, one or more crews of subcontractors. The crew leader directs a crew in the performance of work tasks.

4.1.0 Construction Project Phases, Contracts, and Budgeting

Construction projects consist of three phases: the *development phase*, the *planning phase*, and the *construction phase*. Throughout these phases, the property owner must work directly with engineers and architects to gather data necessary for persuading a contractor to accept the job. Once a job is in progress, the crew leader fills an active role in maintaining the planned budget for the project.

4.1.1 Development Phase

A new building project begins when an owner has decided to build a new facility or add to an existing facility. The development process is the first stage of planning for a new building project. This process involves land research and feasibility studies to ensure that the project has merit. Architects or engineers develop the conceptual drawings that define the project graphically. They then provide the owner with sketches of room layouts and elevations and make suggestions about what construction materials to use.

During the development phase, architects, engineers, and/or the owner develop an estimate for the proposed project and establish a preliminary budget. Once that budget is established, the financing of the project with lending institutions begins. The development team begins preliminary reviews with government agencies. These reviews include zoning, building restrictions, landscape requirements, and environmental impact studies.

The owner must analyze the project's cost and potential **return on investment (ROI)** to ensure that its costs won't exceed its market value and that the project provides a reasonable profit during its existence. If the project passes this test, the responsible architects/engineers will proceed to the planning phase.

4.1.2 Planning Phase

When the architects/engineers begin to develop the project drawings and specifications, they consult with other design professionals such as structural, mechanical, and electrical engineers. They perform the calculations, make a detailed technical analysis, and check details of the project for accuracy.

The design professionals create drawings and specifications. They use these drawings and specifications to communicate the necessary information to the contractors, subcontractors, suppliers, and workers that contribute to a project.

During the planning phase, the owners hold many meetings (*Figure 18*) to refine estimates, adjust plans to conform to regulations, and secure a construction loan. If the project is a condominium, an office building, or a shopping center, then a marketing firm develops a marketing program. In such cases, the selling of the project often starts before actual construction begins.

Figure 18 Architects and clients meet to refine plans.

Next, the design team produce a complete set of drawings, specifications, and bid documents. Then the owner will select the method to obtain contractors. The owner may choose to negotiate with several contractors or select one through competitive bidding. Everyone concerned must also consider safety as part of the planning process. A safety crew leader may walk through the site as part of the pre-bid process.

Contracts for construction projects can take many forms. All types of contracts fall under three basic categories: *firm-fixed-price, cost reimbursable,* and *guaranteed maximum price.*

Firm-fixed-price – In this type of contract, the buyer generally provides detailed drawings and specifications, which the contractor uses to calculate the cost of materials and labor. To these costs, the contractor adds a percentage representing company overhead expenses such as office rent, insurance, and accounting/payroll costs. At the end, the contractor adds a profit factor.

When submitting the bid, the contractor will state very specifically the conditions and assumptions on which the company based the bid. These conditions and assumptions also form the basis from which parties can price allowable changes to the contract. Because contracting parties establish the price in advance, any changes in the job requirements once the job is started will impact the contractor's profit margin.

This is where the crew leader can play an important role by identifying problems that increase the amount of planned labor or materials. By passing this information up the chain of command, the crew leader allows the company to determine if the change is outside the scope of the bid. If so, they can submit a change order request to cover the added cost.

Cost reimbursable contract – In this type of contract, the buyer reimburses the contractor for labor, materials, and other costs encountered in the performance of the contract. Typically, the contractor and buyer agree in advance on hourly or daily labor rates for different categories of worker. These rates include an amount representing the contractor's overhead expense. The buyer also reimburses the contractor for the cost of materials and equipment used on the job.

The buyer and contractor also negotiate a profit margin. On this type of contract, the profit margin is likely to be lower than that of a fixed-price contract because of the significantly-reduced contractor's cost risk. The profit margin is often subject to incentive or penalty clauses that make the amount of profit awarded subject to performance by the contractor. The contract usually ties performance to project schedule milestones.

Guaranteed maximum price (GMP) contract – This form of contract, also called a *not-to-exceed contract,* is common on projects negotiated mainly with the owner. Owner's involvement in the process usually includes preconstruction, and the entire team develops the parameters that define the basis for the work.

With a GMP contract, the owner reimburses the contractor for the actual costs incurred by the contractor. The contract also includes a payment of a fixed fee up to the maximum price allowed in the contract. The contractor bears any cost overruns.

The advantages of the GMP contract vehicle may include the following:

- Reduced design time
- Allows for phased construction
- Uses a team approach to a project
- Reduction in changes related to incomplete drawings

4.1.3 Construction Phase

The designated contractor enlists the help of mechanical, electrical, elevator, and other specialty subcontractors to complete the construction phase. The contractor may perform one or more parts of the construction, and rely on subcontractors for the remainder of the work. However, the general contractor is responsible for managing all the trades necessary to complete the project. *Figure 19* shows the flow of a typical project from beginning to end.

As construction nears completion, the architect/engineer, owner, and government agencies start their final inspections and acceptance of the project. If the general contractor has managed the project, the subcontractors have performed their work, and the architects/engineers have regularly inspected the project to ensure it satisfied the local code, then the inspection process can finish up in a timely manner. This results in a satisfied client and a profitable project for all.

On the other hand, if the inspection reveals faulty workmanship, poor design, incorrect use of materials, or violation of codes, then the inspection and acceptance will become a lengthy battle and may result in a dissatisfied client and an unprofitable project.

The initial set of drawings for a construction project reflects the completed project as conceived by the architect and engineers. During construction, changes are usually necessary because of factors unforeseen during the design phase. For example, when electricians must reroute cabling or conduit, or the installed equipment location is different than shown on the original drawing, such changes must be marked on the drawings. Without this record, technicians called to perform maintenance or modify the equipment later will have trouble locating all the cabling and equipment.

Project supervision must document any changes made during construction or installation on the drawings as the changes occur. Architects usually note changes on hard-copy drawings using a colored pen or pencil, so users can readily spot the change. These marked-up versions are commonly called **redline drawings**. With mobile digital technology, architects and engineers can revise and promulgate the latest drawing versions almost instantly. After the drawings have been revised to reflect the redline changes, the final drawings are called **as-built drawings**, and are so marked. These become the drawings of record for the project.

4.1.4 Project-Delivery Systems

Project-delivery systems are processes for constructing projects, from development through construction. Project delivery systems focus on the following three primary systems, shown in *Figure 20*:

- *General contracting* – The traditional project delivery system uses a general contractor. In this type of project, the owner determines the design of the project, and then solicits proposals from general contractors. After selecting a general contractor, the owner contracts directly with that contractor, who builds the project as the prime, or controlling, contractor.
- *Design-build* – In the design-build system, a single entity manages both the design and construction of a project. Design-build delivery commonly use GMP contracts.
- *Construction management* – The construction management project delivery system uses a construction manager to facilitate the design and construction of a project. Construction managers are very involved in project control; their main concerns are controlling time, cost, and the quality of the project.

4.1.5 Cost Estimating and Budgeting

Before building a project, an estimate must be prepared. Estimating is the process of calculating the cost of a project. There are two types of costs to consider, including direct and indirect costs. Direct costs, also known as general conditions, are those that planners can clearly assigned to a

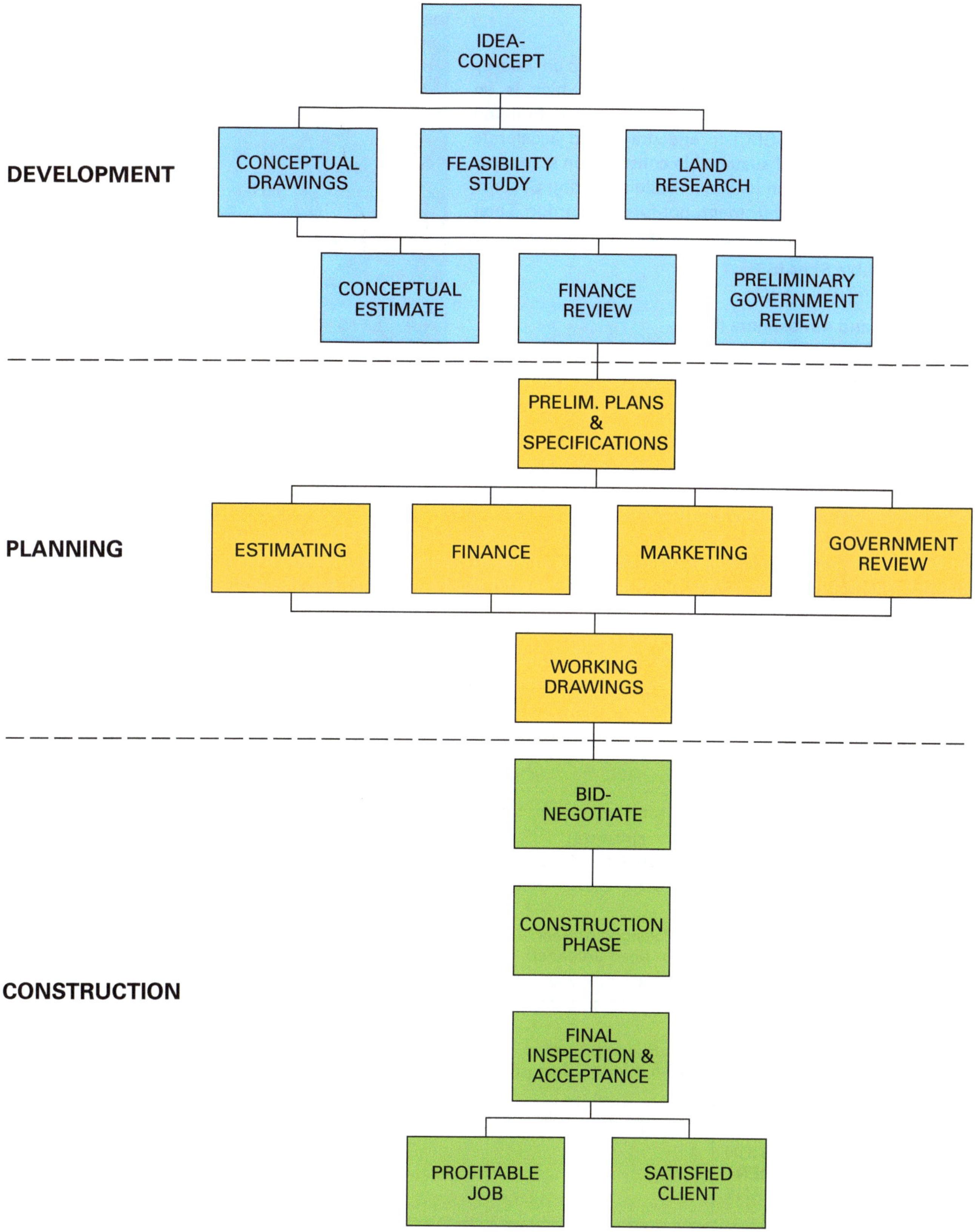

Figure 19 Project flow diagram.

GOING GREEN

LEED

Figure Credit: © Linda Williams/Dreamstime.com

LEED stands for *Leadership in Energy and Environmental Design*, which is an initiative started by the US Green Building Council (USGBC) to encourage and accelerate the adoption of sustainable construction standards worldwide through its Green Building Rating System. USGBC is a non-government, not-for-profit group. Their rating system addresses six categories:

1. Sustainable Sites (SS)
2. Water Efficiency (WE)
3. Energy and Atmosphere (EA)
4. Materials and Resources (MR)
5. Indoor Environmental Quality (EQ)
6. Innovation in Design (ID)

Building owners are the driving force behind the LEED voluntary program. Construction crew leaders may not have input into the decision to seek LEED certification for a project, or what materials are used in the project's construction. However, crew leaders can help to minimize material waste and support recycling efforts, both of which are factors in obtaining LEED certification.

An important question to ask is whether your project is seeking LEED certification. If the project is seeking certification, the next step is to ask what your role will be in getting the certification. If you are procuring materials, what information does the project need and who should receive it? What specifications and requirements do the materials need to meet? If you are working outside the building or inside in a protected area, what do you need to do to protect the work area? How does your crew manage waste? Are there any other special requirements that will be your responsibility? Do you see any opportunities for improvement? LEED principles are described in more detail in NCCER's *Your Role in the Green Environment* (Module ID 70101) and *Sustainable Construction Supervisor* (Module ID 70201).

DELIVERY SYSTEMS

RESPONSIBLE PARTIES	GENERAL CONTRACTING	DESIGN-BUILD	CONSTRUCTION MANAGEMENT
OWNER	Designs project (or hires architect)	Hires general contractor	Hires construction management company
GENERAL CONTRACTOR	Builds project (with owner's design)	Involved in project design, builds project	Builds, may design (hired by construction management company)
CONSTRUCTION MANAGEMENT COMPANY			Hires and manages general contractor and architect

Figure 20 Project-delivery systems.

budget. Indirect costs are overhead costs shared by all projects. Planners generally calculate these costs as an overhead percentage to labor and material costs.

Direct costs include the following:

- Materials
- Labor
- Tools
- Equipment

Indirect costs refer to overhead items such as the following:

- Office rent
- Utilities
- Telecommunications
- Accounting
- Office supplies, signs

The bid price includes the estimated cost of the project as well as the profit. Profit refers to the amount of money that the contractor will make after paying all the direct and indirect costs. If the direct and indirect costs exceed the estimate for the job, the difference between the actual and estimated costs must come out of the company's profit. This reduces what the contractor makes on the job.

Profit is the fuel that powers a business. It allows the business to invest in new equipment and facilities, provide training, and to maintain a reserve fund for times when business is slow. In large companies, profitability attracts investors who provide the capital necessary for the business to grow. For these reasons, contractors can't afford to consistently lose money on projects. If they can't operate profitably, they are forced out of business. Crew leaders can help their companies remain profitable by managing budget, schedule, quality, and safety adhering to the drawings, specifications, and project schedule.

The cost estimate must consider many factors. Many companies employ professional cost estimators to do this work. They also maintain performance data for previous projects. They use this data as a guide in estimating new projects. Development of a complete estimate generally proceeds as follows:

Step 1 Using the drawings and specifications, an estimator records the quantity of the materials needed to construct the job. Construction companies call this step the *quantity takeoff*. The estimator enters the information on a hard-copy or digital takeoff sheet like the one shown in *Figure 21*.

Step 2 The estimator uses the company's productivity rates to calculate the amount of labor required to complete the project. Most companies keep records of these rates for the type and size of the jobs that they perform. The company's estimating department maintains and updates these records.

Step 3 The estimator calculates the labor hours required by dividing the estimated amount of work by the productivity rate.

- For example, if the productivity rate for concrete finishing is 40 square feet per hour, and there are 10,000 square feet of concrete to be finished, then 250 hours of concrete finishing labor is required (10,000 ÷ 4 = 250).
- The estimator multiplies this number by the hourly rate for concrete finishing to determine the cost of that labor category.
- If this work is subcontracted, then the estimator uses the subcontractor's cost estimate, raised by an overhead factor, in place of direct-labor cost.

Step 4 The estimator transfers the total material quantities from the quantity takeoff sheet to a summary or pricing sheet (*Figure 22*). The total cost of materials is calculated after obtaining material prices from local suppliers.

Step 5 Next, the estimator determines the cost of equipment needed for the project. This number could reflect rental cost or a factor applied by the company when they plan to use their own equipment.

Step 6 The estimator totals the cost of all resources on the summary sheet—materials, equipment, tools, and labor. The estimator can also calculate the material unit cost—the total cost divided by the total number of units of listed materials.

Step 7 The estimator adds the cost of taxes, bonds, insurance, subcontractor work, and other indirect costs to the direct costs of the materials, equipment, tools, and labor.

Step 8 A sum of direct and indirect costs yields the total project cost. The contractor adds the expected profit to that total.

WORKSHEET

Takeoff By: ______ DATE ______ PROJECT ______ PAGE #

Checked By: ______ SHEET ____ of ____ ARCHITECT ______

REF.	DESCRIPTION	DIMENSIONS				EXTENSION	QUANTITY	UNIT	TOTAL		REMARKS
		NO	LENGTH	WIDTH	HEIGHT				QUANTITY	UNIT	

Figure 21 Quantity takeoff sheet.

SUMMARY SHEET

By:

DATE ______ PROJECT ______

SHEET ______ of ______ TITLE: ______ WORK ORDER # ______

PAGE #

DESCRIPTION	QUANTITY		MATERIAL COST		LABOR MAN HOURS FACTORS					LABOR COST		ITEM COST	
	TOTAL	UT	PER UNIT	TOTAL	CRAFT	PR UNIT	TOTAL	RATE	COST PR	PER	TOTAL	TOTAL	PER UNIT
				MATERIAL							LABOR	TOTAL	

Figure 22 Summary sheet.

NOTE

There are computer programs and mobile applications available to simplify the cost-estimating process. Many of them are tailored to specific industries, such as construction or manufacturing, as well as specific trades within the industries, such as plumbing, electrical, HVAC, and insulating contractors.

Estimating programs are typically set up to include a takeoff form and a form for estimating labor by category. Most of these programs include a database that contains current prices for labor and materials, so they automatically price the job and produce a bid. The programs can also generate purchase orders for materials on the job.

As a crew leader, you may be required to estimate quantities of materials. You will need a set of construction drawings and specifications to estimate the amount of a certain type of material required to perform a job. You should carefully review the appropriate section of the technical specifications and page(s) of drawings to determine the types and quantities of materials required. Then enter the quantities on the worksheet. For example, you should review the specification section on finished carpentry along with the appropriate pages of drawings before taking off the linear feet of door and window trim.

If insufficient materials are available to complete the job and an estimate is required, the estimator must determine how much more construction work is necessary. Knowing this, the crew leader can then determine the materials needed. You must also reference the construction drawings in this process.

4.2.0 Planning

One definition of planning is determining the methods used and their sequence to carry out the different tasks to complete a project (*Figure 23*). Planning involves the following:

- Determining the best method for performing the job
- Identifying the responsibilities of each person on the work crew
- Determining the duration and sequence of each activity
- Identifying the tools and equipment needed to complete a job
- Ensuring that the required materials are at the work site when needed
- Making sure that heavy construction equipment is available when required

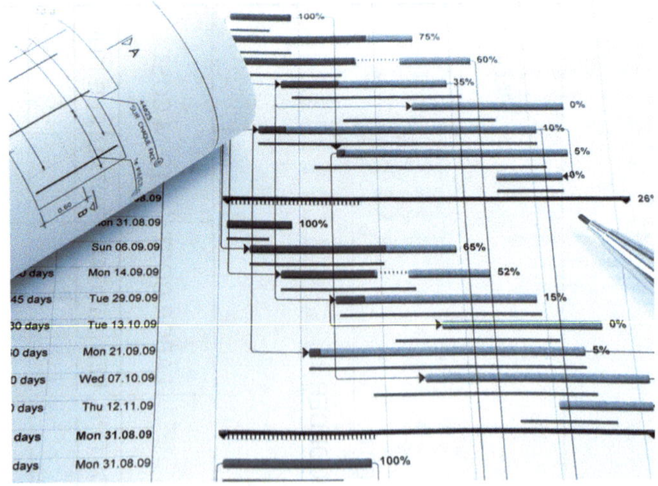

Figure 23 Proper prior planning prevents poor performance.

- Working with other contractors in such a way as to avoid interruptions and delays

Some important reasons for crew-leader planning include the following:

- Controlling the job in a safe manner so that it is built on time and within cost
- Lowering job costs through improved productivity
- Preparing for bad weather or unexpected occurrences
- Promoting and maintaining favorable employee morale
- Determining the best and safest methods for performing the job

With a plan, a crew leader can direct work efforts efficiently and can use resources such as personnel, materials, tools, equipment, work area, and work methods to their full potential.

A proactive crew leader will always have a backup plan in case circumstances prevent the original plan from working. There are many circumstances that can cause a plan to go awry, including adverse weather, equipment failure, absent crew members, and schedule slippage by other crafts.

Project planners establish time and cost limits for the project; the crew leader's planning must fit within those constraints. Therefore, it is important to consider the following factors that may affect the outcome:

- Site and local conditions, such as soil types, accessibility, or available staging areas
- Climate conditions that should be anticipated during the project
- Timing of all phases of work
- Types of materials to be installed and their availability

Participant Exercise E

Assume you are the leader of a crew building footing formwork for the construction shown in the figure below. You have used all of the materials provided for the job, yet you have not completed it. You study the drawings and see that the formwork consists of two side forms, each 12" high. The total length of footing for the entire project is 115'-0". You have completed 88'-0" to date; therefore, you have 27'-0" remaining (115' – 88' = 27').

Your task is to prepare an estimate of materials that you will need to complete the job. In this case, you will estimate only the side form materials (do not consider the miscellaneous materials here).

- Footing length to complete: 27'-0"
- Footing height: 1'-0"

Refer to the worksheet on the next page for a final tabulation of the side forms needed to complete the job.

1. Using the same footing as described in the example above, calculate the quantity (square feet) of formwork needed to finish 203 linear feet of the footing. Place this information directly on the worksheet.
2. You are the crew leader of a carpentry crew whose task is to side a warehouse with plywood sheathing. The wall height is 16', and there is a total of 480 linear feet of wall to side. You have done 360 linear feet of wall and have run out of materials. Calculate how many more feet of plywood you will need to complete the job. If you are using 4 × 8 plywood panels, how many will you need to order to cover the additional work? Write your estimate on the worksheet.

Show your calculations to the instructor.

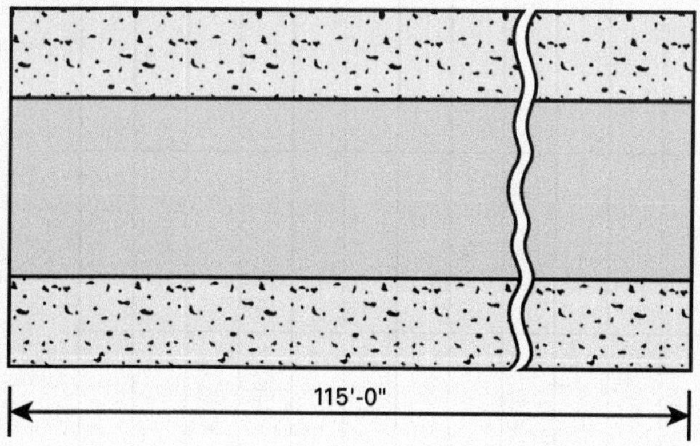

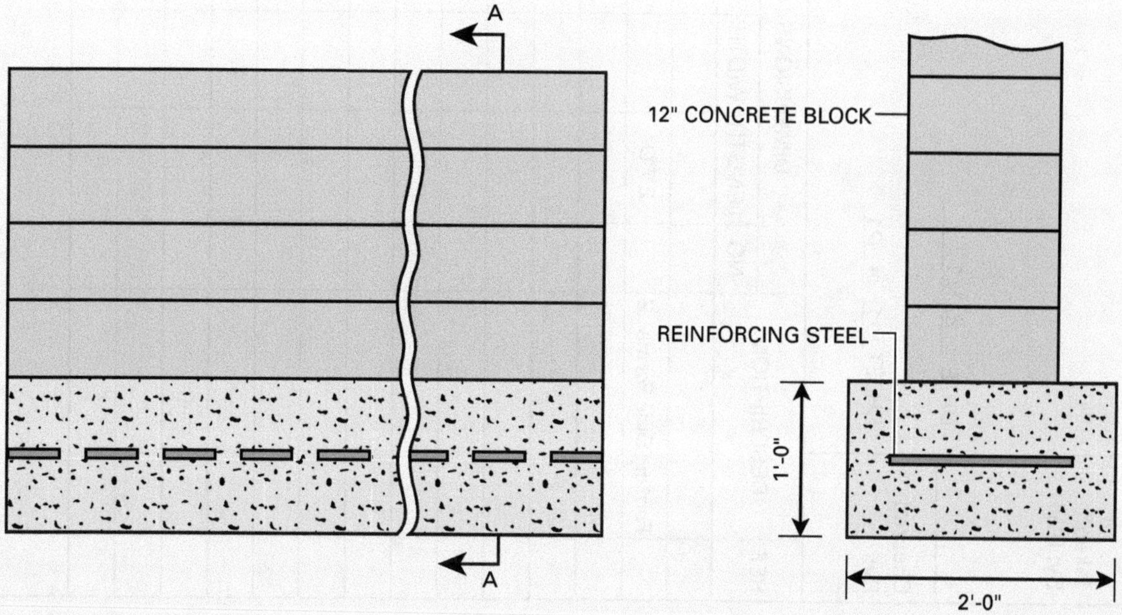

Participant Exercise E Worksheet

WORKSHEET

Takeoff By: RWH	DATE 2/1/17	PROJECT Sam's Diner
Checked By:	SHEET 01 of 01	ARCHITECT 654b

PAGE #1

REF.	DESCRIPTION	DIMENSIONS				EXTENSION	QUANTITY	UNIT	TOTAL		REMARKS
		NO	LENGTH	WIDTH	HEIGHT				QUANTITY	UNIT	
	Footing Side Forms 2		27'0"		1'0"	2x27x1	54	SF	54	SF	

- Equipment and tools required and their availability
- Personnel requirements and availability
- Relationships with the other contractors and their representatives on the job

On a simple job, crews can handle these items almost automatically. However, larger or more complex jobs require the planner to give these factors more formal consideration and study.

4.2.1 Stages of Planning

Formal planning for a construction job occurs at specific times in the project process. The two most important stages of planning occur in the preconstruction phase and during the construction work.

Preconstruction planning – The preconstruction stage of planning occurs before the start of construction. The preconstruction planning process doesn't always involve the crew leader, but it's important to understand what it consists of.

There are two phases of preconstruction planning. The first is developing the proposal, bid, or negotiated price for the job. This is when the estimator, the project manager, and the field superintendent develop a preliminary plan for completing the work. They apply their experience and knowledge from previous projects to develop the plan. The process involves determining what methods, personnel, tools, and equipment the work will require and what level of productivity they can achieve.

The second phase of preconstruction planning occurs after the client awards the contract. This phase requires a thorough knowledge of all project drawings and specifications. During this process, planners select the actual work methods and resources needed to perform the work. Here, crew leaders might get involved, but their planning must adhere to work methods, production rates, and resources that fit within the estimate prepared during contract negotiations. If the project requires a method of construction different

Participant Exercise F

1. In your own words, define planning, and describe how a job can be done better if it is planned. Give an example.

____

____

____

____

2. Consider a job that you recently worked on to answer the following:
   a. List the material(s) used.
   b. List each member of the crew with whom you worked and what each person did.
   c. List the kinds of equipment used.

____

____

____

____

3. List some suggestions for how the job could have been done better, and describe how you would plan for each of the suggestions.

____

____

____

____

from what is normal, planners will usually inform the crew leader of what method to use.

Construction planning – During construction, the crew leader is directly involved in daily planning. This stage of planning consists of selecting methods for completing tasks before beginning work. Effective planning exposes likely difficulties, and enables the crew leader to minimize the unproductive use of personnel and equipment. Proper planning also provides a gauge to measure job progress.

Effective crew leaders develop a tool known as the **look-ahead schedule**. These schedules consider actual circumstances as well as projections two-to-three weeks into the future. Developing a look-ahead schedule helps ensure that all resources are available on the project when needed. Most scheduling apps and programs include this feature that automatically help you focus on this period in the job.

4.2.2 The Planning Process

The planning process consists of the following five steps:

Step 1 Establish a goal.

Step 2 Identify the completed work activities required to achieve the goal.

Step 3 Identify the required tasks to accomplish those activities.

Step 4 Communicate responsibilities.

Step 5 Follow up to verify the goal achievement.

Establishing a goal – The term *goal* has different meanings for different people. In general, a goal is a specific outcome that one works toward. For example, the project superintendent of a home construction project could establish the goal to have a house dried-in by a certain date. (The term *dried-in* means ready for the application of roofing and siding.) To meet that goal, the leader of the framing crew and the superintendent would need to agree to a goal to have the framing completed by a given date. The crew leader would then establish sub-goals (objectives) for the crew to complete each element of the framing (floors, walls, roof) by a set time. The superintendent would need to set similar goals with the crews that install sheathing, building wrap, windows, and exterior doors. However, if the framing crew doesn't meet its goal, that will delay the other crews.

Identifying the required work – The second step in planning is to identify the necessary work to achieve the goal as a series of activities in a certain sequence. You will learn how to break down a job into activities later in this section. At this point, the crew leader should know that, for each activity, one or more objectives must be set.

An objective is a statement of a condition the plan requires to exist or occur at a specific time. An objective must:

- Mean the same thing to everyone involved
- Be measurable, so that everyone knows when it has been reached
- Be achievable with the resources available
- Have everyone's full support

Examples of practical objectives include the following:

- "By 4:30 p.m. today, the crew will have completed installation of the floor joists."
- "By closing time Friday, the roof framing will be complete."

Notice that both examples meet the first three requirements of an objective. Planners assume that everyone involved in completing the task is committed to achieving the objective. The advantage in developing objectives for each work activity is that it allows the crew leader to determine if the crew is following the plan. In addition, objectives serve as sub-goals that are usually under the crew leader's control.

Some construction work activities, such as installing 12" footing forms, are done so often that they require little planning. However, other jobs, such as placing a new type of mechanical equipment, require substantial planning. This type of job demands that the crew leader set specific objectives.

Whenever faced with a new or complex activity, take the time to establish objectives that will serve as guides for accomplishing the job. You can use these guides in the current situation, as well as for similar work in the future.

Identifying the required tasks – To plan effectively, the crew leader must be able to break a work activity assignment down into smaller tasks. Large jobs include a greater number of tasks than small ones, but all jobs can be broken down into manageable components.

When breaking down an assignment into tasks, make each task identifiable and definable. A task is identifiable when one knows the types and amounts of resources it requires. A task is definable if it has a specific duration. For purposes of efficiency, the job breakdown should not be too detailed or complex, unless the job has never been done before or must be performed with strictest efficiency.

For example, a suitable breakdown for the work activity to install square vinyl floor tiles in a cafeteria might be the following:

Step 1 Prepare the floor.

Step 2 Stage the tiles.

Step 3 Spread the adhesive.

Step 4 Lay the tiles.

Step 5 Clean the tiles.

Step 6 Wax the floor.

The crew leader could create even more detail by breaking down any one of the tasks into subtasks. In this case, however, that much detail is unnecessary and wastes the crew leader's time and the project's money.

Planners can divide every work activity into three general parts:

- Preparing
- Performing
- Cleaning up

Some of the most frequent mistakes made in the planning process are forgetting to prepare and to clean up. The crew leader must not overlook preparation and cleanup.

After identifying the various tasks that make up the job and developing an objective for each task, the crew leader must determine what resources the job requires. Resources include labor, equipment, materials, and tools. In most jobs, the job estimate identifies these resources. The crew leader must make sure that these resources are available on the site when needed.

"By failing to prepare, you are preparing to fail."
—Benjamin Franklin

Communicating responsibilities – No supervisor can complete all the activities within a job alone. The crew leader must rely on other people to get everything done. Therefore, most jobs have a crew of people with various experiences and skill levels to assist in the work. The crew leader's job is to draw from this expertise to get the job done well and in a safe and timely manner.

Once the various activities that make up the job have been determined, the crew leader must identify the person or persons responsible for completing each activity. This requires that the crew leader be aware of the skills and abilities of the people on the crew. Then, the crew leader must put this knowledge to work in matching the crew's skills and abilities to specific tasks required to complete the job.

After matching crew members to specific activities, the crew leader must then assign work to the crew. The crew leader normally communicates responsibilities verbally; the crew leader often talks directly to the responsible person for the activity. There may be times when the crew leader assigns work through written instructions or indirectly through someone other than the crew leader. Either way, crew members should know what it is they are responsible for accomplishing on the job.

Following up – Once the crew leader has delegated the activities to the appropriate crew members, there must be follow-up to make sure that the crew has completed them correctly and efficiently. Task follow-up involves being present on the job site to make sure all the resources are available to complete the work, ensuring that the crew members are working on their assigned activities, answering any questions, and helping to resolve any problems that occur while the work is being done. In short, follow-up activity means that the crew leader is aware of what's going on at the job site and is doing whatever is necessary to make sure that the crew completes the work on schedule. *Figure 24* reviews the planning steps.

The crew leader should carry a small note pad or electronic device for planning and taking notes. That way, you can record thoughts about the project as they occur, and you won't forget pertinent details. The crew leader may also choose to use a manual planning form such as the one illustrated in *Figure 25*.

As the job progresses, refer to these resources to see that the tasks are being done according to plan. This is job analysis. Construction projects that don't proceed according to work plans usually end up costing more and taking longer. Therefore, it is important that crew leaders refer to the planning documents periodically.

The crew leader is involved with many activities on a day-to-day basis. Thus, it is easy to forget important events if they aren't recorded. To help keep track of events such as job changes, interruptions, and visits, the crew leader should keep a **job diary**. A job diary is a notebook in which the crew leader records activities or events that take place on the

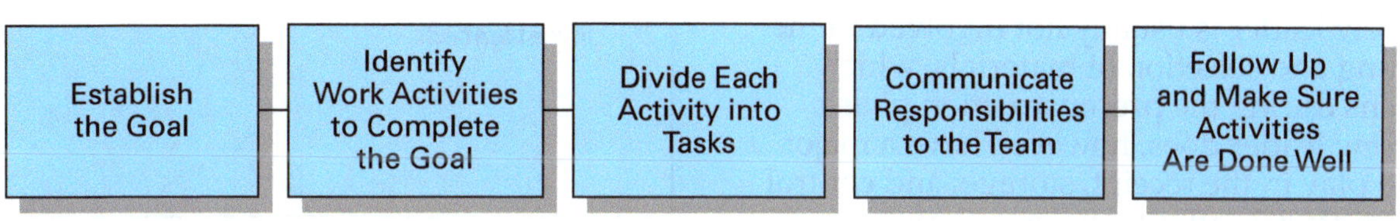

Figure 24 Steps to effective planning.

DAILY WORK PLAN

"PLAN YOUR WORK AND WORK YOUR PLAN = EFFICIENCY"

Plan of ____________________ Date __________

PRIORITY	DESCRIPTION	✓ When Completed X Carried Forward

Figure 25 Planning form.

job site that may be important later. When making entries in a job diary, make sure that the information is accurate, factual, complete, consistent, organized, and up-to-date. This is especially true if documenting personnel problems. If the company requires maintaining a job diary, follow company policy in determining which events and what details you should record. However, if there is a doubt about what to include, it is better to have too much information than too little. *Figure 26* shows a sample page from a job diary.

4.2.3 Planning Resources

Once a job has been broken down into its tasks or activities, the various resources needed to perform them must be determined and accounted for. Resource planning includes the following specific considerations:

- *Safety planning* – Using the company safety manual as a guide, the crew leader must assess the safety issues associated with the job and take necessary measures to minimize any risk to the crew. This may involve working with the company or site safety officer and may require a formal hazard analysis.
- *Materials planning* – Preconstruction planning identifies the materials required for the job and lists them on the job estimate. Companies usually order the materials from suppliers who have previously provided quality materials on schedule and within estimated cost.

  The crew leader is usually not involved in the planning and selection of materials, which happens during the preconstruction phase. The crew leader does, however, have a major role to play in the receipt, storage, and control of the materials after they reach the job site.
- The crew leader is also involved in planning materials for tasks such as job-built formwork and scaffolding. In addition, the crew leader may run out of a specific material, such as fasteners, and need to order more. In such cases, be sure to consult the appropriate supervisor, since most companies have specific purchasing policies and procedures.

July 8, 2017

Weather: Hot and Humid

Project: Company XYZ Building

- The paving contractor crew arrived late (10 am).
- The owner representative inspected the footing foundation at approximately 1 pm.
- The concrete slump test did not pass. Two trucks had to be ordered to return to the plant, causing a delay.
- John Smith had an accident on the second floor. I sent him to the doctor for medical treatment. The cause of the accident is being investigated.

Figure 26 Sample page from a job diary.

- *Site planning* – There are many planning elements involved in site work. The following are some of the key elements:
  - Access roads
  - Emergency procedures
  - Material and equipment storage
  - Material staging
  - Parking
  - Sedimentation control
  - Site security
  - Storm water runoff
- *Equipment planning* – The preconstruction phase addresses much of the planning for use of construction equipment. This planning includes the types of equipment needed, the use of the equipment, and the length of time it will be on the site. The crew leader must work with the main office to make certain that the equipment reaches the job site on time. The crew leader must also ensure that crew equipment operators are properly trained.

  Coordinating the use of the equipment is also very important. Some equipment operates in combination with other equipment. For example, dump trucks are generally required when loaders and excavators are used. The crew leader should also coordinate equipment with other contractors on the job. Sharing equipment can save time and money and avoid duplication of effort.

  The crew leader must reserve time for equipment maintenance to prevent equipment failure. In the event of an equipment failure, the crew leader must know who to contact to resolve the problem. An alternate plan must be ready in case one piece of equipment breaks down, so that the other equipment doesn't sit idle. The crew leader should coordinate these contingency plans with the main office or the crew leader's immediate superior.
- *Tool planning* – A crew leader is responsible for planning tool usage for a job. This task includes the following:
  - Determining the tools required
  - Informing the workers who will provide the tools (company or worker)
  - Making sure the workers are qualified to use the tools safely and effectively
  - Determining what controls to establish for tools
- *Labor planning* – All jobs require some sort of labor because the crew leader can't complete all the work alone. When planning for labor, the crew leader must do the following:
  - Identify the skills needed to perform the work.
  - Determine the number of people having those specific skills that are required.
  - Decide who will be on the crew.

  In many companies, the project manager or job superintendent determines the size and makeup of the crew. Then, supervision expects the crew leader to accomplish the goals and objectives with the crew provided. Even though the crew leader may not have any involvement in staffing the crew, the crew leader is responsible for training the crew members to ensure that they have the skills needed to do the job.
  In addition, the crew leader is responsible for keeping the crew adequately staffed at all times to avoid job delays. This involves dealing with absenteeism and turnover, two common problems discussed in earlier sections.

4.2.4 Scheduling

Planning and scheduling are closely related and are both very important to a successful job. Planning identifies the required activities and how and in what order to complete them. Scheduling involves establishing start and finish times/dates for each activity.

A schedule for a project typically shows the following:

- Operations listed in sequential order
- Units of construction
- Duration of activities
- Estimated date to start and complete each activity
- Quantity of materials to be installed

There are different types of schedules used today. They include the bar chart, the network schedule, also called the critical path method (CPM) or precedence diagram, and the short-term, or look-ahead schedule.

The following is a summary of the steps a crew leader must complete to develop a schedule.

Step 1 Make a list of all the activities that the plan requires to complete the job, including individual work activities and special tasks, such as inspections or the delivery of materials. At this point, the crew leader should just be concerned with generating a list, not with determining how to accomplish the activities, who will perform them, how long they will take, or the necessary sequence to complete them.

Step 2 Use the list of activities created in Step 1 to reorganize the work activities into a logical sequence. When doing this, keep in mind that certain steps can't happen before the completion of others. For example, footing excavation must occur before concrete emplacement.

Step 3 Assign a duration or length of time that it will take to complete each activity and determine the start time for each. Then place each activity into a schedule format. This step is important because it helps the crew leader compare the task time estimates to the scheduled completion date or time.

The crew leader must be able to read and interpret the job schedule. On some jobs, the form provides the beginning and expected end date for each activity, along with the expected crew or worker's production rate. The crew leader can use this information to plan work more effectively, set realistic goals, and compare the starts and completions of tasks to those on the schedule.

Before starting a job, the crew leader must do the following:

- Determine the materials, tools, equipment, and labor needed to complete the job.
- Determine when the various resources are needed.
- Follow up to ensure that the resources are available on the job site when needed.

The crew leader should verify the availability of needed resources three to four working days before the start of the job. This should occur even earlier for larger jobs. Advance preparation will help avoid situations that could potentially delay starting the job or cause it to fall behind schedule.

Supervisors can use bar chart schedules, also known as *Gantt charts*, for both short-term and long-term jobs. However, they are especially helpful for jobs of short duration.

Bar charts provide management with the following:

- A visual presentation of the overall time required to complete the job using a logical method rather than a calculated guess
- A means to review the start and duration of each part of the job
- Timely coordination requirements between crafts
- Alternative sequences of performing the work

A bar chart works as a control device to see whether the job is on schedule. If the job isn't on schedule, supervision can take immediate action in the office and the field to correct the problem and increases the likelihood of completing the activity on schedule. *Figure 27* illustrates a bar or Gantt chart.

Another type of schedule is the network schedule, which shows dependent (critical path) activities and other activities completed in parallel with but not part of the critical path. In *Figure 28*, for example, reinforcing steel can't be set until the concrete forms have been built and placed. Other activities are happening in parallel, but the forms are in the critical path.

When building a house, builders can't install and finish drywall until wiring, plumbing, and HVAC ductwork have been roughed-in. Because other activities, such as painting and trim work, depend on drywall completion, the drywall work is a critical-path operation. In other words, until it is complete, workers can't start the other tasks, and the project itself will likely experience delay by the amount of delay starting any dependent

Did You Know?

Recycling Materials

Habitat for Humanity, a charitable organization that builds homes for disadvantaged families, accepts surplus building materials for use in their projects. In some cities, they have stores called ReStores, which serve as retail outlets for such materials. They may also accept materials salvaged during demolition of a structure. Companies can obtain LEED credits through practices such as salvaging building materials, segregating scrap materials for recycling, and taking steps to minimize waste.

Figure Credit: Sushil Shenoy/Virginia Tech

Did You Know?

Scheduling Apps

Many companies use project scheduling software. Among the top construction project management applications at the time of publication are WorkflowMax, Procore, BuildTools, Co-Construct, and BuilderTREND. Such programs are spreadsheets on steroids, which can respond to schedule changes by altering downstream events and producing a revised schedule. Most of the newest programs are cloud-based, and users can access and update them remotely using mobile, tablet, and computer devices.

activity. Likewise, workers can't even start the drywall installation until the rough-ins are complete. Therefore, the project superintendent is likely to focus on those activities when evaluating schedule performance.

The advantage of a network schedule is that it allows project leaders to see how a schedule change with one activity is likely to affect other activities and the project in general. Planners lay out a network schedule on a timeline and usually show the estimated duration for each activity. Planners use network schedules for complex jobs that take a long time to complete. The PERT (program evaluation and review technique) schedule is a form of network schedule.

Since the crew needs to hold to the job schedule, the crew leader needs to be able to plan daily production. As discussed earlier, short-term scheduling is a method used to do this. *Figure 29* displays an example of a short-term, look-ahead schedule.

The information to support short-term scheduling comes from the estimate or cost breakdown. The schedule helps to translate estimate data and the various job plans into a day-to-day schedule of events. The short-term schedule provides the crew leader with visibility over the immediate future of the project. If actual production begins to slip behind estimated production, the schedule will warn the crew leader that a problem lies ahead and that a schedule slippage is developing.

Crew leaders should use short-term scheduling to set production goals. Generally, workers can improve production when they:

- Know the amount of work to be accomplished
- Know the time available to complete the work
- Can provide input when setting goals

Example:

A carpentry crew on a retaining wall project is about to form and pour catch basins and put up wall forms. The crew has put in several catch basins, so the crew leader is sure that they can perform the work within the estimate. However, the crew leader is concerned about their production of the wall forms. The crew will work on both the basins and the wall forms at the same time. The scheduling process in this scenario could be handled as follows:

1. The crew leader notices the following in the estimate or cost breakdown:
   a. Production factor for wall forms = 16 worker-hours (w-h) per 100 ft^2
   b. Work to be done by measurement = 800 ft^2
   c. Total time: (800 $ft^2 \times 16$ w-h/100 ft^2) = 128 w-h
2. The carpenter crew consists of the following:
   a. One carpenter crew leader
   b. Four carpenters
   c. One laborer
3. The crew leader determines the goal for the job should be set at 128 w-h (from the cost breakdown).
4. If the crew remains the same (six workers), the work should be completed in about 21 crew-hours (128 w-h $\div$ 6 workers/crew = 21.3 crew-hours).
5. The crew leader then discusses the production goal (completing 800 ft^2 in 21 crew-hours) with the crew and encourages them to work together to meet the goal of getting the forms erected within the estimated time.

In this example, the crew leader used the short-term schedule to translate production into work-hours or crew-hours and to schedule work so that the crew can accomplish it within the estimate. In addition, setting production targets provides the motivation to produce more than the estimate requires.

No matter what type of schedule is used, supervision must keep it up to date to be useful to the crew leader. Inaccurate schedules are of no value. The person responsible for scheduling in the office handles the updates. This person uses information gathered from job field reports to do the updates.

The crew leader is usually not directly involved in updating schedules. However, completing field or progress reports used by the company may be a daily responsibility to keep the schedule up to date. It's critical that the crew leader fill out any required forms or reports completely and accurately.

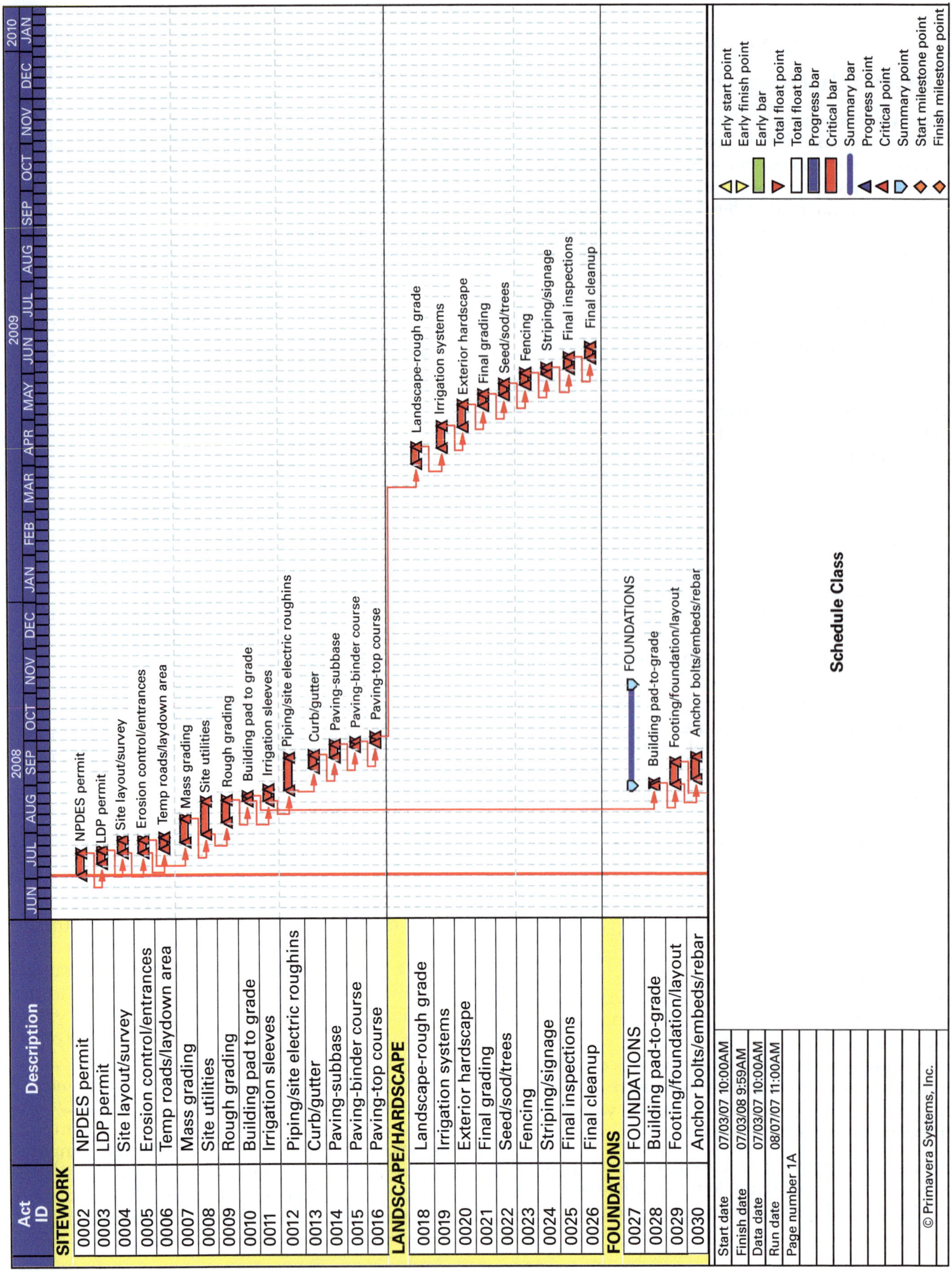

Figure 27 Example of a bar-chart schedule.

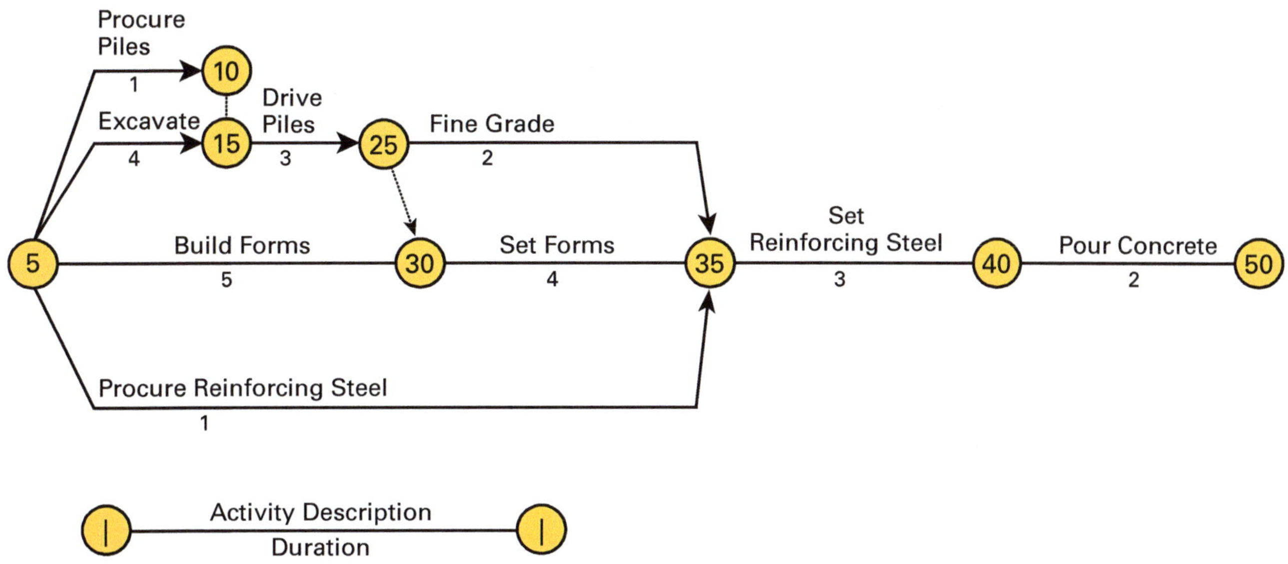

Figure 28 Example of a network schedule.

4.3.0 Cost Control

Being aware of costs and controlling them is the responsibility of every employee on the job. It's the crew leader's job to ensure that employees uphold this responsibility. Control refers to the comparison of estimated performance against actual performance and following up with any needed corrective action. Crew leaders who use cost-control practices are more valuable to the company than those who do not.

On a typical job, many activities are going on at the same time, even within a given crew. This can make it difficult to control the activities involved. The crew leader must be constantly aware of the costs of a project and effectively control the various resources used on the job.

When resources aren't controlled, the cost of the job increases. For example, a plumbing crew of four people is installing soil pipe and runs out of fittings. Three crew members wait (*Figure 30*) while one crew member goes to the plumbing-supply dealer for a part that costs only a few dollars. It takes the crew member an hour to get the part, so four worker-hours of production have been lost. In addition, the total cost of the delay must include the travel costs for retrieving the supplies.

4.3.1 Assessing Cost Performance

Cost performance on a project is determined by comparing actual costs to estimated costs. Regardless of whether the job is a contract-bid project or an in-house project, the company must first establish a budget. In the case of a contract bid, the budget is generally the cost estimate used to bid the job. For an in-house job, participants will submit labor and material forecasts, and someone in authority will authorize a project budget.

It is common to estimate cost by either breaking the job into funded tasks or by forecasting labor and materials expenditures on a timeline. Many companies create a **work breakdown structure (WBS)** for each project. Within the WBS, planners assign each major task a discrete charge number. Anyone working on that task charges that number on their time sheet, so that project managers can readily track cost performance. However, knowing how much money the company is spending doesn't necessarily determine cost performance.

Although financial reports can show that actual expenses are tracking with forecast expenses, they don't show if the work itself is occurring at the required rate. Thus, it is possible to have spent half the budget, but have less than half of the work compete. When the project is broken down into funded tasks related to schedule activities and events, there is far greater control over cost performance.

4.3.2 Field Reporting System

The total estimated cost comes from the job estimates, but managers obtain the actual cost of doing the work from an effective field-reporting system. A field-reporting system consists of a series of forms, which are completed by the crew leader and others. Each company has its own forms and methods for obtaining information. The following paragraphs describe the general information and the process of how they are used.

First, records must document the number of hours each person worked on each task. This

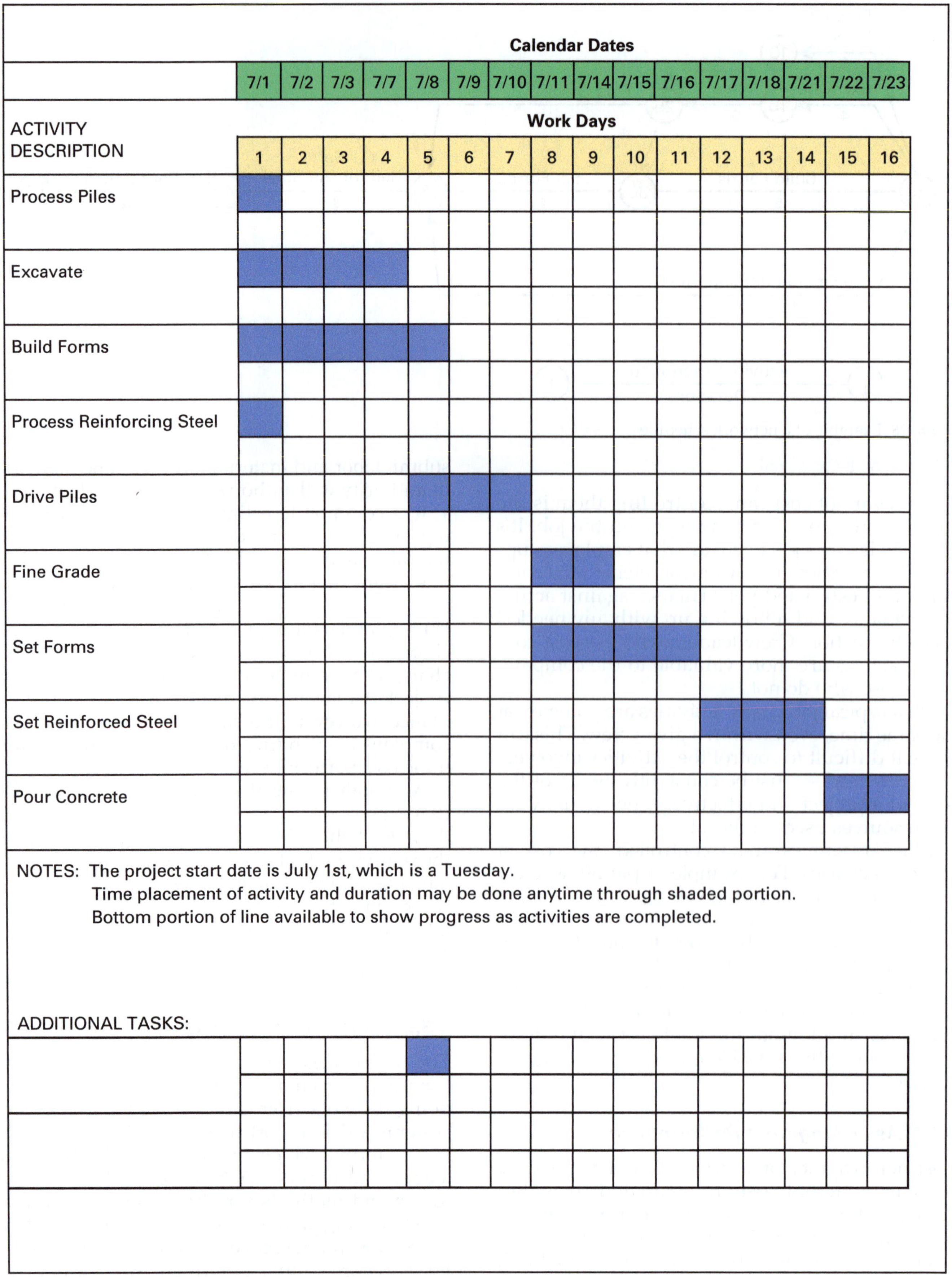

Figure 29 Short-term schedule.

Figure 30 Idle workers cost money.

information comes from daily time cards. Once the accounting department knows how many hours each employee worked on an activity, it can calculate the total cost of the labor by multiplying the number of hours worked by the wage rate for each worker. Managers can then calculate the cost for the labor to do each task as the job progresses. They compare this cost and the estimated costs during the project and at its completion. Managers use a similar process to determine if the costs to operate equipment are comparable to the estimated equipment operating costs.

As the crew places materials, a designated person will measure the quantities from time to time, and send this information to the company office and, possibly, the crew leader. This information, along with the actual cost of the material and the number of hours it took the workers to install it, is compared to the estimated cost. If the cost is greater than the estimate, management and the crew leader must take action to reduce the cost.

For this comparison process to be of use, the information obtained from field personnel must be correct. It is important that the crew leader be accurate in reporting. This is another reason to maintain a daily job diary as discussed earlier. In the event of a legal/contractual conflict with the client, courts can use such diaries as evidence in legal proceedings, and they can be helpful in reaching a settlement.

Example:

You are running a crew of five concrete finishers for a subcontractor. When you and your crew show up to finish a slab, the general contractor (GC) informs you, "We're a day behind on setting the forms, so I need you and your crew to stand down until tomorrow." What do you do?

Solution:

You should first call your office to let them know about the delay. Then, immediately record it in your job diary. A six-man crew for one day represents 48 worker-hours. If your company charges $30 an hour, that's a potential loss of $1,440, which the company would want to recover from the GC. If there is a dispute, your entry in the job diary could result in a favorable decision for your employer.

4.3.3 Crew Leader's Role in Cost Control

The crew leader is often the company's representative in the field, where the work takes place. Therefore, the crew leader contributes a great deal to determining job costs. When supervision assigns work to a crew, the crew leader should receive a budget and schedule for completing the job. It is then up to the crew leader to make sure the crew finishes the job on time and stays within budget. The crew leader achieves this by actively managing the use of labor, materials, tools, and equipment.

> *"Beware of little expenses;*
> *a small leak will sink a great ship."*
> *—Benjamin Franklin*

If the actual costs are at or below the estimated costs, the job is progressing as planned and scheduled, then the company will realize the expected profit. However, if the actual costs exceed the estimated costs, one or more problems may result in the company losing its expected profit, and maybe more. No company can remain in business if it continually loses money.

One of the factors that can increase cost is client-related changes (often called scope creep). The crew leaders must be able to assess the potential impact of such changes and, if necessary, confer with the employer to determine the course of action. If contractor-related losses are occurring, the crew leader and superintendent will need to work together to get the costs back in line.

"There is no such thing as scope creep, only scope gallop."
—Cornelius Fitchner

The following are some causes that can make actual costs exceed estimated costs, along with some potential actions the crew leader can take to bring the costs under control:

> **NOTE**
>
> Before starting any action, the crew leader should check with company supervision to see that the action proposed is acceptable and within the company's policies and procedures.
>
> - *Late delivery of materials, tools, and/or equipment* – Plan ahead to ensure that job resources will be available when needed
> - *Inclement weather* – Work with the superintendent and have alternate plans ready
> - *Unmotivated workers* – Counsel the workers
> - *Accidents* – Enforce the existing safety program (or evaluate the program for adequacy)

There are many other methods to get the job done on time if it gets off schedule. Examples include working overtime, increasing the size of the crew, prefabricating assemblies, or working staggered shifts. However, these examples may increase the cost of the job, so the crew leader should not do them without the approval of the project manager.

4.4.0 Resource Control

The crew leader's job is to complete assigned tasks safely according to the plans and specifications, on schedule, and within the scope of the estimate. To accomplish this, the crew leader must closely control how resources of materials, equipment, tools, and labor are used. The crew must minimize waste whenever possible.

Control involves measuring performance and correcting departures from plans and specifications to accomplish objectives. Control anticipates predictable deviations from plans and specifications based on experience and takes measures to prevent them from occurring.

An effective control process can be broken down into the following steps:

Step 1 Establish standards and divide them into measurable units. For example, a baseline can be created using experience gained on a typical job, where 2,000 linear feet (LF) of 1¼" copper water tube was installed in five days. Thus, dividing the total LF installed by the number of installation days gives the average installation rate. In this case, for 1¼" copper water tube, the standard rate was 400 LF/day.

Step 2 Measure performance against a standard. On another job, a crew placed 300 LF of the same tube during a given day. Thus, this actual production of 300 LF/day did not meet the standard rate of 400 LF/day.

Step 3 Adjust operations to meet the standard. In Step 2, if the plan scheduled the job for five days, the crew leader would have to take action to meet this goal. If 300 LF/day is the actual average daily production rate for that job, the crew must increase their placement rate by 100 LF/day to meet the standard. This increase may come by making the workers place piping faster, adding workers to the crew, or receiving authorization for overtime.

The crew leader's responsibility in materials control depends on the policies and procedures of the company. In general, the crew leader is responsible for ensuring on-time delivery, preventing waste, controlling delivery and storage, and preventing theft of materials.

4.4.1 Ensuring On-Time Delivery

It's essential that the materials required for each day's work be on the job site when needed. The crew leader should confirm in advance the placement of orders for all materials and that they will arrive on schedule. A week or so before the delivery date, follow up to make sure there will be no delayed deliveries or items on backorder.

If other people are responsible for providing the materials for a job, the crew leader must follow up to make sure that the materials are available when needed. Otherwise, delays occur as crew members stand around waiting for the delivery of materials.

4.4.2 Preventing Waste

Waste in construction can add up to loss of critical and costly materials and may result in job delays. The crew leader needs to ensure that every crew member knows how to use the materials efficiently. Crew leaders should monitor their crews to make certain that no materials are wasted.

An example of waste is a carpenter who saws off a piece of lumber from a full-sized piece, when workers could have found the length needed in the lumber scrap pile. Another example of waste

Did You Know?

Just in Time (JIT)

Just-in-time (JIT) delivery is a strategy in which materials arrive at the job site when needed. This means that a crew may install the materials right off the truck. This method reduces the need for on-site storage and staging areas. It also reduces the risk of loss or damage while moving products about the site. Other modern material management methods include the use of radiofrequency identification (RFID) tags that make it easy to inventory and locate material in crowded staging areas. Using an RFID inventory control system requires special scanners.

involves rework due to bad technique or a conflict in craft scheduling requiring removal and reinstallation of finished work. The waste in time spent and replacement material costs occur when redoing the task a second time.

Under LEED, waste control is very important. Companies receive LEED credits for finding ways to reduce waste and for recycling construction waste. Workers should segregate waste materials by type for recycling, if feasible (*Figure 31*).

Figure 31 Waste materials separated for recycling.

4.4.3 Verifying Material Delivery

A crew leader may be responsible for the receipt of materials delivered to the work site. When this happens, the crew leader should require a copy of the shipping invoice or similar document and check each item on the invoice against the actual materials to verify delivery of the correct amounts.

The crew leader should also check the condition of the materials to verify that nothing is defective before signing the invoice. This can be difficult and time consuming because it requires the crew to open cartons and examine their contents. However, this step cannot be overlooked, because a signed invoice indicates that the recipient accepted all listed materials and in an undamaged state. If the crew leader signs for the materials without checking them, and then finds damage, no one will be able to prove that the materials came to the site in that condition.

After checking and signing the shipping invoice, the crew leader should give the original or a copy to the superintendent or project manager. The company then files the invoice for future reference because it serves as the only record the company has to check bills received from the supplier company.

4.4.4 Controlling Delivery and Storage

Another essential element of materials control is the selection of where the materials will be stored on the job site. There are two factors in determining the appropriate storage location. The first is convenience. If possible, the materials should be stored near where the crew will use them. The time and effort saved by not having to carry the materials long distances will greatly reduce the installation costs.

Next, the materials must be stored in a secure area to avoid damage. It is important that the storage area suit the materials being stored. For instance, materials that are sensitive to temperature, such as chemicals or paints, should be stored in climate-controlled areas to prevent waste.

4.4.5 Preventing Theft and Vandalism

Theft and vandalism of construction materials increase costs. The direct and indirect costs of lost materials include obtaining replacement materials, production time lost while the needed materials are missing, and the costs of additional security precautions can add significantly to the overall cost to the company. In addition, the insurance that the contractor purchases will increase in cost as the theft and vandalism rate grows.

The best way to avoid theft and vandalism is a secure job site. At the end of each work day, store unused materials and tools in a secure location, such as a locked construction trailer. If the job site is fenced or the building has lockable accesses, the materials can be stored within. Many sites have security cameras and/or intrusion alarms to help minimize theft and vandalism.

4.4.6 Equipment Control

The crew leader may not be responsible for long-term equipment control. However, the equipment required for a specific job is often the crew leader's responsibility. The first step is to schedule when a crew member must transport the required equipment from the shop or rental yard. The crew leader is responsible for informing the shop of the location of its use and seeing a worker returns it to the shop when finished with it.

It is common for equipment to lay idle at a job site because of a lack of proper planning for the job and the equipment arrived early. For example, if wire-pulling equipment arrives at a job site before the conduit is in place, this equipment will be out of service while awaiting the conduit installation. In addition to the wasted rental cost, damage, loss, or theft can also occur with idle equipment while awaiting use.

The crew leader needs to control equipment use, ensure that the crew operates the equipment according to its instructions, and within time and cost guidelines. The crew leader must also assign or coordinate responsibility for maintaining and repairing equipment as indicated by the applicable preventive maintenance schedule. Delaying maintenance and repairs can lead to costly equipment failures. The crew leader must also ensure that the equipment operators have the necessary credentials to operate the equipment, including applicable licenses.

The crew leader is responsible for the proper operation of all other equipment resources, including cars and trucks. Reckless or unsafe operation of vehicles will likely result in damaged equipment, leading to disciplinary action in the case of the workers, fines and repair costs levied by the vehicle vendors, and a loss of confidence in the crew leader's ability to govern crew-members' actions.

The crew leader should also arrange to secure all equipment at the close of each day's work to prevent theft. If continued use of the equipment is necessary for the job, the crew leader should make sure to lock it in a safe place; otherwise, return it to the shop.

4.4.7 Tool Control

Among companies, various policies govern who provides hand and power tools to employees. Some companies provide all the tools, while others furnish only the larger power tools. The crew leader should be familiar with and enforce any company policies related to tools.

Tool control has two aspects. First, the crew leader must control the issue, use, and maintenance of all tools provided by the company. Second, the crew leader must control how crew members use the tools to do the job. This applies to tools supplied by the company as well as tools that belong to the workers.

Using the proper tools correctly saves time and energy. In addition, proper tool use reduces the chance of damage to the tool during use, as well as injury to the user and to nearby workers. This is especially true of edge tools and those with any form of fixed or rotating blade. A dull tool is far more dangerous than a sharp one.

Tools must be adequately maintained and properly stored. Making sure that tools are cleaned, dried, and lubricated prevents rust and ensures that the tools are in the proper working order. If a tool is damaged, it is essential to repair or replace it promptly. Otherwise, an accident or injury could result.

> **NOTE**
>
> Regardless of whether a worker or the company owns a tool, OSHA holds companies responsible for the consequences of using a tool on a job site. The company is accountable if an employee receives an injury from a defective tool. Therefore, the crew leader needs to be aware of any defects in the tools the crew members are using.

Users should take care of company-issued tools as if they were their own. Workers should not abuse tools simply because they don't their property.

One of the major sources of low productivity on a job is the time spent searching for a tool. To prevent this from occurring, supervision should establish a storage location for company-issued tools and equipment. The crew leader should make sure that crew members return all company-issued tools and equipment to this designated location after use. Similarly, workers should organize their personal toolboxes so that they can readily find the appropriate tools and return their tools to their toolboxes when they are finished using them. This is a matter of professionalism that crew leaders are in an excellent position to develop.

Studies have shown that some keys to an effective tool control system are:

- Limiting the number of people allowed access to stored tools
- Limiting the number of people held responsible for tools and holding them accountable
- Controlling the ways in which a tool can be returned to storage
- Making sure tools are available when needed

4.4.8 Labor Control

Labor typically represents more than half the cost of a project, and therefore has an enormous impact on profitability. For that reason, it's essential to manage a crew and their work environment in a way that maximizes their productivity. One of the ways to do that is to minimize the time spent on unproductive activities such as the following:

- Engaging in bull sessions
- Correcting drawing errors
- Retrieving tools, equipment, and materials
- Waiting for other workers to finish

If crew members are habitually goofing off, it is up to the crew leader to counsel those workers. The crew leader's daily diary should document the counseling. The crew leader will need to refer repeated violations to the attention of higher management as guided by company policy.

All sorts of errors will occur during a project that need correction. Some errors, such as mistakes on drawings, may be outside of the crew leader's control. However, before a drawing error results in performing an action that will need to be corrected later, a careful examination of the drawings by the crew leader may discover the error before work begins, saving time and materials. If crew members are making mistakes due to inexperience, the crew leader can help avoid these errors by providing on-the-spot training and by checking on inexperienced workers more often.

The availability and location of tools, equipment, and materials to a crew can have a profound effect on their productivity. As discussed in the previous section, the key to minimizing such problems is proactive management of these resources.

Delays caused by crews or contractors can be minimized or avoided by carefully tracking the project schedule. In doing so, crew leaders can anticipate delays that will affect the work and either take action to prevent the delay or redirect the crew to another task.

4.4.9 Production and Productivity

Production is the amount of construction materials put in place. It is the quantity of materials installed on a job, such as 1,000 linear feet of waste pipe installed for a task. On the other hand, productivity is a rate, and depends on the level of efficiency of the work. It is the amount of work done per hour or day by one worker or a crew.

Production levels are set during the estimating stage. The estimator determines the total amount of materials placed from the plans and specifications. After the job is complete, supervision can assess the actual amount of materials installed, and can compare the actual production to the estimated production.

Productivity relates to the amount of materials put in place by the crew over a certain period of time. The estimator uses company records during the estimating stage to determine how much time and labor it will take to place a certain quantity of materials. From this information, the estimator calculates the productivity necessary to complete the job on time.

Participant Exercise G

1. List the methods your company uses to minimize waste.

____

____

____

____

____

2. List the methods your company uses to control small tools on the job.

____

____

____

____

____

3. List five ways that you feel your company could control labor to maximize productivity.

____

____

____

____

____

For example, it might take a crew of two people ten days to paint 5,000 square feet. To calculate the required average productivity, divide 5,000 ft^2 by 10 days. The result is 500 ft^2/day. The crew leader can compare the daily production of any crew of two painters doing similar work with this average, as discussed previously.

Planning is essential to productivity. The crew must be available to perform the work and have all the required materials, tools, and equipment in place when the job begins.

The time on the job should be for business, not for taking care of personal problems. Anything not work-related should be handled after hours, away from the job site, if possible. Even the planning of after-work activities, arranging social functions, or running personal errands should occur after work or during breaks. Only very limited and necessary exceptions to these rules should be permitted (e.g., making or meeting medical appointments), and these should be spelled out in company or crew policies.

Organizing field work can save time. The key to effectively using time is to work smarter, not necessarily harder. Working smart can save your crew from doing unnecessary work. For example, most construction projects require that the contractor submit a set of as-built plans at the completion of the work. These plans describe the actual structure and installation of materials. The best way to prepare these plans is to mark up a set of working plans as the work is in progress as described earlier. That way, workers and supervision won't forget pertinent details and waste time trying to reconstruct how the company did the work.

> *"Knowledge is power is time is money."*
> *—Robert Thier, Storm and Silence*

The amount of material used should not exceed the estimated amount. If it does, either the estimator has made a mistake, undocumented changes have occurred, or rework has caused the need for additional materials. Whatever the case, the crew leader should use effective control techniques to ensure the efficient use of materials.

When bidding a job, most companies calculate the cost per worker-hour. For example, a ten-day job might convert to 160 w-h (two painters for ten days at eight hours per day). If the company charges a labor rate of $30/hour, the labor cost would be $4,800. The estimator then adds the cost of materials, equipment, and tools, along with overhead costs and a profit factor, to determine the price of the job.

After completion of a job, information gathered through field reporting allows the company office to calculate actual productivity and compare it to the estimated figures. This helps to identify productivity issues and improves the accuracy of future estimates.

The following labor-related practices can help to ensure productivity:

- Ensure that all workers have the required resources when needed.
- Ensure that all personnel know where to go and what to do after each task is completed.
- Make reassignments as needed.
- Ensure that all workers have completed their work properly.

Additional Resources

Construction Workforce Development Professional, NCCER. 2016. New York, NY: Pearson Education, Inc.

Mentoring for Craft Professionals, NCCER. 2016. New York, NY: Pearson Education, Inc.

Blueprint Reading for Construction, James A. S. Fatzinger. 2003. New York, NY: Pearson Education, Inc.

Construction Leadership from A to Z: 26 Words to Lead By, Wally Adamchik. 2011. Live Oak Book Company.

The following websites offer resources for products and training:

Architecture, Engineering, and Construction Industry (AEC), **www.aecinfo.com**

US Green Building Council (USGBC), **www.usgbc.org/leed**

4.0.0 Section Review

1. Which of these activities occurs during the development phase of a project?
   a. Architect/engineer sketches are prepared and a preliminary budget is developed.
   b. Government agencies give a final inspection of the design, check for adherence to codes, and inspect materials used.
   c. Detailed project drawings and specifications are prepared.
   d. Contracts for the project are awarded.

2. A job diary should typically record any of the following *except* _____.
   a. items such as job interruptions and visits
   b. the exact times of scheduled lunch breaks
   c. changes needed to project drawings
   d. the actual time each job task related to a particular project took to complete

3. Which of the following is a correct statement regarding project cost?
   a. Cost is handled by the accounting department and isn't a concern of the crew leader.
   b. The difference between the estimated cost and the actual cost affects a company's profit.
   c. Wasted material is factored into the estimate and is never a concern.
   d. The contractor's overhead costs aren't included in the cost estimate.

4. To prevent job delays due to late delivery of materials, the crew leader should _____.
   a. demand a discount from the supplier to compensate for the delay
   b. tell a crew member to go look for the delivery truck
   c. refuse the late delivery and re-order the materials from another supplier
   d. check with the supplier in advance of the scheduled delivery

SUMMARY

In construction, the crew leader is the company's supervisor on the ground at the scene where the actual work takes place. The modern crew leader must be prepared to deal with a host of challenges ranging from differences in age, culture, race, gender, language, and educational backgrounds to attitudes regarding authority and personal integrity. Rather than being overwhelmed, the effective and successful crew leader will use these challenges to broaden his or her understanding of people and to grow in character.

Crew leaders must develop an understanding of the company, its policies, and how to become an effective and valuable member of the supervisory team. They must also be able to effectively communicate up and down the chain, to motivate the crew's workers, and develop a sense of ownership within the crew so that they have the confidence and will to do the jobs.

Safety is an overriding concern for the company, and crew leaders need to make safe work practices and conditions the number-one priority. Once a worker has lost an eye, a limb, or life itself, it's too late to reconsider how the task might have been done safer. Workers are a company's most valuable asset and the crew leader is in the position of being the best example of safety to the crew.

While crew leaders view their role as mainly people and work managers, these tasks gain a greater sense of importance when crew leaders understand their responsibilities to the company for whom they work. Effectively controlling worker productivity, resources, schedules, and other aspects of the project directly contribute to cost control, which reflects in greater profits for the company. Companies recognize effective crew leaders and they can expect long and successful careers as a result.

Review Questions

1. Younger-generation workers may have a better perception of success if _____.
   a. crew leaders assign their projects in smaller, well-defined units
   b. they receive closer supervision
   c. crew leaders leave them alone
   d. they are given the perception that they are in charge

2. Companies can help prevent sexual harassment in the workplace by _____.
   a. requiring employee training that avoids the potentially offensive subject of stereotypes
   b. developing a consistent policy with appropriate consequences for engaging in sexual harassment
   c. communicating to workers that, legally, the victim of sexual harassment is only the one being directly harassed
   d. establishing an effective complaint or grievance process for victims of harassment

3. Which of the following statements about a formal organization is *true*?
   a. It is a group of members created mainly to share professional information.
   b. It is a group established for developing industry standards.
   c. It is a group of individuals led by someone all working toward the same goal.
   d. It is a group that is identified by its overall function rather than by key positions.

4. Which of these departments in a large company would likely arrange equipment rentals?
   a. Accounting
   b. Estimating
   c. Human Resources
   d. Purchasing

5. When a person is promoted to crew leader, the amount of time they spend on craft work _____.
   a. increases
   b. decreases
   c. stays the same
   d. doubles

6. Which of the following is a *correct* statement regarding annual crew evaluations?
   a. Their results should never come as a surprise to the crew member.
   b. Formal evaluations should be conducted every three months.
   c. Comments on performance should be withheld until the formal evaluation.
   d. Advise the crew members to improve their performance just before the formal evaluation is scheduled.

7. As a crew leader, what should you avoid when giving recognition to crew members?
   a. Looking for good work and knowing what good work is
   b. Being available at the job site to witness good work
   c. Giving praise for poor work to make the worker feel better
   d. Encouraging work improvement through your confidence in a worker

8. Within the construction industry, who normally assigns a crew to the crew leader?
   a. The US Department of Labor
   b. The company
   c. The crew leader
   d. The crew members

9. Some decisions are difficult to make because you _____.
   a. have no experience dealing with the circumstances requiring a decision
   b. don't want to make a decision
   c. prefer to think it's someone else's responsibility to make a decision
   d. don't recognize the need for a decision

10. The first step in problem solving should be _____.
    a. choosing a solution to the problem
    b. evaluating the solution to the problem
    c. defining the problem
    d. testing the solution to the problem

11. The leading cause of fatalities in the construction industry is _____.
    a. electrocution
    b. asphyxiation
    c. struck by something
    d. falls

12. OSHA inspection of a business or job site _____.
    a. can be done only by invitation
    b. can be conducted at random
    c. is done only after an accident
    d. is conducted only if a safety violation occurs

13. To proactively identify and assess hazards, it is essential that employers _____.
    a. ensure that crew leaders are present and available on the job site
    b. conduct safety inspections of contractor offices
    c. design safety logs to document safety violations
    d. conduct accident investigations

14. Employers who are subject to OSHA record-keeping requirements must maintain a log of recordable occupational injuries and illnesses for _____.
    a. three years
    b. five years
    c. seven years
    d. as long as the company is in business

15. Which of the following statements regarding safety training sessions is *correct*?
    a. The project manager usually holds them.
    b. They are held only for new employees.
    c. They should be conducted frequently by the crew leader.
    d. They are required only after an accident has occurred.

16. The type of contract in which the client pays the contractor for their actual labor and material expenses they incur is known as a _____.
    a. firm fixed-price contract
    b. time-spent contract
    c. performance-based contract
    d. cost-reimbursable contract

17. On a design-build project, _____.
    a. the same contractor is responsible for both design and construction
    b. the owner is responsible for providing the design
    c. the architect does the design and the general contractor builds the project
    d. a construction manager is hired to oversee the project

18. When planning labor for onsite work, the crew leader must _____.
    a. factor in the absenteeism and turnover rates
    b. factor in time for training classes
    c. select the material to be used by each crew member
    d. identify the skills needed to perform the work

19. Which of the following is *not* a source of data in the field-reporting system?
    a. The cost-estimate for the project
    b. Crew leader's daily diary
    c. Onsite inventory of placed materials
    d. Daily time cards

20. The following duties are important in work site equipment control *except* _____.
    a. scheduling the pickup and return of a piece of equipment
    b. ensuring the equipment is operated and maintained correctly
    c. monitoring safe and responsible operation of vehicles
    d. selection and purchase of equipment to be stocked at the work site equipment room

Trade Terms Introduced in This Module

Autonomy: The condition of having complete control over one's actions, and being free from the control of another. To be independent.

Bias: A preconceived inclination against or in favor of something.

Cloud-based applications: Mobile and desktop digital programs that can connect to files and data located in distributed storage locations on the Internet ("the Cloud"). Such applications make it possible for authorized users to create, edit, and distribute content from any location where Internet access is possible.

Craft professionals: Workers who are properly trained and work in a particular construction trade or craft.

Crew leader: The immediate supervisor of a crew or team of craft professionals and other assigned persons.

Critical path: In manufacturing, construction, and other types of creative processes, the required sequence of tasks that directly controls the ultimate completion date of the project.

Demographics: Social characteristics and other factors, such as language, economics, education, culture, and age, that define a statistical group of individuals. An individual can be a member of more than one demographic.

Ethics: The moral principles that guides an individual's or organization's actions when dealing with others. Also refers to the study of moral principles.

Infer: To reach a conclusion using a method of reasoning that starts with an assumption and considers a set of logically-related events, conditions, or statements.

Intangible: Not touchable, material, or measureable; lacking a physical presence.

Job description: A description of the scope and responsibilities of a worker's job so that the individual and others understand what the job entails.

Job diary: A written record that a supervisor maintains periodically (usually daily) of the events, communications, observations, and decisions made during the course of a project.

Legend: In maps, plans, and diagrams, an explanatory table defining all symbolic information contained in the document.

Lethargy: Sluggishness, slow motion, lack of activity, or a lack of enthusiasm.

Letter of instruction (LOI): A written communication from a supervisor to a subordinate that informs the latter of some inadequacy in the individual's performance, and provides a list of actions that the individual must satisfactorily complete to remediate the problem. Usually a key step in a series of disciplinary actions.

Local area networks (LAN): Communication networks that link computers, printers, and servers within a small, defined location, such as a building or office, via hard-wired or wireless connections.

Lockout/tagout (LOTO): A system of safety procedures for securing electrical and mechanical equipment during repairs or construction, consisting of warning tags and physical locking or restraining devices applied to controls to prevent the accidental or purposeful operation of the equipment, endangering workers, equipment, or facilities.

Look-ahead schedule: A manual- or software-scheduling tool that looks several weeks into the future at the planned project events; used for anticipating material, labor, tool, and other resource requirements, as well as identifying potential schedule conflicts or other problems.

Negligence: Lack of appropriate care when doing something, or the failure to do something, usually resulting in injury to an individual or damage to equipment.

Organizational chart: A diagram that shows how the various management and operational responsibilities relate to each other within an organization. Named positions appear in ranked levels, top to bottom, from those functional units with the most and broadest authority to those with the least, with lines connecting positions indicating chains of authority and other relationships.

Paraphrase: Rewording a written or verbal statement in one's own words in such a way that the intent of the original statement is retained.

Pragmatic: Sensible, practical, and realistic.

Proactive: To anticipate and take action in the present to deal with potential future events or outcomes based on what one knows about current events or conditions.

Project manager: In construction, the individual who has overall responsibility for one or more construction projects; also called the general superintendent.

Return on investment (ROI): A measure of the gain or loss of money resulting from an investment; normally measured as a percentage of the original investment.

Safety data sheets (SDS): Documents listing information about a material or substance that includes common and proper names, chemical composition, physical forms and properties, hazards, flammability, handling, and emergency response in accordance with national and international hazard communication standards. Also called material safety data sheets (MSDS).

Sexual harassment: Any unwelcome verbal or nonverbal form of communication or action construed by an individual to be of a sexual or gender-related nature.

Smartphone: The most common type of cellular telephone that combines many other digital functions within a single mobile device. Most important of these, besides the wireless telephone, are high-resolution still and video cameras, text and email messaging, and GPS-enabled features.

Superintendent: In construction, the individual who is the on-site supervisor in charge of a given construction project.

Synergy: Any type of cooperation between organizations, individuals, or other entities where the combined effect is greater than the sum of the individual efforts.

Textspeak: A form of written language characteristic of text messaging on mobile devices and text-based social media, which usually consists of acronyms, abbreviations, and minimal punctuation.

Wide area networks (WAN): Dedicated communication networks of computers and related hardware that serve a given geographic area, such as a work site, campus, city, or a larger but distinct area. Connectivity is by wired and wireless means, and may use the Internet as well.

Wi-Fi: The technology allowing communications via radio signals over a LAN or WAN equipped with a wireless access point or the Internet. (Wi-Fi stands for wireless fidelity.) Many types of mobile, portable, and desktop devices can communicate via Wi-Fi connections.

Work breakdown structure (WBS): A diagrammatic and conceptual method for subdividing a complex project, concept, or other thing into its various functional and organizational parts so that planners can analyze and plan for each part in detail.

OSHA Forms for Safety Records

Employers who are subject to the recordkeeping requirements of the Occupational Safety and Health Act of 1970 must maintain records of all recordable occupational injuries and illnesses. The following are three important OSHA forms used for this recordkeeping (these forms can be accessed at **www.osha.gov**):

- OSHA Form 300, *Log of Work-Related Injuries and Illnesses*
- OSHA Form 300A, *Summary of Work-Related Injuries and Illnesses*
- OSHA Form 301, *Injury and Illness Incident Report*

OSHA's Form 300 (Rev. 01/2004)

Log of Work-Related Injuries and Illnesses

Attention: This form contains information relating to employee health and must be used in a manner that protects the confidentiality of employees to the extent possible while the information is being used for occupational safety and health purposes.

Year ________

U.S. Department of Labor
Occupational Safety and Health Administration

Form approved OMB no. 1218-0176

You must record information about every work-related injury or illness that involves loss of consciousness, restricted work activity or job transfer, days away from work, or medical treatment beyond first aid. You must also record significant work-related injuries and illnesses that are diagnosed by a physician or licensed health care professional. You must also record work-related injuries and illnesses that meet any of the specific recording criteria listed in 29 CFR 1904.8 through 1904.12. Feel free to use two lines for a single case if you need to. You must complete an injury and illness incident report (OSHA Form 301) or equivalent form for each injury or illness recorded on this form. If you're not sure whether a case is recordable, call your local OSHA office for help.

Establishment name ____________________

City ____________________ State ____________________

Identify the person			Describe the case			Classify the case										
(A) Case No.	(B) Employee's Name	(C) Job Title (e.g., Welder)	(D) Date of injury or onset of illness (mo./day)	(E) Where the event occurred (e.g. Loading dock north end)	(F) Describe injury or illness, parts of body affected, and object/substance that directly injured or made person ill (e.g. Second degree burns on right forearm from acetylene torch)	CHECK ONLY ONE box for each case based on the most serious outcome for that case:			Enter the number of days the injured or ill worker was:		Check the "injury" column or choose one type of illness: (M)					
						Days away from work	Remained at work									
							Job transfer or restriction	Other record-able cases	Away From Work (days)	On job transfer or restriction (days)	Injury	Skin Disorder	Respiratory Condition	Poisoning	Hearing Loss	All other illnesses
						(H)	(I)	(J)	(K)	(L)	(1)	(2)	(3)	(4)	(5)	(6)
					Page totals	0	0	0	0	0	0	0	0	0	0	0
											Injury (1)	Skin Disorder (2)	Respiratory Condition (3)	Poisoning (4)	Hearing Loss (5)	All other illnesses (6)

Be sure to transfer these totals to the Summary page (Form 300A) before you post it.

Public reporting burden for this collection of information is estimated to average 14 minutes per response, including time to review the instruction, search and gather the data needed, and complete and review the collection of information. Persons are not required to respond to the collection of information unless it displays a currently valid OMB control number. If you have any comments about these estimates or any aspects of this data collection, contact: US Department of Labor, OSHA Office of Statistics, Room N-3644, 200 Constitution Ave, NW, Washington, DC 20210. Do not send the completed forms to this office.

Page 1 of 1

Figure A01 OSHA Form 300

OSHA's Form 300A (Rev. 01/2004)

Summary of Work-Related Injuries and Illnesses

Year ______

U.S. Department of Labor
Occupational Safety and Health Administration

Form approved OMB no. 1218-0176

All establishments covered by Part 1904 must complete this Summary page, even if no injuries or illnesses occurred during the year. Remember to review the Log to verify that the entries are complete and accurate before

Using the Log, count the individual entries you made for each category. Then write the totals below, making sure you've added the entries from every page of the log. If you had no cases write "0."

Employees former employees, and their representatives have the right to review the OSHA Form 300 in its entirety. They also have limited access to the OSHA Form 301 or its equivalent. See 29 CFR 1904.35, in OSHA's Recordkeeping rule, for further details on the access provisions for these forms.

Number of Cases

Total number of deaths	Total number of cases with days away from work	Total number of cases with job transfer or restriction	Total number of other recordable cases
0	0	0	0
(G)	(H)	(I)	(J)

Number of Days

Total number of days away from work	Total number of days of job transfer or restriction
0	0
(K)	(L)

Injury and Illness Types

Total number of...
(M)

(1) Injury	0	(4) Poisoning	0
(2) Skin Disorder	0	(5) Hearing Loss	0
(3) Respiratory Condition	0	(6) All Other Illnesses	0

Post this Summary page from February 1 to April 30 of the year following the year covered by the form

Public reporting burden for this collection of information is estimated to average 58 minutes per response, including time to review the instruction, search and gather the data needed, and complete and review the collection of information. Persons are not required to respond to the collection of information unless it displays a currently valid OMB control number. If you have any comments about these estimates or any aspects of this data collection, contact: US Department of Labor, OSHA Office of Statistics, Room N-3644, 200 Constitution Ave, NW, Washington, DC 20210. Do not send the completed forms to this office.

Establishment information

Your establishment name ______________________

Street ______________________

City ____________ State ____________ Zip ________

Industry description (e.g., Manufacture of motor truck trailers)

____

Standard Industrial Classification (SIC), if known (e.g., SIC 3715)

__ __ __ __

OR North American Industrial Classification (NAICS), if known (e.g., 336212)

__ __ __ __ __ __

Employment information

Annual average number of employees ________

Total hours worked by all employees last year ________

Sign here

Knowingly falsifying this document may result in a fine.

I certify that I have examined this document and that to the best of my knowledge the entries are true, accurate, and complete.

______________________ Company executive ______________ Title

______________________ Phone ______________ Date

Figure A02 OSHA Form 300A

OSHA's Form 301
Injuries and Illnesses Incident Report

Attention: This form contains information relating to employee health and must be used in a manner that protects the confidentiality of employees to the extent possible while the information is being used for occupational safety and health purposes.

U.S. Department of Labor
Occupational Safety and Health Administration

Form approved OMB no. 1218-0176

This *Injury and Illness Incident Report* is one of the first forms you must fill out when a recordable work-related injury or illness has occurred. Together with the *Log of Work-Related Injuries and Illnesses* and the accompanying *Summary*, these forms help the employer and OSHA develop a picture of the extent and severity of work-related incidents.

Within 7 calendar days after you receive information that a recordable work-related injury or illness has occurred, you must fill out this form or an equivalent. Some state workers' compensation, insurance, or other reports may be acceptable substitutes. To be considered an equivalent form, any substitute must contain all the information asked for on this form.

According to Public Law 91-596 and 29 CFR 1904, OSHA's recordkeeping rule, you must keep this form on file for 5 years following the year to which it pertains

If you need additional copies of this form, you may photocopy and use as many as you need.

Completed by ____________

Title ____________

Phone ____________ Date ____________

Information about the employee

1) Full Name ____________

2) Street ____________

City ____________ State ______ Zip ______

3) Date of birth ____________

4) Date hired ____________

5) ☐ Male
☐ Female

Information about the physician or other health care professional

6) Name of physician or other health care professional ____________

7) If treatment was given away from the worksite, where was it given?

Facility ____________

Street ____________

City ____________ State ______ Zip ______

8) Was employee treated in an emergency room?
☐ Yes
☐ No

9) Was employee hospitalized overnight as an in-patient?
☐ Yes
☐ No

Information about the case

10) Case number from the Log ______ *(Transfer the case number from the Log after you record the case.)*

11) Date of injury or illness ______

12) Time employee began work ______ AM/PM

13) Time of event ______ AM/PM ☐ Check if time cannot be determined

14) **What was the employee doing just before the incident occurred?** Describe the activity, as well as the tools, equipment or material the employee was using. Be specific. Examples: "climbing a ladder while carrying roofing materials"; "spraying chlorine from hand sprayer"; "daily computer key-entry."

15) **What happened?** Tell us how the injury occurred. Examples: "When ladder slipped on wet floor, worker fell 20 feet"; "Worker was sprayed with chlorine when gasket broke during replacement"; "Worker developed soreness in wrist over time."

16) **What was the injury or illness?** Tell us the part of the body that was affected and how it was affected; be more specific than "hurt", "pain", or "sore." Examples: "strained back"; "chemical burn, hand"; "carpal tunnel syndrome."

17) **What object or substance directly harmed the employee?** Examples: "concrete floor"; "chlorine"; "radial arm saw." If this question does not apply to the incident, leave it blank.

18) **If the employee died, when did death occur?** Date of death

Public reporting burden for this collection of information is estimated to average 22 minutes per response, including time for reviewing instructions, searching existing data sources, gathering and maintaining the data needed, and completing and reviewing the collection of information. Persons are not required to respond to the collection of information unless it displays a current valid OMB control number. If you have any comments about this estimate or any other aspects of this data collection, including suggestions for reducing this burden, contact: US Department of Labor, OSHA Office of Statistics, Room N-3644, 200 Constitution Ave, NW, Washington, DC 20210. Do not send the completed forms to this office.

Figure A03 OSHA Form 301

Additional Resources

This module presents thorough resources for task training. The following reference material is recommended for further study.

Construction Workforce Development Professional, NCCER. 2016. New York, NY: Pearson Education, Inc.

Mentoring for Craft Professionals, NCCER. 2016. New York, NY: Pearson Education, Inc.

Blueprint Reading for Construction, James A. S. Fatzinger. 2003. New York, NY: Pearson Education, Inc.

Construction Leadership from A to Z: 26 Words to Lead By, Wally Adamchik. 2011. Live Oak Book Company.

Generational Cohorts and their Attitudes Toward Work Related Issues in Central Kentucky, Frank Fletcher, et al. 2009. Midway College, Midway, KY. **www.kentucky.com.**

It's Your Ship: Management Techniques from the Best Damn Ship in the Navy, Captain D. Michael Abrashoff, USN. 2012. New York City, NY: Grand Central Publishing.

Survival of the Fittest, Mark Breslin. 2005. McNally International Press.

The Definitive Book of Body Language: The Hidden Meaning Behind People's Gestures and Expressions, Barbara Pease and Allan Pease. 2006. New York City, NY: Random House / Bantam Books.

The Young Person's Guide to Wisdom, Power, and Life Success: Making Smart Choices, Brian Gahran, PhD. 2014. San Diego, CA: Young Persons Press. **www.WPGBlog.com**.

The following websites offer resources for products and training:

Aging Workforce News, **www.agingworkforcenews.com**.

American Society for Training and Development (ASTD), **www.astd.org**.

Architecture, Engineering, and Construction Industry (AEC), **www.aecinfo.com**.

Equal Employment Opportunity Commission (EEOC), **www.eeoc.gov**.

National Association of Women in Construction (NAWIC), **www.nawic.org**.

National Census of Fatal Occupational Injuries (NCFOI), **www.bls.gov**.

National Institute of Occupational Safety and Health (NIOSH), **www.cdc.gov/niosh**.

National Safety Council, **www.nsc.org**.

Occupational Safety and Health Administration (OSHA), **www.osha.gov**.

Society for Human Resources Management (SHRM), **www.shrm.org**.

United States Census Bureau, **www.census.gov**.

United States Department of Labor, **www.dol.gov**.

US Green Building Council (USGBC), **www.usgbc.org/leed**.

Wi-Fi® is a registered trademark of the Wi-Fi Alliance, **www.wi-fi.org**

Figure Credits

© Photographerlondon/Dreamstime.com, Module Opener
© iStockphoto.com/shironosov, Figure 1
© iStockphoto.com/kali9, Figures 2, 10
© iStockphoto.com/nanmulti, Figure 3
© iStockphoto.com/Halfpoint, Figure 13
© iStock.com/vejaa, Figure 16
DIYawards, Figure 17
© iStockphoto.com/Tashi-Delek, Figure 18
© iStockphoto.com/kemaltaner, Figure 23
Sushil Shenoy/Virginia Tech, Figure 31
John Ambrosia, Figure 27
© iStockphoto.com/izustun, Figure 30

Section Review Answer Key

Answer	Section Reference	Objective
Section One		
1. a	1.1.0	1a
2. b	1.2.2	1b
3. b	1.3.0	1c
Section Two		
1. b	2.0.0	2a
2. c	2.2.1	2b
3. a	2.3.1	2c
4. c	2.4.1	2d
5. a	2.5.1	2e
Section Three		
1. a	3.0.0	3a
2. d	3.2.0; Table 1	3b
3. a	3.3.0	3c
4. b	3.4.0	3d
Section Four		
1. a	4.1.1	4a
2. b	4.2.2	4b
3. b	4.3.3	4c
4. d	4.4.1	4d

NCCER CURRICULA — USER UPDATE

NCCER makes every effort to keep its textbooks up-to-date and free of technical errors. We appreciate your help in this process. If you find an error, a typographical mistake, or an inaccuracy in NCCER's curricula, please fill out this form (or a photocopy), or complete the online form at **www.nccer.org/olf**. Be sure to include the exact module ID number, page number, a detailed description, and your recommended correction. Your input will be brought to the attention of the Authoring Team. Thank you for your assistance.

Instructors – If you have an idea for improving this textbook, or have found that additional materials were necessary to teach this module effectively, please let us know so that we may present your suggestions to the Authoring Team.

NCCER Product Development and Revision

13614 Progress Blvd., Alachua, FL 32615

Email: curriculum@nccer.org

Online: www.nccer.org/olf

❑ Trainee Guide ❑ Lesson Plans ❑ Exam ❑ PowerPoints Other ____________

Craft / Level: Copyright Date:

Module ID Number / Title:

Section Number(s):

Description:

Recommended Correction:

Your Name:

Address:

Email: Phone:

Glossary

Abrasion: The roughening or wearing-away of a surface by rubbing or by the impact of particles. Sand and other materials can abrade surfaces as they move past them.

Amorphous silica: A term identifying especially pure, noncrystalline (that is, amorphous) artificial mineral wool fibers made from molten silica.

Autonomy: The condition of having complete control over one's actions, and being free from the control of another. To be independent.

Batten: In layout work, a long, thin, flexible strip of wood or other material that can evenly bend without kinks. Used for creating smooth, sweeping single- and compound-curves.

Bead: A raised ridge or groove along an edge of sheet metal.

Beauty band: A narrow strip of sheet-metal jacket that is installed over the rough edges of adjacent jacket to provide a finished appearance. Often supplied with beading, crimped edges, or both.

Bias: A preconceived inclination against or in favor of something.

Billets: A block of material usually built up by cementing thinner layers together.

Bisector: In geometry, a line that divides an angle in half or bisects it.

Block: A usually rigid form of insulating material where its thickness dimension relative to its width and length give it a blocky appearance.

Board: A usually semi-rigid form of insulation typically having an appearance more like a rectangular panel than a block.

Buns: A large uniform block of material from which thinner boards or smaller blocks are cut.

Cathodic protection: A method of preventing underground metal objects from corrosion aided by exposure to ground water. This may be done by either attaching a more easily-corrodible metal to the object or, in the case of underground pipes, supplying a low-voltage direct current through the metal that blocks the corrosive effect.

Center cap: A circular or cone-shaped piece of jacket sheet metal installed at the center of a vessel head to finish off the jacket installation there.

Centerpiece: A disk of jacket material fastened to the insulation at the center of a vessel's head to anchor the inner ends of the gore segments during installation.

Circumference: The distance around the perimeter of a circle or a circular object. Measured in units of length, such as inches, feet, centimeters, or meters.

Cloud-based applications: Mobile and desktop digital programs that can connect to files and data located in distributed storage locations on the Internet ("the Cloud"). Such applications make it possible for authorized users to create, edit, and distribute content from any location where Internet access is possible.

Concentric: The condition where two or more circles or circular objects have the same center point or lie on the same axis.

Convex: A surface that appears to be outwardly curved or domed.

Corrosion under insulation (CUI): Corrosion occurring out of sight under pipe and equipment insulation due to the failure of the jacket, the vapor retarder boundary, or both, allowing water to penetrate the insulation and contact the metal surface.

Course: As with masonry and similar trades, refers to one row of many rows of sections of insulation.

Craft professionals: Workers who are properly trained and work in a particular construction trade or craft.

Crew leader: The immediate supervisor of a crew or team of craft professionals and other assigned persons.

Crimp: A treatment of an edge of sheet metal to improve its rigidity, consisting of either many peaks and valleys along the edge, like gear teeth, or a single shallow fold parallel to the edge.

Critical path: In manufacturing, construction, and other types of creative processes, the required sequence of tasks that directly controls the ultimate completion date of the project.

Cryogenic: For mechanical insulating purposes, any application or system operating at temperatures below –50°F (–46°C). However, the National Institute of Standards and Technology defines the upper end of the cryogenic band at –292°F (–180°C) for scientific purposes.

Demographics: Social characteristics and other factors, such as language, economics, education, culture, and age, that define a statistical group of individuals. An individual can be a member of more than one demographic.

Develop: As used in geometric layouts, to draw the true shape of a three-dimensional surface in two dimensions. It is an important part of fabricating jacket for three-dimensional vessels and equipment using two-dimensional sheet metal.

Dished top: When referring to a vessel head, it is a shape similar to an upside-down bowl; a low, domed appearance compared to a flat-topped vessel.

E-glass: One of several chemical categories of glass. E-glass is most commonly used in fiberglass and glass-reinforced products.

Eccentric: The condition where two circles, circular objects, or rotating objects do not have the same center point or center of rotation.

Elliptical: Shaped like an ellipse, which is a closed shape likened to a flattened circle. Curved lines and three-dimensional shapes formed from a portion of an ellipse are also referred to as being elliptical.

Emissivity: A measure of how rapidly a material radiates heat away from its surface.

Encasement: The fabric that encloses the insulation in a reusable insulation cover.

Endothermic: Any process that gains thermal energy (absorbs heat) from it surroundings.

Equal tee: Identifies a piping tee where the main and branch pipes have the same diameters.

Equipment: In the context used here, the term generally refers to pumps, vessels, and irregularly-shaped mechanical devices.

Ethics: The moral principles that guides an individual's or organization's actions when dealing with others. Also refers to the study of moral principles.

Fire-resistance rating: The time that a material by itself can resist the effects of a fire or continue to perform its function as measured by a standardized fire-resistance test.

Firestopping: A material or construction method that interferes with the entry or movement of fire through a fire-rated barrier.

Flashing: Narrow strips of sheet metal fabricated to protect joints in a covering as part of a weatherproofing system that may include caulking and neoprene washers for screw fasteners.

Floating ring: In the insulating trade, a ring formed of twisted wires, cable, or thin rod that is not attached to the underlying surface (i.e., it's floating) and is used to anchor insulation securement bands. Also called *chokers*, they are installed around manway covers, pipe flanges, and other obstructions on the surface to be insulated.

Footings: The underlying supports for a column or wall.

Friable: The property of a brittle material that tends to crumble into small particles when crushed, rubbed, or scraped.

Galvanic corrosion: A form of corrosion (rusting) resulting from the creation of an electric current flowing between dissimilar metals in contact with each other.

Ganister: A hard, fine-grained quartzite rock (naturally-cemented quartz sandstone) used in high-temperature insulation.

Gores: Sections of a curved surface, such as the surface of a sphere, that when flattened out, have little or no change of shape. The surface maps of model globes of the earth are created from gores glued to a ball.

Hemmed: The folded-over edge of a sheet of material, as in a fabric hem. A hemmed edge is smooth and eliminates a safety hazard.

Hog rings: Open, c-shaped rings with both ends beveled so that they interlock when clamped into place. Special pliers or pneumatic tools are used to install and remove them. Also called *C-rings*.

Hook-and-loop: A material used to temporarily and repeatedly attach fabrics and other kinds of objects together. One surface is covered with a material containing tiny hooks and the other surface has a material containing tiny loops. When brought together, the hooks and loops engage. Originally manufactured under the Velcro® brand name.

Infer: To reach a conclusion using a method of reasoning that starts with an assumption and considers a set of logically-related events, conditions, or statements.

Infrared (IR): A form of light not visible to humans that effectively transfers thermal energy from one object to any other that absorbs it.

Intangible: Not touchable, material, or measureable; lacking a physical presence.

Intumescent: The property of a material that swells when heated, increasing its volume and decreasing its density, usually by the production of gas voids inside the material. The insulating property of the voids slows the transmission of heat through the material.

Job description: A description of the scope and responsibilities of a worker's job so that the individual and others understand what the job entails.

Job diary: A written record that a supervisor maintains periodically (usually daily) of the events, communications, observations, and decisions made during the course of a project.

Juncture: A line formed by the intersection of two surfaces. An edge of a cube is the juncture of two of its faces.

Knuckle: In general, an area of sharper curvature in a surface compared to the surrounding surfaces. For vessel heads, the surface near the head perimeter that is more sharply curved than the central dish-part of the head or its perimeter.

Knurled: Having narrow milled ridges on a shaft or knob that are used to aid against slippage in a shaft hole or when gripped.

Lagging: With reference to insulation, heavy sheet metal sheathing over insulation, usually ribbed or corrugated, as weatherproofing and to protect insulation against industrial washdown operations.

Lags: Narrow sections of insulation cut from rigid board or block material whose long sides have been beveled. The beveled sides minimize the air spaces between similar lags next to it when installed on a curved surface.

Laser cutter: A device that uses a form of high-energy light, usually of a single color and produced in a tight, intense beam, to cut materials. Laser cutters can be designed for extremely precise shape fabrication.

Lateral length: For a conical fitting, the top-to-bottom length of the fitting as measured along its side (or lateral) surface.

Legend: In maps, plans, and diagrams, an explanatory table defining all symbolic information contained in the document.

Lethargy: Sluggishness, slow motion, lack of activity, or a lack of enthusiasm.

Letter of instruction (LOI): A written communication from a supervisor to a subordinate that informs the latter of some inadequacy in the individual's performance, and provides a list of actions that the individual must satisfactorily complete to remediate the problem. Usually a key step in a series of disciplinary actions.

Loadbearing: The function of any structural member, such as a wall, that supports part of the structure's weight.

Local area networks (LAN): Communication networks that link computers, printers, and servers within a small, defined location, such as a building or office, via hard-wired or wireless connections.

Lockout/tagout (LOTO): A system of safety procedures for securing electrical and mechanical equipment during repairs or construction, consisting of warning tags and physical locking or restraining devices applied to controls to prevent the accidental or purposeful operation of the equipment, endangering workers, equipment, or facilities.

Look-ahead schedule: A manual- or software-scheduling tool that looks several weeks into the future at the planned project events; used for anticipating material, labor, tool, and other resource requirements, as well as identifying potential schedule conflicts or other problems.

Needled: A term describing a manufacturing method for nonwoven fabrics with randomly-oriented fibers. These fabrics are created by rapidly and repeatedly inserting numerous barbed needles into a mass of fibers, during which they are pulled into a tightly tangled, cloth-like mass.

Negligence: Lack of appropriate care when doing something, or the failure to do something, usually resulting in injury to an individual or damage to equipment.

Noise abatement: Noise reduction; in this case, by use of insulating material.

Non-conducting: The property of a substance that resists or prevents the movement of heat or electricity.

Offset compound shears: A type of hand shears with jaws designed to permit cutting small-radius curves in sheet metal without crimping the material and keeping the worker's hand clear of the cut. Most manufacturers supply color-coded tools to identify the designed direction of the cut: green is clockwise cutting; red is counterclockwise cutting; yellow is straight cutting. Also called *aviation snips*.

Organizational chart: A diagram that shows how the various management and operational responsibilities relate to each other within an organization. Named positions appear in ranked levels, top to bottom, from those functional units with the most and broadest authority to those with the least, with lines connecting positions indicating chains of authority and other relationships.

Paraphrase: Rewording a written or verbal statement in one's own words in such a way that the intent of the original statement is retained.

Perimeter: The outside boundary of any object or region, especially one that encloses a space inside. The line forming a circle and the sidewall of a vessel are examples of perimeters.

Pittsburgh seam: A seam between two sheet-metal edges consisting of a right-angle flanged male tab on one edge and a folded female pocket on the other edge. The flange is inserted into the pocket and the side extension of the female pocket is folded over to lock in the male flange. The various folds and closures can be manually- or machine-formed in the sheet-metal edges.

Plasma cutter: A device that uses a high-velocity, high-temperature jet of ionized air or other gas emitted from a nozzle to cut and shape materials.

Pragmatic: Sensible, practical, and realistic.

Pressed cover: A type of preformed insulation jacket cover for elbows and tees available from many insulation distributors. The cover is made by shaping sheet metal in a form press and can be identified by the crimped fingers in the throat of the cover.

Radial lines: In geometry, straight lines drawn outward from the center of a circle, like a radius.

Refractory insulation: Temperature-resistant insulation that can retain its properties above 1,000°F (538°C). Typically installed inside of furnaces and other high-temperature apparatus.

Rolled angle irons: Curved sections of standard angle iron that has been shaped in an industrial rolling mill. The rolled angle iron is welded to the curved sidewall of a vessel to form a horizontal shelf or ledge.

Scored: A type of grooving in rigid board insulation consisting of vertical slots or notches. The scoring gives some flexibility to the otherwise inflexible insulation to minimize fracturing of the material when bent. Compare with the term *V-groove*.

Sector: In geometry, an area contained inside a circle's perimeter between two radial lines.

Self-drilling: Describes a fastener (screw) that is equipped with a special drill-bit-like point that drills through the metalwork to be fastened and taps its own threaded hole. Compare with the term *self-tapping*.

Self-tapping: Describes a fastener equipped with a notched, tapered point that allows the screw to create its own threads in metals softer than the screw. These screws require a predrilled hole. Compare with the term *self-drilling*.

Sheet-metal brake: A heavily-constructed machine designed to create straight folds in sheet metal. The process of bending metal into a corner or fold is called *braking*.

Sound isolation: Any material or construction method that interferes with the movement of sound (noise) from its source to somewhere else.

Spheroidal: Shaped like a sphere (or a part of a sphere). Spheroidal forms include spheres, hemispheres, and ellipsoids. In the insulating trade, it describes many types of cylindrical vessel heads.

Standing seam: A method for connecting two pieces of sheet metal. Squared flanges are formed in the adjacent edges of the sheet-metal sections and brought together. At the top of the flanges, the pieces are folded together, locking them into a tight joint.

Stiffeners: A piece of heavy formed steel, often with a T, H, I, or other cross section, welded to a flat surface to limit flexing. Also an angled ridge formed in sheet metal for the same purpose.

Sub-girts: In industrial construction, the term *girt* refers to a secondary support iron (T-, H-, or L-bar) attached to primary structural components and surfaces for stiffening, reinforcement, or to support other equipment. A sub-girt is a similar bar attached to the underlying structure that supports an overlying cover or sheathing.

Sweep elbow: A type of pipe elbow that has a longer radius than the standard short- or long-radius elbows.

Takeoff: The process of surveying, measuring, itemizing, and counting all materials and equipment needed for a construction project, as indicated by the drawings.

Tan-to-tan: A mechanical trade term referring to the distance between the tangent lines at opposite ends of a double-headed cylindrical vessel. Refer to the term *tangent line*.

Tangent lines: When referring to cylindrical vessels, the lines where the vessel head begins to curve away from the straight vessel sidewall. On most vessels, these lines are visible as either weld beads or flanges.

Tanks: Fully- or partially-enclosed vessels mainly used for storing liquids or gases. Compare with the term *vessel*.

Thermograph: An image created by detecting and recording infrared light. Colored thermographs show the intensity of infrared by colors, which indicate heat sources.

Trapezoids: Four-sided figures in which two opposite sides are parallel and one of the parallel sides is shorter than the other.

Ultraviolet (UV): A form of light invisible to humans that is more energetic than visible light, and can cause damaging chemical reactions when absorbed by skin or materials.

Unequal tee: A type of piping tee where the main and branch pipes are of unequal size. The branch pipe may have either a smaller or larger diameter than the main pipe.

V-grooved: A term describing thicker and typically rigid board insulation that has manufactured or fabricated V-shaped slots cut into the material. These grooves allow the board to be bent around curved surfaces with minimal and allowable cracking.

Ventilation: The natural or artificial entry and exchange of fresh air within a structure.

Vessels: In the context of mechanical insulating, any fully- or partially-enclosed structures that can be used for storing or processing any type of substance, usually at pressures above or below atmospheric pressure. Vessels are used for many purposes, including containing chemical reactions, in addition to storing or transferring liquids and gases.

Wind forces: Also called aerodynamic forces, they are the forces generated by air flowing over a surface, like the lifting forces on an airplane wing or wind on a sail.

Wing seal: A type of metal banding seal or buckle equipped with two tabs or wings that are bent over the folded end of the band to secure it.

X-Y controller: A computer-controlled device that precisely positions a machine or tool in the front-to-back and side-to-side directions in a plane.

Proactive: To anticipate and take action in the present to deal with potential future events or outcomes based on what one knows about current events or conditions.

Project manager: In construction, the individual who has overall responsibility for one or more construction projects; also called the *general superintendent*.

Return on investment (ROI): A measure of the gain or loss of money resulting from an investment; normally measured as a percentage of the original investment.

Safety data sheets (SDS): Documents listing information about a material or substance that includes common and proper names, chemical composition, physical forms and properties, hazards, flammability, handling, and emergency response in accordance with national and international hazard communication standards. Also called *material safety data sheets* (*MSDS*).

Sexual harassment: Any unwelcome verbal or nonverbal form of communication or action construed by an individual to be of a sexual or gender-related nature.

Smartphone: The most common type of cellular telephone that combines many other digital functions within a single mobile device. Most important of these, besides the wireless telephone, are high-resolution still and video cameras, text and email messaging, and GPS-enabled features.

Superintendent: In construction, the individual who is the on-site supervisor in charge of a given construction project.

Synergy: Any type of cooperation between organizations, individuals, or other entities where the combined effect is greater than the sum of the individual efforts.

Textspeak: A form of written language characteristic of text messaging on mobile devices and text-based social media, which usually consists of acronyms, abbreviations, and minimal punctuation.

Wide area networks (WAN): Dedicated communication networks of computers and related hardware that serve a given geographic area, such as a work site, campus, city, or a larger but distinct area. Connectivity is by wired and wireless means, and may use the Internet as well.

Wi-Fi: The technology allowing communications via radio signals over a LAN or WAN equipped with a wireless access point or the Internet. (Wi-Fi stands for *wireless fidelity*.) Many types of mobile, portable, and desktop devices can communicate via Wi-Fi connections.

Work breakdown structure (WBS): A diagrammatic and conceptual method for subdividing a complex project, concept, or other thing into its various functional and organizational parts so that planners can analyze and plan for each part in detail.

Index

D

E

F

T

U

V

W

X

Z